BIOGAS: FROM WASTE TO FUEL

BIOGAS: FROM WASTE TO FUEL

Edited by:

Navodita Bhatnagar

ARCLER
PRESS

www.arclerpress.com

Biogas: From Waste to Fuel

Navodita Bhatnagar

Arcler Press
2010 Winston Park Drive,
2nd Floor
Oakville, ON L6H 5R7
Canada
www.arclerpress.com
Tel: 001-289-291-7705
001-905-616-2116
Fax: 001-289-291-7601
Email: orders@arclereducation.com

ISBN: 978-1-77407-362-9 (Hardcover)

DECLARATION

ABOUT THE EDITOR

Navodita Bhatnagar finished her masters in technology and currently pursuing her PhD in environmental sciences from Institute of Technology, Carlow, Ireland. She's a keen enthusiast of environment protection and management. An alumunus of Indo-German centre for sustainibility which is essentially a collaborative interface to gather scientists in the field of environment sciences to come together and work for the common cause which is environment protection.

TABLE OF CONTENTS

LIST OF CONTRIBUTORS

Navodita Bhatnagar

Ilona Sárvári Horváth
Swedish Centre for Resource Recovery, University of Borås, 501 90 Borås, Sweden

Meisam Tabatabaei
Microbial Biotechnology Department, Agricultural Biotechnology Research Institute of Iran (ABRII), AREEO, Karaj, Iran
Biofuel Research Team (BRTeam), Karaj, Iran

Keikhosro Karimi
Department of Chemical Engineering, Isfahan University of Technology, Isfahan 84156-83111, Iran
Microbial Industrial Biotechnology Group, Institute of Biotechnology and Bioengineering, Isfahan University of Technology, Isfahan 84156-83111, Iran

Rajeev Kumar
Center for Environmental Research and Technology (CE-CERT), Bourns College of Engineering, University of California, Riverside, California, USA

Yitayal Addis Alemayehu
Department of Environmental science; Faculty of natural and computational sciences; Kotebe University College, Addis Ababa, Ethiopia

Franziska Enzmann
DECHEMA Research Institute, Industrial Biotechnology, Theodor-Heuss-Allee 25, 60486 Frankfurt am Main, Germany

Florian Mayer
DECHEMA Research Institute, Industrial Biotechnology, Theodor-Heuss-Allee 25, 60486 Frankfurt am Main, Germany

Michael Rother
Technische Universität Dresden, Institut für Mikrobiologie, Zellescher Weg 20b, 01217 Dresden, Germany

Dirk Holtmann
DECHEMA Research Institute, Industrial Biotechnology, Theodor-Heuss-Allee 25, 60486 Frankfurt am Main, Germany

Robbert Kleerebezem
Department of Biotechnology, Delft University of Technology, Julianalaan 67, 2628BC Delft, The Netherlands

Bart Joosse
Department of Biotechnology, Delft University of Technology, Julianalaan 67, 2628BC Delft, The Netherlands

Rene Rozendal
Paques BV, T. de Boerstraat 24, 8561EL Balk, The Netherlands

Mark C. M. Van Loosdrecht
Department of Biotechnology, Delft University of Technology, Julianalaan 67, 2628BC Delft, The Netherlands

Andreas Otto Wagner
Department of Microbiology, Universität Innsbruck, Technikerstraße 25d, A-6020 Innsbruck, Austria

Nina Lackner
Department of Microbiology, Universität Innsbruck, Technikerstraße 25d, A-6020 Innsbruck, Austria

Mira Mutschlechner
Department of Microbiology, Universität Innsbruck, Technikerstraße 25d, A-6020 Innsbruck, Austria

Eva Maria Prem
Department of Microbiology, Universität Innsbruck, Technikerstraße 25d, A-6020 Innsbruck, Austria

Rudolf Markt
Department of Microbiology, Universität Innsbruck, Technikerstraße 25d, A-6020 Innsbruck, Austria

Paul Illmer
Department of Microbiology, Universität Innsbruck, Technikerstraße 25d, A-6020 Innsbruck, Austria

Alessandra Cesaro
Sanitary Environmental Engineering Division (SEED), Department of Civil Engineering, University of Salerno, via Giovanni Paolo II, 84084 Fisciano (SA), Italy

Vincenzo Belgiorno
Sanitary Environmental Engineering Division (SEED), Department of Civil Engineering, University of Salerno, via Giovanni Paolo II, 84084 Fisciano (SA), Italy

Federica Cucchiella
Department of Industrial and Information Engineering and Economics, University of L'Aquila, Via G. Gronchi 18, 67100 L'Aquila, Italy

Idiano D'Adamo
Department of Industrial and Information Engineering and Economics, University of L'Aquila, Via G. Gronchi 18, 67100 L'Aquila, Italy

Massimo Gastaldi
Department of Industrial and Information Engineering and Economics, University of L'Aquila, Via G. Gronchi 18, 67100 L'Aquila, Italy

Susanne Theuerl
Leibniz Institute for Agricultural Engineering and Bioeconomy, Max-Eyth-Allee 100, 14469 Potsdam, Germany

Johanna Klang
Leibniz Institute for Agricultural Engineering and Bioeconomy, Max-Eyth-Allee 100, 14469 Potsdam, Germany

Annette Prochnow
Leibniz Institute for Agricultural Engineering and Bioeconomy, Max-Eyth-Allee 100, 14469 Potsdam, Germany
Humboldt Universität zu Berlin, Albrecht Daniel Thaer Institute for Agricultural and Horticultural Sciences, Hinter der Reinhardtstr. 6–8, 10115 Berlin, Germany

M. S. Shah
Department of Petroleum and Mining Engineering, Jessore University of Science and Technology, Jessore 7408, Bangladesh

P. K. Halder
Department of Industrial and Production Engineering, Jessore University of Science and Technology, Jessore 7408, Bangladesh
School of Engineering, Royal Melbourne Institute of Technology University, Melbourne, VIC 3001, Australia

A. S. M. Shamsuzzaman
Department of Industrial and Production Engineering, Jessore University of Science and Technology, Jessore 7408, Bangladesh

M. S. Hossain
Department of Petroleum and Mining Engineering, Jessore University of Science and Technology, Jessore 7408, Bangladesh

S. K. Pal
Department of Petroleum and Mining Engineering, Jessore University of Science and Technology, Jessore 7408, Bangladesh

E. Sarker
Hajee Mohammad Danesh Science and Technology University, Dinajpur 5200, Bangladesh

LIST OF FIGURES

PART II

Chapter 6

Chapter 8

LIST OF FIGURES

PART II

Chapter 6

Chapter 8

LIST OF TABLES

LIST OF ABBREVIATIONS

BES	bioelectrochemical system
CoA	coenzyme A
CoB	coenzyme B
CoM	coenzyme M
CODH	CO dehydrogenase
CSTRs	continuously stirred tank reactors
DET	direct electron transfer
DIET	direct interspecies electron transfer
Ech	energy converting hydrogenase
Fd	ferredoxin
Fdox	oxidized ferredoxin
Fdred	reduced ferredoxin
Fdh	formate dehydrogenase
Fpo	F420H2 dehydrogenase
Hbd	3-hydroxybutyryl-CoA dehydrogenase
Hdr	heterodisulfide reductase
H4MTP	tetrahydromethanopterin
H4SPT	tetrahydrosarcinapterin
IET	indirect electron transfer
MES	microbial electrosynthesis
MET	mediated electron transfer
MFR	methanofuran
MPh	methanophenazine
MPhH2	reduced methanophenazine
Mtr	methyl-H4MPT:coenzyme M methyltransferase
Mvh-Hdr	(methyl viologen-reducing) hydrogenase and heterodisulfide reductase
PHB	polyhydroxybutyric acid
SHE	standard hydrogen electrode

SLP	substrate level phosphorylation
UASB	upflow anaerobic sludge blanket
Vho	(F420 non-reducing) hydrogenase

PREFACE

Sustainable development has become a household term in today's day and age where efficient planning of resources, their utilization and economic efficiency have become the key drivers of a country's growth and development.

World is facing environmental problems ranging from global warming, acid rains, glaciers melting, pollution, resource exploitation, loss of non-renewable sources of energy, deforestation and so on and so forth.

This not only affects the environment but also the socio-economic status of a country.

Both developed and developing countries are struggling with their own version of environmental pollution.

- Decline in global mortality rate, longevity and population growth have all led to tremendous exploitation of natural resources while the rate of replenishment is rather slow.
- One of the key issues that has been globally agreed upon to be focused at from an environmental perspective is waste management, treatment and resource recovery.
- One of the most important concerns that developing countries have to deal with is the struggle to feed growing population and meet the fuel demands while also taking care of forest cover and keeping the environment clean.
- Rural development is another factor that needs development and improvement in the living conditions as otherwise because of low economic flexibility they will need to turn to the resources that need to be preserved.
- In the area of resource conservation, anaerobic digestion and biogas production can become highly beneficial.
- Biogas- waste to fuel covers various aspects of biogas production technology. It discusses at length about the anaerobic digestion process that results in biogas production and also the play of physiological parameters in this process.

- Biogas can be upgraded to almost pure methane and the product thus obtained is called as biomethane.

Recently, a lot of research has been focused toward finding cost effective methods for biogas upgradation and it has been discussed in detail in the book.

Globally, waste disposal and renewable energy crisis have become burning issues, especially, in the developing countries.

It has become imperative to have honest conversations about tackling these issues and focusing research towards sustainable solutions to these problems.

Biogas technology has twofold benefits in this scenario by providing a sustainable solution for disposing organic waste and also converting it into biogas and fertilizer. Scientists are also looking at novel substrates for anaerobic digestion in order to increase the scope of this technology.

These include paper and pulp industry, agriculture sector, municipal sewage waste, food waste, slaughterhouse waste, animal manure and so on. Many EU nations have become world leaders in biogas production and using it as fuel for various purposes while there are others which are developing their AD sector rapidly. However, a large part of the world still remains untouched and needs to strengthen their AD sector.

All of the aforementioned aspects of biogas production and implementation have been discussed in this book and hopefully it will provide the readers a fresh perspective on anaerobic digestion and biogas.

The book is divided into two parts. Part I discusses the basic aspects of biogas technology beginning from the basics of anaerobic digestion to various types of wastes that can be used for biogas production. Various methods involved in purifying and upgrading biogas to biomethane are also discussed.

Algal biomass has emerged as a novel feedstock for biogas production and has a huge potential for commercial purposed for which the fourth chapter is dedicated. Finally, the last chapter discusses various positive and negative impacts of AD on the society and the world.

Part II of the book includes various peer reviewed articles that discuss the applications of biogas technology in different countries. These articles will help the reader understand the current scenario of application of AD and biogas based electricity and fuel in households and for transportation.

The first article discusses recent trends in the field of AD and biogas followed by development of biogas technology in rural areas. The articles also concern involvement of methanogens in the process of AD and the possibility of AD without biogas.

This is followed by discussion of various pretreatment strategies for recalcitrant feedstocks like lignocellulosic material, opportunities and challenges for biogas combined with bioethanol, application of biomethane in the transport section and closing with microbiome involved in AD and using biogas as bio-CNG in Bangladesh.

Hope the readers get an insight into the biogas and biomethane industry at the basic and at an advanced level.

PART – I
ANAEROBIC DIGESTION AND BIOGAS OVERVIEW

1

Introduction to Anaerobic Digestion (AD)

Navodita Bhatnagar

Biogas is a product of a naturally occurring process in which microorganisms break down organic matter into simpler products. This happens in the absence of oxygen and the process is called as anaerobic digestion (AD). Biogas is a mixture of different gases in various amounts but majority of it is constituted by methane which is a fuel gas and can be used for various purposes. Biogas also consists of up to 30% of carbon dioxide and a trace amounts of gases like hydrogen sulphide, nitrous oxide, water vapor and so on. Alexander Volta first discovered, in 1776, the presence of flammable gas - methane above marsh lands which may have resulted from the dead and decaying matter. This was further consolidated in 1804, when Dalton discovered the molecular formula of methane molecule.

INTRODUCTION

First AD plant that was built was in Bombay during the British India in 1859 and it functioned on the waste coming from a hospital.

- It then reached to Britain in 1859 where waste from a sewage treatment plant was used and the gas generated was used for lighting street-lights. However, the underlying biology behind AD was still unknown.
- First large- scale AD plant was built in Birmingham in 1911 for treatment of the sewage waste in the city. It was not until the 1900s that the microbiological aspects of AD were properly understood.

- Buswell laid out the basics behind AD in 1930s, highlighting the microbiology involved and also discovered what is famously known as the Buswell equation now.

World war II saw an increased emphasis on production of biogas from waste, mainly agricultural, as many different European nations faced an energy crisis owing to the war. Since then, EU nations like Germany, France, Sweden and Denmark have pioneered in biogas based renewable energy.

This development was not only confined to Europe, a lot of attention was gained in biogas technologies form south Asian countries, especially, India and China. While Europe focuses on a centralized approach, countries like India and China employed the decentralized approach resulting in millions of small-scale AD reactors that now stand throughout the countries.

Readers might be familiar with the terms compressed natural gas (CNG) and liquefied petroleum gas (LPG). These are the two different forms in which natural gas is used in transportation sector and household purposes.

Natural gas forms after millions of years just like all other petroleum products and its reserves are also limited in number making it a non-renewable resource. But the interesting fact here is that both CNG and LPG have methane as the main constituent and also the fuel that burns to produce energy.

Biogas production takes a few days and can be scaled up or scaled down as per requirement. Biogas also has the same constituent gas that makes it a fuel which is methane just like CNG and LPG. Moreover, biogas is renewable and as long as there exists biomass, biogas can be produced repeatedly.

Below is a list of recent developments which have led to the growing popularity that has been gained by AD in the recent times

- Not too long ago, AD was used to treat sewage waste and now it has grown into a popular technology in terms of the understanding of the processes and various methods to enhance biogas production by it.
- Environmental benefits of AD have attracted interest from both the developed and the developing world thus leading to installation of number of AD plants throughout the world. These may either be centralized supplying energy to the power grid or decentralized for use within the farm.
- Not only that, AD has also gained popularity by providing more

economic treatment for different types of waste like municipal waste, agricultural and industrial waste along with production of energy.

- This is much more efficient than chemical or thermal processes requiring input of strong chemicals or heat.
- Moreover, the liquid digestate which is a by-product of this process is hugely beneficial as an organic fertilizer for agricultural crops.
- This can help recycle nutrients by putting them back into the soil through the digestate which works better than raw manure as the nutrients are more readily available for uptake by the plants.
- Due to anaerobic conditions, many pathogens found in organic waste do not survive the process that in turn adds on another advantage to the already beneficial AD process.

Despite the many advantages, AD has still not grown to its full capacity in many countries, especially those where agriculture is one of the major industries for the economy. This could be due to a lot of issues including the climate in these countries, operational costs, government regulations and subsidies for renewable energy.

1.1 ANAEROBIC DIGESTION PROCESS

Biogas is formed during a natural process which is called as anaerobic digestion (AD). It is carried out by specialized microorganisms through a series of steps namely- hydrolysis, acidogenesis, acetogenesis and methanogenesis.

Steps involved in the process of AD are described in the flow diagram in figure 1.

The final product is production of biogas which is a mixture of methane (50-70%), carbon dioxide (20-30%) and other gases in trace quantities (water vapor, hydrogen sulfide, nitrous oxide hydrogen and nitrogen).

AD is a biological process and thus is dependent on many physiological factors required for the optimal functioning of the microbes and their enzymes.

These factors include temperature, pH, alkalinity, agitation, salt concentration and inhibitions.

1.1.1 Hydrolysis

In this step, microbes break down complex polymers into smaller compounds which are more accessible to the microbes for further breakdown. Hydrolytic bacteria convert carbohydrates, fats and protein into sugars, long chain fatty acids and amino acids respectively with the help of their enzymes.

It is difficult for the hydrolytic bacteria to break down lignocellulosic materials due to their structure and thus, may require additional enzymes to be added externally to facilitate the hydrolysis process.

As a result, hydrolysis can often be the rate limiting step as all the subsequent steps are dependent on the products of hydrolysis.

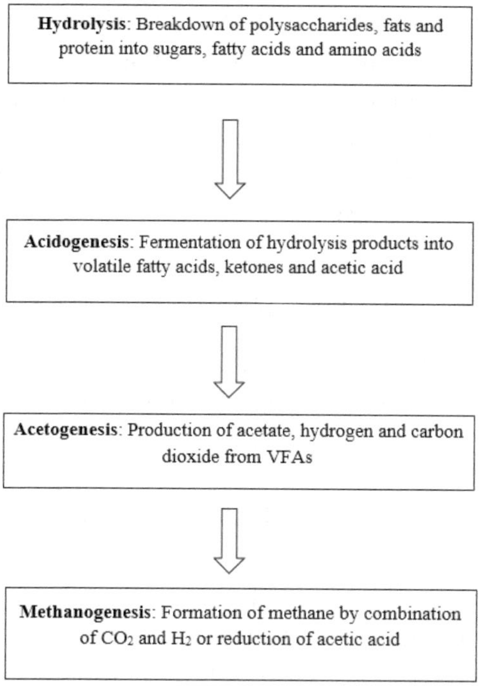

Figure 1: Steps involved in anaerobic digestion.

1.1.2 Acidogenesis

The products of hydrolysis are further broken down into short chain fatty acids like acetic acid, propionic or butyric acid by the fermentative or acidogenic bacteria. These acids are also called as volatile fatty acids or

VFAs and are produced by bacteria like Lactobacillus, Clostridium and Streptococcus.

Along with VFAs, small amounts of hydrogen gas, carbon dioxide and alcohols may also be produced during acidogenesis.

1.1.3 Acetogenesis

As the name suggests, acetogenesis involves production of acetate. This step is carried out by acetogenic bacteria like Syntrophobacter and Syntrophomonas.

The VFAs produced in the previous step are converted into acetate along with the production of hydrogen gas.

These are strictly anerobic bacteria which function through the acetyl CoA pathway and are sensitive to oxygen.

The hydrogen generated in this step is consumed during the methanogenesis step thus maintaining the partial pressure of hydrogen in the reaction and not letting it overshoot as that may be inhibitory to the acetogenic microbes (Caruso et al., 2019)

1.1.4 Methanogenesis

In the final step, methanogenic bacteria and archaea like Methanobacte and methanococcus produce methane.

The methanogenic archaea are obligatory anaerobic and are thus very sensitive to the presence of oxygen.

There are two pathways through which methane production occurs in AD which are the hydrogenotrophic and Acetoclastic pathways.

In hydrogenotrophic pathway, CO_2 and H_2 combine to form methane by the hydrogenotrophic methanogens.

While Acetoclastic methanogenesis occurs by reduction of acetate into methane.

Approximately two thirds of the methane is formed via Acetoclastic methanogenesis while the remaining one-third is formed by hydrogenotrophic methanogenesis.

1.2 FACTORS AFFECTING AD

It is important to understand the process of AD and the factors that affect it to truly understand the different aspects of biogas production.

Microorganisms involved in AD are affected by various physiological factors that are discussed below:

1.2.1 pH

pH is a very important parameter in determining the rate of chemical or biological processes. Most living organisms need a near neutral pH for their optimum functioning.

The same is true for AD where the process has a narrow range of optimum pH. Researchers have reported the process to be stable at other pH ranges but the range of 6.3-7.8 has been reported to be the most ideal for AD (Sarker et al., 2019).

However, pH does change throughout the process and at different stages of AD. For instance, during the hydrolysis stage, optimum pH is between 5-6 as the acidogenic bacteria involved in this stage require slightly acidic pH.

Moreover, the pH goes up to 6.5-8 during the methanogenesis stage.

It should also be kept in mind that during the later stages of the process pH needs to be above 6, otherwise, it may lead to accumulation of volatile fatty acids thereby further dropping the pH and thus inhibiting the methanogenic population.

Temperature also affects the pH by affecting solubility of carbon dioxide in the reaction, thus, at higher temperature the pH will be higher since the carbon dioxide solubility is low and lower formation of carbonic acid.

There may also be a possibility of increased pH due to ammonia production as a result of protein degradation, thus, it is essential to maintain a balance between the two which may be achieved by introducing a buffer like bicarbonate.

Hydrolysis and acidogenesis are less sensitive to pH and can operate within a range of 5.5-6.5, however they may be inhibited in extremely acidic or extremely basic pH range.

In commercial plants, automatic pH controllers can maintain the pH of the reactor in the desired pH range while in laboratory systems buffers are used, for instance, sodium carbonate and sodium bicarbonate buffers.

Indirectly, pH can also be controlled by other process parameters such as hydraulic retention time (HRT) or organic loading rate (OLR).

1.2.2 Temperature

Temperature is another key physiological factor in any biological process. It is important for the microbes to not only survive but also be able to multiply and produce enzymes necessary for the process of AD at any given temperature.

It also plays an important role in maintaining kinetic stability and maximise methane yield in the process. Most commonly, mesophilic temperature is considered ideal for the anaerobic microorganisms.

Even a fluctuation of a few degrees in the temperature can result in significant variation in the methane yield as the microbes that have been acclimatized to a temperature will need to readjust to a new temperature which may also result in changes in their structure.

Different microbial populations, for instance methanogens and acidogenic bacteria, have different sensitivities towards temperature fluctuation.

It was reported that thermophilic temperature resulted in increased rate of hydrolysis of recalcitrant compounds thus resulting in higher biogas yield (Sarker et al., 2019).

Temperatures lower than the mesophilic range may slow down the process by decreasing microbial growth thereby resulting in lower biogas production levels.

It is not impossible to carry out AD at the psychrophilic range (4-25 °C), but it is certainly more difficult and may require additional resources to get optimum microbial growth and biogas productivity.

However, if the temperature is too high (above 40 °C), there may be an increase in the rate of degradation of substrate but also an increased production of volatiles that may have an inhibitory effect on methanogenic microorganisms.

Another reason for not using thermophilic temperature is the possible denaturation of bacterial enzymes, that are pivotal to this process, at a very high temperature.

Most commonly, temperature between 35-37 °C are preferred in lab scale and commercial AD plants (Meegoda et al., 2018).

Thermophilic temperature also allows the HRT to be shorter thus helping reduce the size of the reactor, but the process is generally unstable and more sensitive to inhibition as compared to mesophilic AD.

1.2.3 Mixing

Mixing is required to enhance the contact between the biomass and substrate constantly, provide uniform temperature distribution throughout the reactor and avoid formation of dead zones in the reactor.

It is also important to constantly stir the reaction mixture to avoid formation of scum layer and to maintain homogeneity throughout the reactor.

- It is important to maintain mixing speed however as too low speed may prove inefficient while a very high speed may lead to syntrophic relationships within the microbial communities to be broken.
- Mixing can be achieved through various different methods including mechanical stirrers, internal gas mixing or by purging the slurry back into the system continuously.
- In case of use of an external pump, large quantities of digestate may be pumped out and then back into the reactor from the top or the bottom.
- However, it should be made sure that the recirculation rate us high enough in order to achieve complete mixing. In mechanical blade assisted stirring systems, generally, low speed blades are used where the slurry is mixed with a rotating impeller.
- Gas mixing systems are a bit different from the mechanical ones wherein the gas is released into the digester and mixes the slurry through diffusion rather than physical mixing.

Mixing also reduced the incidence of foaming and sedimentation by either mechanical or hydraulic methods. It also improves mass transfer between the gas and the liquid phases thus helping in increasing gas yields.

It should, however, be note that the intensity and speed of mixing are very important in determining the overall yields as high speed stirring may lead to formation of dead zones. Another consideration while designing an AD reactor is the cost effects.

Mixing may improve the yield to an extent but if the input costs and energy in employing mechanical stirrers is higher than the output generated, then it may be reconsidered.

New technologies like gas bubbling or re-purging may be more efficient for certain feedstocks as they do not have the same cost and energy requirements as mechanical stirring.

1.2.4 Substrate Characteristics

Substrate characterisation is an important parameter in determining overall design of the operation strategy for AD.

It is critical to identify the physical and chemical characteristics of the feedstock that will eventually be degraded via AD as it would help reveal information regarding the reactor design, process stability and operating parameters.

These characteristics include

- total and volatile solids content of the substrate,
- moisture content,
- nutrients and organic matter,
- lipid,
- protein,
- ammonia,
- particle size,
- volatile fatty acids,
- carbohydrates,
- lignin and
- biodegradability.

Generally, a substrate with high volatile solids (VS) is preferred as VS represents the organic fraction of any material, thus, higher VS may result in higher biogas yields.

Organic waste with high fat content is a desirable feedstock for AD, however, studies have reported problems like low hydrolysis rate.

Methane yields are generally higher for high lipid substrates, but they also have a quite long retention time. High carbohydrate and protein materials are faster to produce methane but also pose a risk of process failure due to accumulation of VFAs and ammonia in the system (Mir et al., 2016)

Particle size is mainly important in the hydrolysis step as smaller particle size facilitates a bigger surface area for contact between the substrate and biomass/ enzyme reaction. Substrate complexity and availability also affects, directly or indirectly, the rate of AD.

It may also be estimated from the source of the waste, animal manure for instance is generally rich in protein while waste from paper industry would be carbohydrate rich.

High ammonia and sulfide levels in waste such as slaughterhouse waste or MSW which has high protein and fats can be toxic to the methanogens.

High lipid waste can also lead to problems like sludge floating, blockage of pipeline and transport, and also clogging of the gas collector systems.

Since lipid rich waste also promises high methane production, these issues can be overcome by co-digesting such waste with high carbohydrate waste like paper waste or rice husk.

Problems posed by high protein feedstock may be overcome by various approaches as well which include reduction in the particle size, longer HRT, ammonia stripping, mesophilic temperature range and using buffer to maintain the reactor pH (Sarker et al., 2019).

1.2.5 Carbon to Nitrogen Ratio

Substrate characteristics are also indicative of the carbon to nitrogen (C/N) ratio which yet another crucial parameter in designing of an AD process. C/N ratio is the ratio of organic carbon to the nitrogen present in any particular substrate.

Studies have shown that a C/N ratio between 25-30 is considered to be ideal for the AD process.

If the ratio is higher than 30, then there are chances of ammonia accumulation and consequent inhibition of the microbes as ammonia is toxic to the methanogens.

However, in case of lower C/N ratios below 25, the microbes may experience a deficiency of Nitrogen and a decreased pH due to accumulation of VFAs.

C/N ratio also represents a macronutrient balance failing which the process of AD may be inhibited.

Co-digestion of different feedstocks is one such strategy that may be used to improve C/N ratios of a substrate.

For instance, animal manure generally has a low C/N ratio which can be improves by co-digesting it with plant biomass.

Methane production can increase by co-digestion of animal manure with energy crops, agricultural waste and food waste by up to 65% (Sarker et al., 2019).

AD is a process of synergistic interaction of microbes forming a consortium and carrying out a series of steps.

Due to lack of proper buffering, each of these steps can be inhibited as they all need a specific pH range, for instance, methanogenesis occurs at around neutral to basic pH and if there is accumulation of VFAs thus lowering the pH during acidogenesis, the final step of methanogenesis would be inhibited.

Thus, will result in lower methane yields and thus lower process efficiency thus making pH balance important.

This is where C/N ratio becomes very important as higher ratio may inhibit methanogenesis while a very low C/N ratio may lead to ammonia accumulation thus increasing the pH and leading to low methane yields.

Co-digestion can improve the C/N ratio and provide macronutrient balance (Mir et al., 2016).

1.2.6 Organic Loading Rate (OLR)

Organic loading rate or OLR is a very important parameter for both batch and continuously fed AD systems.

It is the amount of substrate fed into the reactor per unit volume of the inoculum and generally expressed in kg/m^3.

If the OLR is increased, biogas and methane yield will increase up to an extent beyond which the relation won't remain linear. This happens as a result of disruption of the equilibrium between the substrate and the inoculum.

While if the OLR is too low it may lead to starvation of the microbes, a very high OLR may lead to inhibition of microbial activity.

High OLR may lead to increased rate of hydrolysis as compared to the rate of methanogenesis which may result in increased VFA production that will cause failure of the reactor due to low pH.

Thus, it is important to work out the range of OLR within which there are least chances of process failure (Meegoda et al., 2018).

High OLRs are generally preferred as they enable treating higher amounts of waste within the same reactor size, reduce the HRT and thus make the process more cost effective.

However, low HRT can also lead to VFA accumulation, washout of the microbial population, and lower reactor stability.

Thus, research has been invested in optimizing OLR best suited to produce maximum biogas yield in a given reactor configuration.

These strategies involve pretreatment of the substrate, co-digestion, adding trace elements and other additives to maintain micronutrient balance and also operating at higher temperature to allow higher OLR and shorter HRT(Mir et al., 2016).

1.2.7 Hydraulic Retention Time (HRT)

Hydraulic retention time or HRT is the product of the biological volume of a reactor and the flow rate. In a more general sense, retention time is indicative of the time for which the liquids continue to remain inside the reactor.

HRT is dependent on various factors including the physiological factors and microbial growth rate.

- OLR and HRT can be related in the way that at a higher OLR the corresponding HRT will be shorter.
- As a result of the short HRT, accumulation of VFAs may occur thus leading to inhibition of the process, however, shorter HRT is preferred to decrease input costs and to obtain a higher efficiency in general.
- Complex wastes like lignocellulosic or high fiber wastes may require longer HRT as compared to simpler easily degradable waste (Meegoda et al., 2018).

HRT also depends on the operating parameters like reactor volumes, temperature and, flow rate in case of a continuous system. Higher HRT is required for complex substrates like high lipid or lignocellulosic substrates, while those rich in carbohydrates do not require long HRT.

Also, at higher temperature, rate of hydrolysis is faster and thus shorter retention times are required as compared to mesophilic temperature range.

Optimal HRT lies between 10-25 days as in case of very short HRT there are chances of VFA build up while for very long HRT reactor size requirement would increase.

Longer HRT may also be required if the reactor is operating in the psychrophilic temperature range as the methanogens grow slower than acidogens in this case (Sarker et al., 2019) .

1.3 TYPES OF ANAEROBIC REACTORS

Anaerobic reactors can operate on many different configurations based on the process requirements, type of feedstock and physiochemical conditions.

Based on the mode of feeding the substrate to the reactor, there are three different modes of operation namely batch, fed batch and continuous.

The moisture content may also vary and thus gives us two categories of AD reactors namely wet and dry. Temperature of operation is also different for different processes thus three possible modes of operation namely mesophilic, psychrophilic and thermophilic.

1.3.1 Wet Reactors

A wet digestion process is one in which the dry matter concentrations are less than 15% (Lissens et al., 2001). In this instance, substrates are conditioned to a desired consistency using water (typically 10% dry matter) prior to addition into the system.

In addition, "dry" substrates require constant mixing in order to prevent precipitation of the suspended solids in the feedstock. In the case of low dry matter waste such as food waste or farm waste in the form of slurry, processing in wet digesters is desirable.

Wet digestion is advantageous in the sense that it allows dilution of inhibitory substances by water and requires less complicated machinery although the consumption of water and energy for heating for the process remains high.

1.3.2 Dry Reactors

Dry digesters operate on feedstock which has a total solid content of 25-40%. In such systems, there is a lower requirement of fresh water as compared to wet AD process and the consistency of the feed is much thick.

As a result, these systems require constant mixing to avoid substrate settling and the maintenance of flow throughout the reactor. Dry systems are suited for feed stocks with high dry matter, for example lawn cuttings or energy crops.

Dry digesters require less complicated pre-treatment compared to wet systems and also require less fresh water, however, they require more sophisticated machinery including, impellors, propellers, agitators and baffles, to facilitate appropriate mixing of solids.

1.3.3 Batch Reactors

Based on the mode of feeding, batch and continuous are the two types of configurations that are used.

In batch mode, the reactor is essentially a closed system in which substrate and nutrients are fed once at the start of the process with growth occurring according to the logarithmic growth curve of bacteria (Fig 2a).

Once all the nutrients are exhausted, the microbial population enters death phase and the effluent is collected by opening the system.

1.3.4 Continuous Reactors

In continuous systems, the feedstock is fed continuously and an equal amount of digestate is removed from the system via displacement (Fig 2b).

This is essentially an open system and the bacterial growth in this configuration doesn't cease as long as the reactor operates.

Continuous reactors are more suited to commercial scale applications such as AD of food waste or AD of municipal waste where the influx of waste is done on a regular basis and large volumes of waste need to be dealt with on a day to day basis.

(a)

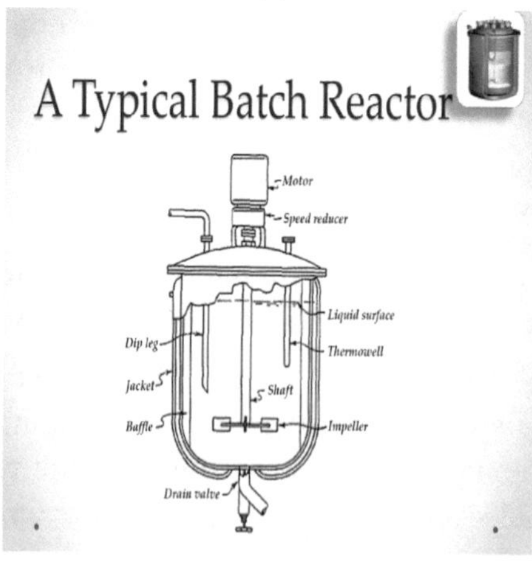

(b)

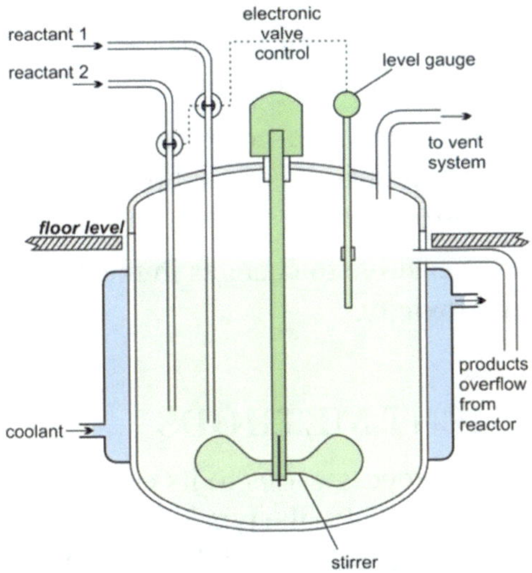

Figure 2: (a) Batch reactor configuration (b) continuously stirred tank reactor (source: google images).

1.3.5 Psychrophilic Reactors

Psychrophilic reactors operate at low temperatures, ranging from 5-20°C. Since the temperature is low, the SRTs are generally higher in psychrophilic anaerobic digestion (PAD) systems.

Currently, there are no industrial reactors operating in the psychrophilic temperature ranges.

However, studies have proven that psychrophilic anaerobic reactors can give comparable yields when compared to mesophilic or thermophilic systems.

1.3.6 Mesophilic Reactors

Mesophilic AD systems operate in the moderate temperature ranges typically from 30-35°C.

It is often advantageous over psychrophilic and thermophilic ranges as it is more economical to operate the reactor in a mesophilic temperature range owing to lower energy requirements.

Such reactors are used for treatment of municipal and industrial wastewater.

1.3.7 Thermophilic Reactors

In the thermophilic range, the reactor operates at higher temperatures of up to 50-60°C which results faster digestion process and pathogen reduction.

Unfortunately, a major drawback of this operating temperature is the low number of naturally occurring thermophilic methane forming bacteria coupled with their slow growth rates.

They are also more sensitive to changes in the digester temperature and also changes in pH and toxicity.

1.4 PRETREATMENT METHODS

The feedstock for an AD process varies in its composition depending upon its source, its solid content and also the type of organic material present in it. Some wastes are highly biodegradable while some others are not.

While some of the feedstocks are easily digestible, some are hard to breakdown, also known as recalcitrant, and thus need pretreatment to enhance their hydrolysis and thus further breakdown during the AD process.

Crops residues are one of the most readily available biomasses around the world and these include cereal husk, straw and so on. In principle, wastes rich in lignin and cellulose, for e.g. wood, grass cuttings and straw are not easily hydrolysed which makes the AD process inefficient.

Enhancing the hydrolysis process during AD can make it more economically favorable and lead to reduced HRT.

The straw or the husk is rich in lignocellulose which is itself composed of lignin, cellulose and hemicellulose.

Hemicellulose surrounds the cellulosic skeleton and lignin is present in this structure for protection bonded through crosslinks and covalent bonds. As a result, the polysaccharides become quite inaccessible and this type of feedstock is called as lignocellulosic feedstock.

While it's not impossible to unlink these polysaccharide-lignin linkages, it is definitely difficult to degrade them fully by AD alone and thus pretreatment offers itself as a way of improving this process.

Pretreatment is aimed at disrupting the structure of lignocellulosic bonds to expose the underlying cellulose fiber thus facilitating its participation in the AD process. Thus, can be done through physical, chemical, biological or radiation treatments of the feedstock either individually or in combination with each other.

Studies testing various pretreatment methods that can be applied to disintegrate the feedstock and make it more bioavailable for microbial degradation have been done.

With recent advances in research on AD, variety of pretreatment methods are now available including thermal, chemical, mechanical and biological treatment methods.

Pretreatment can increase biodegradability of substrates by improving their bioavailability for the microbial enzymes involved in AD.

In studies focusing on highly lignified substrates i.e. farm waste and grass silage which are difficult to degrade. Hydrolysis being the rate limiting step, removal of lignin and hemicelluloses in such wastes can be considerably enhanced through various pretreatment methods.

Besides the lignocellulosic substrates, there are other types of feedstocks which may be rich in fat or fibrous which need to be treated before AD in order to enhance the rate of hydrolysis which is usually the rate limiting step.

1.4.1 Physical Pretreatment

Disintegration of substrate which doesn't involve use of microbes or chemicals but instead the use of physical forces like mechanical or thermal are clubbed under physical pretreatment methods.

Mechanical method involves disintegration of solid particles via grinding/ crushing to make the polymeric units break down into smaller easily accessible oligomers and also increase the surface area of the substrate particles for better contact with the microbes.

Milling is a commonly used mechanical method that includes collioid mill, fibrillator and dissolver for wet materials and roller mill, hammer mill for dry materials. Wet milling is preferred over dry milling as it consumes less energy and also possesses high pulverization properties.

Thermal pretreatment is a routinely used method at industrial scale that includes exposing the substrate to a high temperature generally ranging from 50-200°C and also for various time intervals.

Higher temperatures lead to pathogen removal and reduction in viscosity of the digestate which are favorable for the subsequent stages of the process.

A very well established thermal treatment method is steam explosion which is essentially heating the susbtrate with saturated steam at high pressure for a short span of time followed by sudden reduction of pressure that leads to explosive decompression of the substrate.

1.4.2 Chemical Pretreatmetn

Chemical treatment involves use of string chemicals like acids, bases or oxidizing agents to breakdown complex polymers into simpler units.

Treatment with alkalis involves use of chemicals like sodium hydroxide, calcium hydroxide, or ammonia which help removing lignin and hemicellulose from the substrate thereby increasing availability of substrate to enzyme.

It is seen that alkali treatment is better for agricultural waste instead of woody substrates. It is also observed that alkali pretreatment is more effective than acid treatment or oxidative reagents as it breaks the ester bonds in lignin, hemicellulose and cellulose better than the other chemicals available.

Acid treatment is quite effective in hydrolyzing hemicellulose and lignin but it could also result in formation of inhibitory by-products like furfural and hydroxymethylfurfural.

It may also lead to loss of fermentable sugar and demands neutralization of acidic conditions before moving forward to the process of AD thus making acid treatment less suitable form of chemical pretreatment.

Besides the above mentioned, oxidizing agents can be used too, ozonation being one of them. Ozonation is preferable as it does not leave chemical residues or cause an increase in salt concentration as compared to other methods. Besides, it also indirectly leads to pathogen removal by disinfection.

1.4.3 Biological: Enzyme Pretreatment

Biological treatment uses microbes or enzymes isolated from them for hydrolysis of polymers aerobically as well as anaerobically using fungi, microbial consortia and isolated enzymes. Aerobic pretreatment involves composting or micro-aeration that aids in higher production of hydrolytic enzymes. Studies revealed that aerobic pretreatment led to higher VFA formation owing to increased activities of hydrolytic and acidogenic bacteria.

Fungal pretreatment relies on evaluation of fungi that have the ability to selectively degrade lignin and hemicellulose with minimal utilization of cellulose.

For fungal pretreatment, certain fungal species have been used including the brown, white or soft rot fungi out of which white was found to the most effective.

It was found suitable to carry out the treatment for two to eight weeks at mesophilic temperature for lignin rich substrates like leaves and grass. Figure 3 shows a schematic of pretreatment of lignocellulosic material.

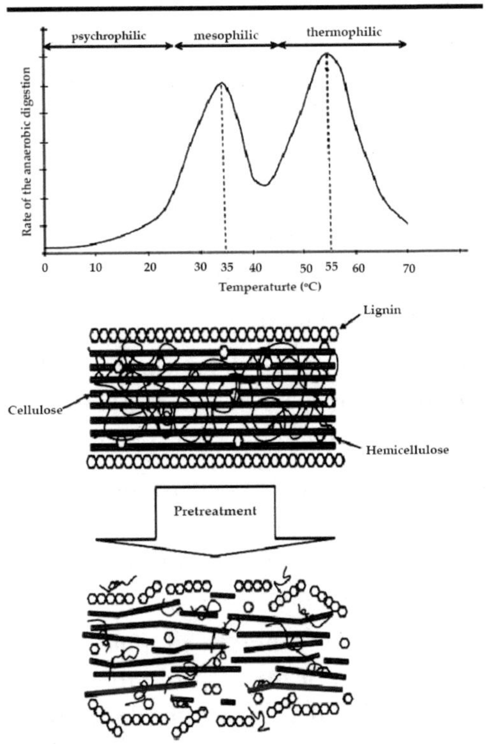

Figure 3: Pretreatment of lignocellulosic biomass (source: Armah and Armah 2017).

Studies have been done which focus on using consortia of microbes incorporating microbes screened from the environment for treatment of substrates.

Unlike fungal treatment, here the target is cellulose and hemicellulose instead of lignin.

This method was found to be suitable for cellulose rich substrates like corn straw, cassava residues and manure biofibres.

Enzymatic hydrolysis is more competitive than other methods as it consumes low water and energy, offers lower costs of waste utilization and also avoids the problems associated with corrosion of equipment.

The enzymes used are largely dependent on the composition of the different substrates and include cellulases, hemicellulases, proteases, lipases and amylases.

It was reported that solid-state fermentation of orange peels by fungal strains of Sporotrichum, Aspergillus, Fusarium and Penicillum enhanced substrate availability along with reduction in the level of antimicrobial substances.

Biological pretreatment offers advantages over other treatment methods in terms of lower energy requirement, absence of chemicals and carrying out the process under mild environmental conditions.

The major drawback, however, is that the rate of reaction is quite low when compared with rest of the methods.

A very important factor to consider here is the loss of carbohydrates during the treatment which must be minimized and the removal efficiency of lignin which must be maximized in order to achieve higher biogas yields.

Enzyme treatment is also a more eco-friendly approach as it does not involve the use of toxic chemicals and thus the residues after the process can be released into the environment without requirement of sophisticated technologies to remove any chemicals.

Biological pretreatment may not need pure enzymes necessarily as fungal mash, bacterial isolates or even whole microorganisms can be used which will remove the additional steps required for purification and can thus be cost effective as compared to using pure enzyme formulations.

1.5 BIOMETHANE POTENTIAL ASSAYS

A critical parameter to evaluate an AD process is measuring the amount of biogas and methane being produced by different substrates.

In order to scale up a process under specific physio chemical conditions, use of AD additives, and a particular or combination of substrates, it is important to measure the amount of biogas formed under each condition at laboratory scale first.

Biochemical Methane Potential tests (BMP) are intended to determine the anaerobic degradability of various substrates.

They provide a simple, repeatable and inexpensive method for prediction of the methane and biogas productivities in order to design a large-scale anaerobic reactor for biogas production.

They can help determine the amount of organic carbon a substrate contains that is essentially converted to methane with minimal requirement of labour and establishment costs (Moody et al., 2009).

There are various protocols for variety of substrates that are being used by researchers and they all differ in terms of experimental design, physiochemical conditions and equipment used.

Thus, there is no universal protocol that works for all different kind of substrates and protocols from existing literature can be modified as suited.

Nevertheless, BMP assays are used widely for determining potential efficiency of an anaerobic process and measuring concentration of organics in the waste.

Angelidaki et al., 2009 proposed an experimental guideline for assessment of biodegradability of any defined or undefined material to produce methane in order to establish a reliable and repeatable BMP assay protocol (fig 3).

The focus of this group in their protocol was on solid organic biowaste, agricultural waste, sludge and similar substrates.

Figure 4: Biomethane potential assay set up (source: Angelidaki et al., 2009).

1.6 PATHOGEN REMOVAL BY AD

Environmental pollution has become a serious concern as to massive industrialization and population growth along with advent of pesticides, chemical fertilizers and the curse of the century- plastics. This has polluted soil, water bodies leading to shortage in the supply of potable water.

- AD does not directly affect pollution but indirectly, it can provide resources such as organic fertilizer to replace harsh chemical fertilizers thus having less harmful substances that may leach into the soil and water bodies.
- Wastewater from different industries can also be treated using AD in conjunction with other feedstocks.
- Water and soil are polluted not only by chemical but also biological contaminants, like, pathogens that are harmful to human and animal health.
- There is a requirement for decontamination organic wastes are exceedingly loaded with pathogenic microorganisms which include bacteria, virus, parasites and fungi and can therefore be harmful for humans and animals.
- Decontamination of these pathogens becomes of utmost importance to avoid risking human and animal life and for this purpose, AD offers itself as an attractive economically efficient solution.

2

Waste to Biogas

Navodita Bhatnagar

It is not uncommon to read about the rising environmental concerns owing to the incessant use of fossil base fuels and the repercussions of relying heavily on non-renewable sources of energy. This not only leads to depletion of limited natural resources, but the fossil-based fuels are also not the best for the environment as they lead to present day problems like climate change and global warming. Due to an increased demand for renewable energy throughout the world and also the rising environmental concerns like global warming and climate change, global focus has shifted from fossil-based fuels to alternate energy sources.

INTRODUCTION

Recently, biomass derived energy has gained a lot of importance throughout the world as it is not only sustainable but also cost friendly, carbon neutral, greenhouse gas neutral and easily available in abundance (fig. 2). Theoretically, any material that is organic in nature can be used for production of biogas by AD.

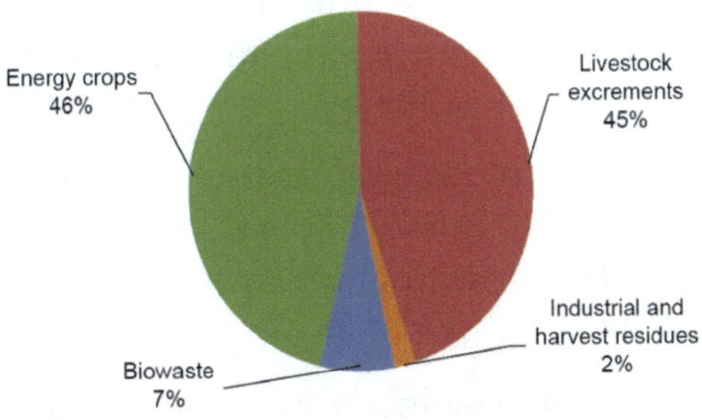

Figure 5: Typical feedstocks in biogas plants in 2010 in Germany (adapted from Achinas et al., 2017).

These include animal and plant waste, industrial waste, pulp and paper, slaughterhouse waste, municipal sewage waste and solid organic waste. Biogas is not only a renewable source of energy but is also carbon neutral. Besides, there are added advantages like reduction of waste going to landfills which further leads to reduction in landfill gas emissions.

Methane is the major component of landfill gas and is a much more potent GHG than carbon dioxide. Through AD, this methane is not only saved from being released into the atmosphere but is also harnessed to be used as a fuel.

Not only this but the by-product of this process which is liquid in nature and termed as digestate is also rich in inorganic nutrients. This makes the digestate a desirable choice for an organic fertilizer.

In the sections below, various types of feedstocks and their contribution to biogas production has been discussed.

2.1 FOOD WASTE

Food waste (FW) includes leftover food from canteens and restaurants, from household and industries and is mostly thrown away. FW also entails the food that may be contaminated or degraded thus making it unfit for human consumption. On an average, approximately 33% of the food that is produced across the world is wasted and needs sustainable disposal methods.

The food and Agricultural organization (FAO) reported that in 2009, approximately 32% of the food produced across the globe annually is either lost or wasted (Morales-Polo et al., 2018).

World hunger and malnourishment are not the only problems faced due to wastage of food. The waste thus generated in huge quantities also needs to be safely disposed and the most commonly used method to achieve that is by landfilling. This may not be the ideal strategy as FW is rich in nutrients and may lead to soil and water pollution.

AD is a promising technology to dispose FW and also generate energy from it. While designing the reactor for an AD operation, it is important to consider the physiochemical properties of the FW as it is important in determining the overall biogas and methane yield.

Traditionally, FW is sent to incineration plants which is not very desirable as this is a high moisture waste which can lead to production of dioxins if burnt along with low moisture wastes.

Furthermore, FW is rich in carbohydrates, proteins and lipids which makes it an attractive feedstock for microbes. Many countries have suffered from serious environmental risk due to improper disposal measures for FW.

As discussed above, landfilling and incineration are not the most environmentally sustainable solutions to the FW disposal issue as they contribute to either soil or air pollution which has also led many countries to ban these practices.

Figure 6: Food waste (FW).

(source:http://theconversation.com/campaigns-urging-us-to-care-more-about-food-waste-miss-the-point-80197)

There also stands the risk of spread of diseases in cases where food waste is used as an animal feed. Thus, there is a dire need for implementation of laws which discourage wastage of food and ensuring food safety.

Incineration of FW was commonly used for generation of energy, but it wasn't the most efficient use of FW as it is rich in organic and inorganics compounds which are destroyed during the combustion process.

There are many processes for which FW can be an ideal substrate which includes production or ethanol and hydrogen. However, AD of FW has been the most researched area in the recent times and that is because of the maximum efficiency of this process.

Table 1: Characteristics of food waste (adapted from (Morales-Polo et al., 2018)

Type and origin	TS%	VS%	Protein %	Lipids %	Carbohydrates %	C/N ratio
Household individual sorted materials						
Meat and bone			70-75	23-30	1	
Fish			75.6	20.2		
Eggs			35	32	2	
Dairy			25-35	20-45	53	
Fruit			4	2	83	
Vegetables			27	1.4	27	
Individual fractions from OFMSC						
Animal KW		33.3	54.4	35.7	9.9	
Vegetable KW		13.4	21.6	19.4	57.6	
Raw animal waste		38.6	59.8	27.2	13	
Raw vegetable waste		10.8	19.4	11.1	69.4	
OFMSC in Sweden		90.8	18.2	20	29.4	17.85
Fractions of OFMSC in Denmark						
Animal food waste	41	84	12	25	52	
Vegetable food waste	24	93	5	14	53	
Food waste		29.3	26.6	35	32.5	12.7
Food waste	18.1	17.1		23.3	61.9	
Food waste	23.1	21				24.5
Food waste for AD	30.9	26.35				13.2
Kitchen Waste	24	23.2	15	23.9	55.2	22.5
Food waste	23.2	21.7	29	6.5	13.7	
OFMSC	20	18	30	4.4	10.7	17
Food waste	16.7	15.3	24	1.4		11.5

Theoretical methane production potential from AD of FW can be quite high thus making it a popular choice for industrial scale AD plants in both developed and developing countries.

However, it is still difficult to operate these large scale reactors with high loading rates of FW as they are usually working at a low OLR. This leads to a reduced process efficiency and also a lower economic feasibility of the process.

Thus, research focusing on increasing the maximum OLR that can be loaded and enhancing gas production while trying to achieve this would be helpful in commercial feasibility of the process.

There are a few typical characteristics of FW which are challenging for the AD process including high salt, oil and protein concentration, low C/N ratio and lack of proper amounts of trace elements.

These characteristics may lead to problems like acidification, ammonia inhibition, trace element deficiency and salt inhibition (Table 1).

This is not good from a commercial aspect and thus overcoming these problems is an important aspect of making AD of FW an economically viable process. The sections below discuss the three main approaches to enhance this process and overcome the barriers as discussed.

2.1.1 Co-digestion

Anaerobic co-digestion s the process of digesting two or more different feedstocks within one system. FW is one such substrate that may mutually benefit from co-digestion with other substrates.

This can also assist in tackling the inhibition during long-term continuous operation of reactors for AD of FW. Different substrates that may be used for this purpose are cattle manure, crop waste, industrial waste and municipal waste.

Co-digestion helps maintain macronutrient and micronutrient balance. Food waste has a low C/N ratio which can be improved by co-digestion with high C/N ratio waste.

It was also reported that there was an improvement in the buffering capacity when FW was co-digested with cattle manure. There is also trace element supplementation when different wastes are co-digested and thus the performance of AD is enhanced as the microbes are provided with balanced micronutrients (Morales-Polo et al., 2018).

Furthermore, co-digestion of FW to maintain trace element balance could be a better alternative than adding synthetic compounds.

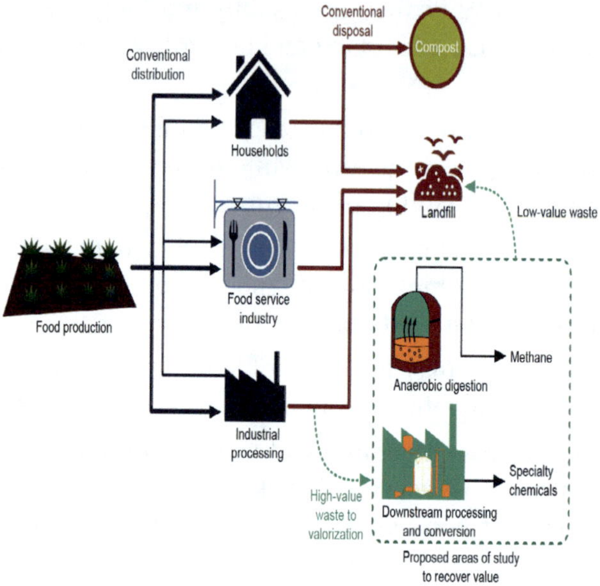

Figure 7: Simplified flow diagram of food from production to disposal (source: RedCorn et al., 2018).

Researchers have also tested adding landfill leachate (LFL) to FW to improve stability of the system and also a way to treat the LFL while harnessing the energy and nutrients it contains. LFL is a rich source of trace metals and thus when added to FW and simultaneously digested, it enhances microbial activity and thus gas yields.

2.1.2 Enzyme Treatment

Enzymes are biological catalysts that accelerate the rate of biological reactions. There are enzymes that can breakdown specific biological compounds and thus help in solubilization of recalcitrant organic waste like lignocellulosic waste or high protein high lipid waste.

FW is composed of different proportions of carbohydrates, lipids and proteins and they can be broken down into sugars, fatty acids and amino acids respectively by the specific enzymes (amylase, lipase and protease).

Approximately 60% of the FW is starch which can be converted into glucose thus making it an important part of the FW as the glucose thus generated during hydrolysis can be easily converted into methane.

Thus, enzymes that can assist in starch breakdown into glucose, like amylase and carbohydrases, can improve the hydrolysis and the methane yield.

Protease can improve the hydrolysis of protein molecules into amino acids by targeting the peptide bonds in the protein structure. The amino acids can be further broken down into ammonia and fatty acids which can then produce methane.

Moreover, proteins are also the sole nitrogen source for the methanogens, thus, any improvement in the hydrolysis of proteins would prove beneficial to the overall process efficiency and gas yield.

Lipids are essentially hydrocarbons and thus have a high theoretical potential to produce methane and biogas.

However, their hydrolysis is slow thus making them a rate limiting component causing lag for the rate of hydrolysis.

Thus, enzymes like lipases may come in quite useful for lipid rich waste which may otherwise be difficult to hydrolyse during AD.

2.1.3 Trace Element Addition

Trace elements or micronutrients are a common limiting factor in the AD of any organic waste. They are important in maintaining optimal microbial activity as they are involved in various metabolic pathways, enzyme co-factors, and thus their concentration in the reaction becomes very important.

Figure 8: Trace elements (source: https://www.fas.scot/article/trace-elements/).

It has also been reported that VFA accumulation and ammonia inhibition may also be tackled by appropriate micronutrient balance and supplementing trace metals.

There are five important trace metals that are required for metalloenzyme functioning and these are Fe, Ni, Co, Mo and Se.

These metalloenzymes are hydrogenases which require Fe and Ni, and formate dehydrogenase which requires Fe, Se and Mo.

Cobalt plays an important role in the function of methyl transferase while Fe is a component of ferredoxin which is a vital part of the electron transport system.

Fe is also found to be the most effective element when it comes to stabilization of FW. Similarly, Ni is a stimulatory in the production of biogas and is also a part of carbon monoxide dehydrogenase therefore making it important for the Acetoclastic and acetogenic reactions (Ye et al., 2018).

Usually, elements like Ca, Mg and K are present in abundance in FW but the ones discussed above are mostly not sufficient for AD. Waste from municipal treatment plants contain the trace elements in sufficient quantities thereby assuring the stability in the AD process.

Trace elements are needed by microbes not only to support enzyme activities but also as building blocks for general growth and maintenance.

- Even the addition of Ca and Mg which are generally available in FW in abundance, can improve methane production by preventing foaming in the reaction.
- Another important trace metal is Tungsten which is important in syntrophic methanogenesis and also in propionate degradation.
- It has also been found out recently that higher concentration of ammonia can be loaded in the reactor with the addition of Se and Co to maintain stability of the AD of FW.
- Trace metal addition depends on various parameters including characteristics of the feedstock, internal levels of trace metals in the FW, microbial community structure and most importantly the operational parameters.
- Moreover, the bioavailability of these trace metals is closely linked with their precipitation and solubilization in the AD system. Free ions can easily form ionic bonds with the anionic carbonate or phosphates thereby resulting in precipitation.

Chelating agents like EDTA may be used in such cases to enhance the bioavailability of the trace metals to the microbes in the system (Mao et al., 2015; Ye et al., 2018).

Studies have also shown that Se and Co play a significant role in propionate oxidation carried out by syntrophic hydrogen transfer pathway by the interspecies in the AD of FW when the concentration of ammonia was high.

In conclusion, it can be stated that addition of trace elements based on their inherent levels in the FW have the potential to improve microbial metalloenzyme function and can thus improve the stability and performance of the AD system at high OLR and ammonia concentrations.

2.2 MUNICIPAL SOLID WASTE

Rapid growth in population, economy and standard of living has also leads to an increase in the generation of municipal solid waste (MSW).

MSW consists of the everyday items that are discarded by the public comprising of food waste, paper, plastic, metals, medical waste and so on (figure 6).

A fraction of this waste is recyclable, another fraction is non-recyclable and non-biodegradable. However, the biodegradable fraction of MSW which typically consists of organic waste is termed as organic fraction of municipal solid waste or OFMSW which also stands out as a good substrate for AD.

Generally, the MSW is sent either to landfills or to incineration plants for its disposal.

While this is a common practice in most countries up until now, it is not the most environment friendly method of disposal.

Landfilling has become a huge problem due to leaching of minerals and toxic chemicals into the soil and also due to lack of space to accommodate the waste.

Landfill leachate which is the liquid produced from leaching of chemicals due to rain or other water sources is another type of waste and is usually sent to wastewater treatment plants for treatment.

Furthermore, landfill gas, which mainly consists of methane, is released into the environment thus leading to GHG emissions (Di Matteo et al., 2017).

Incineration on the other hand leads to production of harmful gases, moreover, the nutrients and organic matter that could have been recovered is destroyed during the process.

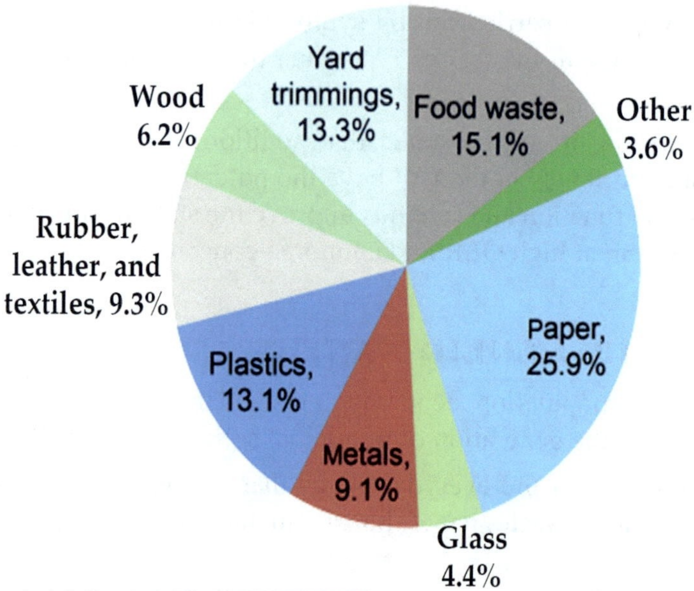

Figure 9: MSW composition generated annually in the U.S.

(source: http://css.umich.edu/factsheets/municipal-solid-waste-factsheet)

With a waste to energy approach on MSW, AD of the OFMSW is considered to be an environmentally sustainable approach while also focusing towards eliminating heavy dependency on fossil based fuels.

In 2011, approximately two billion tons of MSW was generated and it is estimated that this figure will only go up in the coming years. (Amoo and Fagbenle, 2013).

Management of MSW is done differently in different countries, especially based on whether they are developed or developing nations.

Usually, developed countries employ the sorting at the source approach as a result of which most of their waste is already sorted into recyclable, non-recyclable and organic waste even before it is received at the collection centers.

This adds an additional level of difficulty for most developing countries as the waste is generally unsorted and thus more difficult to be disposed. In most developing nations, the waste is disposed by either uncontrolled burning or dumping in open areas or landfills.

These practices contribute to pollution of neighboring water bodies and the land covered by it.

AD offers itself as a viable solution to solid waste management in terms of not only reduction of waste volume but also avoiding soil and water pollution, reducing load on landfills, helping curb the GHG emissions and most importantly, generating renewable energy.

MSW is a clean source of renewable energy by the US environmental pollution agency (USEPA) considerations. It was also concluded after several analysis that waste to energy conversions help reduce up to 36 million tons of GHG emissions annually (Amoo and Fagbenle, 2013).

2.3 ANIMAL WASTE (AW)

Animal farms are installed all over the world as almost 75% of the world population is non-vegetarian and relies on animal meat for their diet (Wikipedia).

As a result, huge quantities of animal waste are also produced by these farms thus putting the environment at a serious risk.

AW has also been found to be one of the major sources contributing to GHG emissions, almost 40% of the methane produced globally comes from agriculture and livestock by products and 18% from disposal of waste (Sorathiya et al., 2014).

Thus, it becomes imperative to search for viable alternatives for AW disposal and treatment. AW is also rich in nutrients and organic matter which may be recovered and reused in the form of manure and applied to agricultural lands for increasing soil fertility (figure 9).

Animal manure generally contains high amount of nitrogen and phosphorous which makes the disposal of this waste a serious concern. If such a waste is not treated properly before releasing it into the environment, it can lead to soil and water pollution of the neighboring areas.

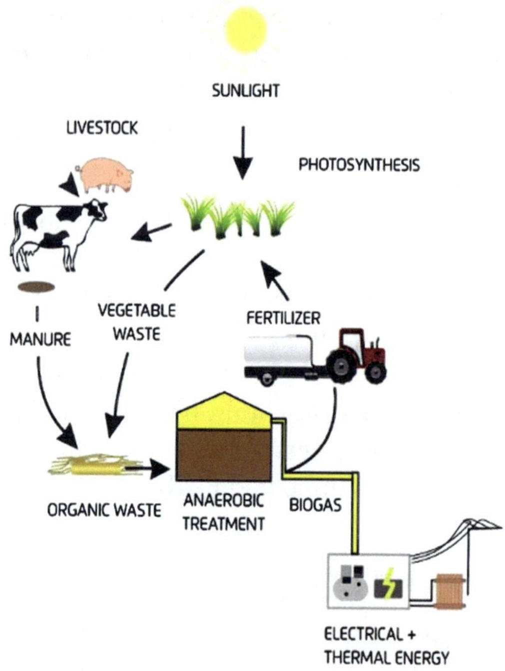

Figure 10: Biogas and fertilizer from animal waste- schematic.

(source: http://www.solution2pollution.com/green-bio-gas-animal.html)

vestock and poultry industries produce meat, milk and egg, and also generate large volumes of waste water and solid wastes that could be beneficial or harmful to the environment. The waste products which includes livestock or poultry excreta and associated feed losses, beddings, wash -water and other such waste materials represent a valuable resource that if used wisely, can replace significant amounts of inorganic fertilizers (Bouwman Contents lists available at Sjournals Journal homepage: www. Sjournals.com Review article I.P. Ogbuewu et al. / Scientific Journal of Review (2012) 1(2) 17-32 18 and Booij, 1988; Leha, 1998; Smith and Vanduk, 1987), conventional feedstuffs (El Boushy and Vander Poel, 2000) and cooking gas (Henuk, 2001), but may be a direct threat to human and animal health (Taiganides, 2002).

Animal wastes in the form of manures are valuable sources of nutrients and organic matter for use in the maintenance of soil fertility and crop production. Studies with animals have shown that 55–90% of the nitrogen and phosphorus content of animal feed is excreted in faeces and urine (Tamminga et al., 2000) normally used as manure. Poultry and swine manure collection in confinement feeding facilities have been recovered for re-feeding to beef cattle, dairy cattle and sheep (Bell, 2002) and have been found to present no serious health hazards to ruminants and poultry or posed any negative effects on meat, egg or milk quality (Mcllroy and Martz, 1978). Animal waste has been traditionally used in the production of biogas in Asia, particularly in tropical areas such as Indonesia, India and Vietnam (Henuk, 2001). However, careless dumping of livestock waste on farm lands and direct discharge to waterways and percolation to groundwater, usually in by-pass flow via cracks and fissures, is a great risk to human and animal health because livestock waste contain myriads of pathogens (Davies, 1997; Dizer et al., 1984), some of which may be zoonotic and can cause systemic or local infections (Dizer et al., 1984; Mackenzie et al., 1994; Davies, 1997; Stanley et al., 1998; Fischer et al., 2000; Cameron et al., 2000). Transmission of disease causing microbe is enhanced by mismanagement of animal waste and may be reduced by proper waste handling methods (Mackenzie et al., 1994). Highly contagious and pathogenic diseases, such as Foot and Mouth Disease and Swine Fever may spread with animal effluent through waterways and, when one farm is infected with the disease, farms downstream will be at considerable risk of infection (Cameron et al., 2000). Livestock waste produces ammonia that can be a potential pollutant causing serious eutrophication of rivers and lakes, characterized by a high concentration of nutrients that creates an ecological imbalance in the water system that support abnormally high levels of algae and aquatic plant growths (Burton and Turner, 2003; IAEA/FAO, 2008). This reduces oxygen levels in the water and has serious implications on the survival of aquatic organisms and, consequently, on food supply and biodiversity (IAEA/FAO, 2008). Livestock wastes are sources of malodours originating from livestock buildings, storage and field application of animal manures. The intensity of malodours is often unacceptable, especially for neighbours in surrounding residential areas. Globally, the concentration of the greenhouse gas methane (CH_4) in the atmosphere has increased by 45% since 1850 (Lelieveld et al., 1998). Increases in livestock production have contributed significantly to this increase and it has been estimated that enteric fermentation of ruminants contributes some 13–15% and livestock waste 5% to the total emission of CH_4 in the 1990s (Hogan et al., 1991; Lelieveld et al., 1998). Agriculture was estimated to have contributed almost 80% to the anthropogenic emissions of N_2O in the 1990s (Khalil and Rasmussen, 1992) and further emission inventories show that livestock production contributed 70–80% of the anthropogenic NH_3 emission in Denmark and Europe (Hutchings et al., 2001).

Animal wastes are generally associated with health risk to humans and animals if not properly managed. There is a pressing need therefore, for a holistic research into effective strategies and techniques in utilizing livestock waste tailored towards the development of sustainable environmental friendly livestock production systems. It is believed that such system(s) will ensure its sustainable use as organic manure, non-conventional feedstuff, a source of biogas as well as reduce its environmental impacts (air and water pollution, ammonia and greenhouse gas emissions) on human and animal health. The objective of this paper was to review the animal production trend, animal waste management and utilization, and its environmental/ health implications.

High nitrogen waste like animal manure may lead to eutrophication of water bodies wherein there is overgrowth of algae and other aquatic plants in these water bodies which leads to a drop in the oxygen available for the other aquatic animals and plants, lack of food and eventually loss of aquatic flora and fauna.

AW may also contain pathogens which may pose a risk to human health. There are many diseases including swine flu, foot and mouth disease which can propagate through water streams and mismanagement of animal waste and affect humans (Ogbuewu et al., 2012)

Considering the multiple advantages offered, AD emerges as one of the most viable technologies for the treatment and disposal of AW. The main product of this process is methane which is a gaseous fuel with the liquid digestate being the by-product.

This not only reduces the GHG emissions by harnessing the methane but also helps reduce soil and water pollution as the digestate can be used as a fertilizer for crops. Besides, this process is also economically beneficial as the feedstock being used is waste and thus cost effective.

AW is generally low in C/N ratio (between 6-8) as the nitrogen content is quite high in most types of animal excreta (Toma et al., 2016).

This is where anaerobic co-digestion comes into play especially with plant biomass as it is a carbon rich substrate and can thus help balance the C/N ratio for AW.

The section below discusses various aspects of AD for different types of AW:

2.3.1 Fish Waste

Fish waste management is important to maintain the marine environment up to the standards as fish farming ends up having severe effects on the marine life.

Commonly, fish farming is carried out in ponds or inside cages which requires constant deployment of artificial feed and drugs which are in turn quite harmful for the environment as it leads to nutrients and chemicals getting released into the marine environment.

Large scale rearing of aquatic animals is called as aquaculture and that of fish is called pisciculture. Aquaculture, in general, can be quite detrimental to the environment as it can affect not only the marine environment but also the aquatic ecosystem.

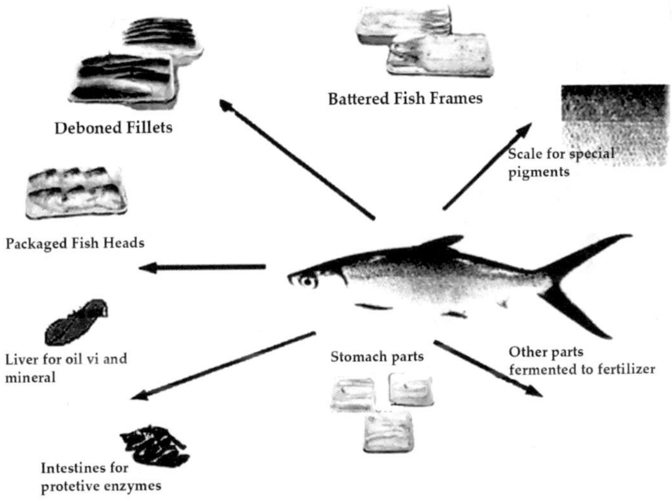

Figure 11: Different uses of fish waste (source: https://www.slideshare.net/omar-alajil/utilization-of-fish-wastes).

As a result, aquatic waste in general and fish waste in particular need better and more environment friendly alternatives for their disposal.

Conventionally, AW like cattle manure, MSW, agro-waste have been quite commonly used in AD plants while there were not many reports of fish waste being disposed the same way.

Researchers evaluated the potential of fish waste for AD and obtained considerable amounts of biogas and methane yield (Arvanitoyannis and Kassaveti, 2008).

Fish waste is not used as an AD feedstock currently, but it holds a huge potential as it is rich in organic matter also in nitrogen.

The digestate from AD of fish waste could be used a fertilizer for the agriculture industry.

2.3.2 Cattle Manure

Cattle are ruminant animals and contain microbes known as methanogens in their digestive system. AD is a naturally occurring process in the guts of cattle like cows and goats. That is also a reason these animals are a major source of GHG emissions into the environment.

Moreover, it is also important to carefully dispose the waste produced by cattle as their excreta also contains the microorganisms necessary to carry out the process of AD thereby leading to GHG emissions in form of methane and carbon dioxide.

Besides, cattle manure is also rich in nutrients and thus the AD by-product i.e. the digestate is an excellent fertilizer.

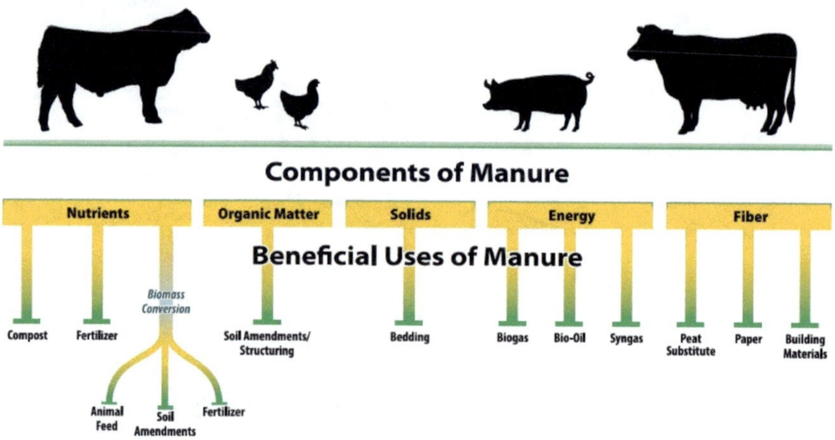

Figure 12: Uses of cattle manure (source: http://www.thecattlesite.com/focus/5m/2311/beneficial-uses-of-manure-and-environmental-protection).

At present, cattle manure is applied directly to the agricultural land as fertilizer to enhance soil fertility, but it comes with added risks of nutrient leaching and also release of harmful pathogens into the environment that can infect humans.

Considering these issues, AD not only seems to take care of the GH emissions and prevent nutrient leaching, but it also minimizes the odor of raw manure and also reduces the pathogen load in it (Atandi and Rahman, 2012).

2.3.3 Poultry Litter

Poultry industry is one of the most flourishing meat industries worldwide and generates huge quantities of waste annually. This waste consists of a lot of mixed components including feathers, bedding material, broken shells and poultry excreta.

Poultry litter (PL) is highly rich in nitrogen due to its high protein content and may cause pollution of neighboring water bodies due to surface run off. Conventionally, PL was either land spread directly on to the field, land filled or incinerated.

Direct land spreading of the PL can increase the nitrogen content of the soil and also lead to release of harmful gases like ammonia and nitrous oxides into the atmosphere. Moreover, many human pathogens like Salmonella and Listeria may also be present in raw untreated PL.

Pasteurizing PL before applying it to the land is one of the approaches that is currently being followed to avoid the human pathogens entering the food chain.

AD of PL is not one of the commonly used disposal methods but it one of the most efficient and sustainable methods for various reasons.

There is reduction in the nitrogen content as can be seen in the digestate, pathogen load is reduced and most importantly, there is production of gaseous fuel which can be used in the farm for generating electricity for various purposes.

Every kg of organic matter will yield 0,5 m3 of biogas. • Biogas contains between 50 – 60% CH4 and has a heating value of 5 to 6 kW. • For comparison: 1 litre of diesel has a heating value of 10.4 kW. • 1 m3 of Natural Gas has a heating value of 9.3 kW. • 1000 kg of Chicken litter (55% DM , containing 42% OM) will yield approx. 200 m3 biogas. 1000 kg of Chicken litter (50% DM) contains approx 27 kg of Nitrogen • with 10 kg in soluble form NH3-N. • NH3-N inhibits the anaerobic digestion when >4kg/m3. • Solution: -Dilution with 5 m3 water • - Removal of ammonia through stripping. • Biogas from chicken litter will produce biogas with high H2S levels (>2000 ppm). • This is corrosive and needs to be removed in a scrubber. • It takes about 30 days for complete digestion of chicken litter. • A digester tank of 2000 m3 can handle approx. 17 ton of chicken litter/day. • This would give you 3400 m3 biogas. • 1 m3 of biogas in a biogas engine can produce approx. 2.1 kWhe of electricity and 2.5 kWth heat ivestock and poultry industries produce meat, milk and egg, and also generate large volumes of waste water and solid wastes that could be beneficial or harmful to the environment. The waste products which includes livestock or poultry excreta and associated feed losses, beddings, wash -water and other such waste materials represent a valuable resource that if used wisely, can replace significant amounts of inorganic fertilizers (Bouwman Contents lists available at Sjournals Journal homepage: www.Sjournals.com Review article I.P. Ogbuewu et al. / Scientific Journal of Review (2012) 1(2) 17-32 18 and Booij, 1988; Leha, 1998; Smith and Vanduk, 1987), conventional feedstuffs (El Boushy and Vander Poel, 2000) and cooking gas (Henuk, 2001), but may be a direct threat to human and animal health (Taiganides, 2002). Animal wastes in the form of manures are valuable sources of nutrients and organic matter for use in the maintenance of soil fertility and crop production. Studies with animals have shown that 55–90% of the nitrogen and phosphorus content of animal feed is excreted in faeces and urine (Tamminga et al., 2000) normally used as manure. Poultry and swine manure collection in confinement feeding facilities have been recovered for re-feeding to beef cattle, dairy cattle and sheep (Bell, 2002) and have been found to present no serious health hazards to ruminants and poultry or posed any negative effects on meat, egg or milk quality (Mcllroy and Martz, 1978). Animal waste has been traditionally used in the production of biogas in Asia, particularly in tropical areas such as Indonesia, India and Vietnam (Henuk, 2001). However, careless dumping of livestock waste on farm lands and direct discharge to waterways and percolation to groundwater, usually in by-pass flow via cracks and fissures, is a great risk to human and animal health because livestock waste contain myriads of pathogens (Davies, 1997; Dizer et al., 1984), some of which may be zoonotic and can cause systemic or local infections (Dizer et al., 1984; Mackenzie et al., 1994; Davies, 1997; Stanley et al., 1998; Fischer et al., 2000; Cameron et al., 2000). Transmission of disease causing microbe is enhanced by mismanagement of animal waste and may be reduced by proper waste handling methods (Mackenzie et al., 1994). Highly contagious and pathogenic diseases, such as Foot and Mouth Disease and Swine Fever may spread with animal effluent through waterways and, when one farm is infected with the disease, farms downstream will be at considerable risk of infection (Cameron et al., 2000). Livestock waste produces ammonia that can be a potential pollutant causing serious eutrophication of rivers and lakes, characterized by a high concentration of nutrients that creates an ecological imbalance in the water system that support abnormally high levels of algae and aquatic plant growths (Burton and Turner, 2003; IAEA/FAO, 2008). This reduces oxygen levels in the water and has serious implications on the survival of aquatic organisms and, consequently, on food supply and biodiversity (IAEA/FAO, 2008). Livestock wastes are sources of malodours originating from livestock buildings, storage and field application of animal manures. The intensity of malodours is often unacceptable, especially for neighbours in surrounding residential areas. Globally, the concentration of the greenhouse gas methane (CH_4) in the atmosphere has increased by 45% since 1850 (Lelieveld et al., 1998). Increases in livestock production have contributed significantly to this increase and it has been estimated that enteric fermentation of ruminants contributes some 13–15% and livestock waste 5% to the total emission of CH_4 in the 1990s (Hogan et al., 1991; Lelieveld et al., 1998). Agriculture was estimated to have contributed almost 80% to the anthropogenic emissions of N_2O in the 1990s (Khalil and Rasmussen, 1992) and further emission inventories show that livestock production contributed 70–80% of the anthropogenic NH_3 emission in Denmark and Europe (Hutchings et al., 2001). Animal wastes are generally associated with health risk to humans and animals if not properly managed. There is a pressing need therefore, for a holistic research into effective strategies and techniques in utilizing livestock waste tailored towards the development of sustainable environmental friendly livestock production systems. It is believed that such system(s) will ensure its sustainable use as organic manure, non-conventional feedstuff, a source of biogas as well as reduce its environmental impacts (air and water pollution, ammonia and greenhouse gas emissions) on human and animal health. The objective of this paper was to review the animal production trend, animal waste management and utilization, and its environmental/health implications.

2.4 INDUSTRIAL WASTE

A comparison between oxygen consuming and oxygen consuming techniques demonstrates that oxygen consuming treatment isn't much favorable for the treatment of squander water.

The aerobic digestion requires higher capital taken a toll for ventilation equipment, progressed working taken a toll (especially energy for pumps or aerators), higher upkeep requirements and attainably checking necessities for checking the dissolved oxygen level within the fluid.

Oxygen consuming digestion technology moreover requests a vitality input to carry out the prepare and requires a few information approximately time.

The working costs are ordinarily much more prominent for aerobic absorption than for anaerobic absorption because of vitality utilized by the blowers, pumps and motors needed to include oxygen to the method as the aerobic technology requires parcel of oxygen for high-impact bacteria, which increments the cost by utilizing the bulk quantity of chemicals (Appels et al., 2008).

One of the greatest disadvantages is that it does not discharge renewable energy asset that's methane gas but discharges more heat and carbon dioxide that influences the environment.

The truth that anaerobic assimilation produces biogas which burns cleanly, which biogas plants make more energy than they devour to function.

The vitality created by aerobic digestion is very much lower than that produced by anaerobic assimilation.

This can be for the most part the calculate that makes anaerobic assimilation the more maintainable option than oxygen consuming assimilation (Stuart, 2006).

2.5 KITCHEN WASTE

Kitchen waste is a very suitable waste for AD as compared to other types of organic wastes as discussed above. It is rich in energy and research directed towards kitchen waste as the sole feedstock for anaerobic digesters would be very beneficial.

Nevertheless, there is high protein and high fat content in this type od waste as a result of which there maybe be ammonia inhibition and VFA accumulation in the system.

Despite the fact that it has a potential for failure of the digester, it has a high calorific value w hich is quite favorable for biogas production by AD.

This also indicates the potential for high methane yield without the requirement for very large sized reactor.

In most developing countries, kitchen waste is usually landfilled or discarded in open dump sites which can pose a threat to public health by incidence of diseases like typhoid and malaria (Pravinkumar et al., 2019).

2.6 ANAEROBIC DIGESTATE

AD is very important in terms of reducing the amount of organic waste by decomposing it to its constituent matter along with generation of energy. As has been discussed earlier, these organic wastes can range from household waste to industrial waste and crop waste to animal waste.

Pathogen removal is another important benefit to the environment from the use of AD.

Anerobic digestate is a rich source of both organic and inorganic nutrients. It is formed as a by-product of the AD process while biogas is the main product formed.

The digestate may either be solid or liquid depending on the water content of the AD process or whether it was dry or wet AD.

It has advantage over chemical fertilizers currently dominating the market as it has the nutrients available in a more easily absorbable form by the plants.

Since it is mostly derived from plants and animals themselves, it has all the essential micronutrients and macronutrients that may be required by the plants for their optimal growth.

Along with nutrients, the digestate also has the organic matter that can add .to the other physical characteristics to the soil like texture and humus

Digestate is a by-product of AD of organic matter to produce biogas. Depending upon whether the AD process was wet or dry, the digestate can be solid or liquid in nature.

Digestate is rich in nutrients like mineral nitrogen which is present in very high concentration in the form of ammonium. It also contains other important macronutrients and trace elements that are required by plants for their optimal growth.

As a result, research has been focused towards identifying optimum concentrations of digestate required for its function as an organic fertilizer for growth of crops.

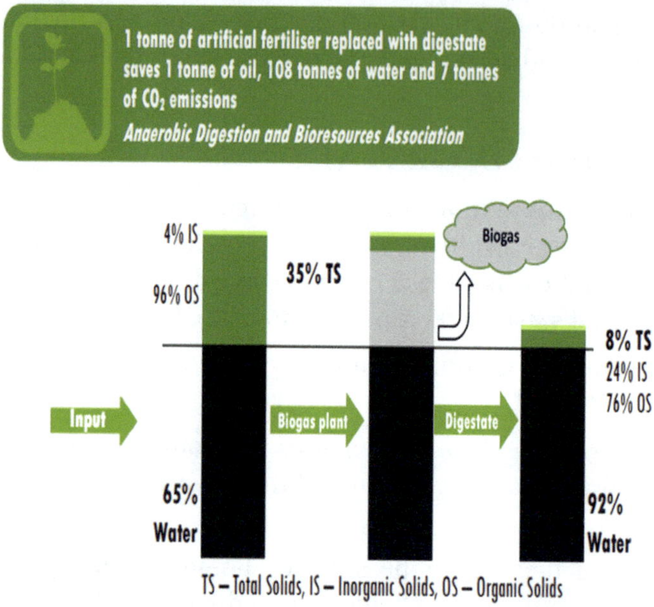

Figure 13: Environmental benefits of digestate (source: http://www.organics-recycling.org.uk/dmdocuments/digestate%20paper%20final%20small.pdf).

Not only this, the digestate can also enrich the soil my contributing to the organic matter or humus in the soil thereby influencing the physical and chemical characteristics, biological make up and texture of the soil.

Recently, research has also been focusing on utilizing the digestate as solid fuel or to be purged back into the reactor for methane production.

Different organic materials can be used in AD individually or in combination like slurries, waste from slaughterhouses, crop waste, municipal waste and so on.

As a result, the quality of fertilizer derived from digestate would depend upon its source along with other physiochemical parameters during operation of the digester.

For instance, organic content in the digestate is inversely proportional to the retention time, the longer the retention times lower would be the organic content of the digestate.

Moreover, at elevated temperatures, there is higher elimination of pathogens and thus thermophilic digestate would have fewer harmful pathogens as compared to mesophilic digestate which may need additional steps of sterilization before land application.

2.6.3 Composition of the Digestate (Macronutrients)

Digestate pH mainly lies in the alkaline range and its exact pH depends on the quality of the feedstock in the AD process. Alkaline pH is also favorable due to the existing problem of soil acidification and thus can be neutralized by the digestate.

The pH can vary throughout the process of Ad and thus it is important to monitor the pH if it will be applied as a fertilizer.

It is also rich in nitrogen in the form of ammonia and nitrogen is an important macronutrient required by plants for their growth. Digestate is beneficial in the sense that it allows enrichment of the soil with nutrients in crop rotation format.

It also provides higher amount of nitrogen as compared to compost. Generally, ammonia nitrogen concentration is higher in feedstocks with higher protein content which includes dairy or meat products.

During AD, the organic nitrogen is converted into ammonia nitrogen which is readily available for uptake by plants.

Ammonia volatilization, however, is an issue that needs to be minimized and can be reduced by changing the injection techniques and the depth of application of the fertilizer.

Digestate is also rich in other macronutrients like potassium and phosphorous which can complement the missing nutrients in the soil and thus make digestate a better supplement than compost.

2.6.2 Composition of the Digestate (Macronutrients)

Trace elements or micronutrients are also required by both plants and animals for maintaining enzyme structures and active sites and other biochemical functions. These trace elements include nickel, cobalt, copper, zinc, aluminum, cadmium, selenium, and so on.

There are certain heavy metals as well which may be added to the system through anthropogenic sources like human or sewage waste, animal feed, sludge and domestic waste.

Digestate is also rich in organic matter which is reflected in its volatile solid content and is part of the feedstock that is actually converted into methane.

The organic matter which is not converted and remains in the digestate is stabilized by molecules like lignin and some other complex molecules that provide stability to the organic matter.

2.6.3 Effect of Digestate on Crop Yield

Certain plants life sunflower, soybean and alfalfa are sensitive to the application of digestate, especially, depending on their life stage. It requires careful process of recovery in case the plants have been exposed to digestate while they are sensitive.

In case of the non-sensitive plants, the digestate can be applied at any stage of development. Phosphorous and nitrogen are generally the main components that are taken as the basis of calculation for the application rate of the digestate.

It has also been noticed that digestate from co-digestion produces more effective results compared to the mono-digested one. This is because combination of feedstocks leads to balancing of the carbon to nitrogen ratios and also the trace element composition to some extent.

3

Biogas Upgradation to Biomethane

Navodita Bhatnagar

Biogas is mainly composed of methane along with 30-50% CO_2 and a few other gases present in trace amounts which include water vapor, nitrogen, hydrogen sulfide, oxygen and hydrocarbons. Thus, to use this gas for commercial purposes and injection into the natural gas grid, it needs to be cleaned and upgraded into a higher percentage of methane. This upgradation results in what is commonly known as biomethane. This can be done either by removing the CO_2 and trace gases or by maintaining the calorific value. There are various different methods available to achieve this which include condensation, absorption and adsorption based approaches. These methods vary in their functioning, their pros and cons and most importantly their efficiency.

INTRODUCTION

The gaseous products apart from methane which include carbon dioxide, hydrogen sulfide, siloxanes, ammonia and hydrogen have the tendency to inhibit the upgradation process by causing corrosion in the pipelines for transporting the final product.

Table 2 discusses these problems for each of the trace compound present in biogas that needs to be filtered out.

Research has established that membrane technology is a promising technology for separation of CO_2 from the biogas and discusses use of polymeric membranes for this purpose.

Table 2: Biogas impurities and their consequences. (source: Ryckebosch et al., 2011)

Water	Corrosion in compressors, gas storage tanks and engines due to reaction with H2S, NH3 and CO2 to form acids; Accumulation of water in pipes; Condensation and/or freezing due to high pressure
Dust	Dust Clogging due to deposition in compressors, gas storage tanks
H_2S	H2S Corrosion in compressors, gas storage tanks and engines; Toxic concentrations of H2S (> 5 cm3 m-3) remain in the biogas; SO2 and SO3 are formed due to combustion, which are more toxic than H2S and cause corrosion with water
CO_2	CO2 Low calorific value
Siloxanes	Siloxanes Formation of SiO2 and microcrystalline quartz due to combustion; deposition at spark plugs, valves and cylinder heads abrading the surface
Hydrocarbons	Hydrocarbons Corrosion in engines due to combustion
Ammonia	NH3 Corrosion when dissolved in water
Air	O2/air Explosive mixtures due to high concentrations of O2 in biogas
Chloride	Cl Corrosion in combustion engines
Flouride	F Corrosion in combustion engines

These polymeric membranes are economically viable in removing both CO_2 and H_2S when compared to the conventional technologies except the fact that polymeric membranes are prone to corrosion by ammonia and other biogas components.

Biogas is a source of renewable energy either on its own or indirectly as a starting material for production of various types of chemicals and synthesis of gas.

Composition of biogas varies depending upon the source and so does the percentage of the impurities present in it.

Landfill biogas contains a mixture of components which includes hydrogen, carbon monoxide, ammonia, volatile organics, siloxanes and halogenated hydrocarbons.

Most importantly, removal of hydrogen sulfide and carbon dioxide is essential in improving the standard of biogas quality and use for energy production.

Carbon dioxide is present in much larger amount compared to hydrogen sulfide, yet it is less harmful than the corrosive hydrogen sulfide which can destroy the metallic components like storage tanks, tubes, valves and pumps.

Therefore, it is essential to remove these gases before the upgraded biogas can be utilized.

There is a plethora of technologies available these days which can assist in biogas purification based on the needs depending on the source and utility.

The upgradation involves two major steps which are firstly the removal of contaminants and secondly the upgradation of the calorific value of resulting biogas by adjustment of carbon dioxide content.

Upgradation may be done through physical, chemical or biological treatment depending on the costs and efficiency of each method. Biomethane obtained after biogas purification can be 95-99% pure and may contain 1-5% carbon dioxide (Awe et al., 2017).

3.1 PHYSICAL/ CHEMICAL REMOVAL OF CARBON DIOXIDE (CO_2)

CO_2 removal is required to enhance the density of biomethane, and various physical and chemical technologies are available for this including physical/ chemical absorption, membrane separation, water and organic solvent scrubbing and membrane separation.

These technologies not only help in separation of CO_2 from biogas but also help in its bioconversion into commercially valuable products.

3.1.1 Water Scrubbing

The principle behind water scrubbing for CO_2 separation is its solubility in water which is approximately 26 times higher than methane as a result of which its concentration is higher in the liquid phase, which is water, while methane concentration is higher in the gas phase.

Besides there are other contaminants which also dissolve in water which include hydrogen sulfide, hydrogen and ammonia, thus, leaving mostly methane in the water which flows out of the absorption column.

Temperature is also important in this context as CO_2 solubility decrease with increasing temperature (Chandra et al., 2012).

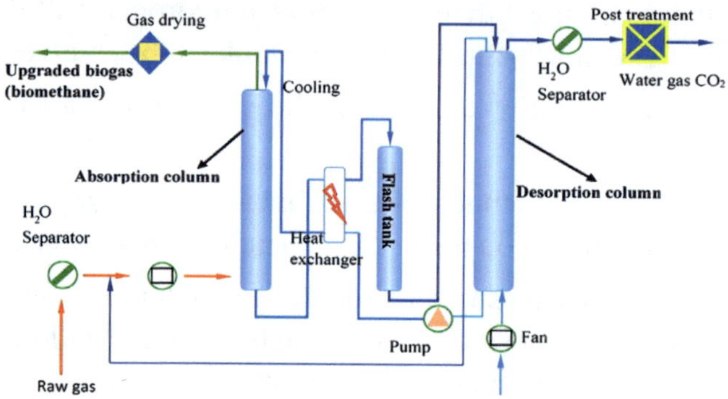

Figure 14: Water scrubbing for CO_2 removal from biogas (source: Awe et al., 2017).

Water scrubbing is one of the most commonly used methods for biogas upgradation accounting for 41% of the global share (Muñoz et al., 2015). This is mostly because of the economic viability of the process and simplicity of its operation.

The packed absorption column is fed with biogas from the bottom while water is fed through spraying from above the column. The water leaving the system contains dissolved contaminants which are easy enough to dispose thus making this process quite straightforward.

The effluent water is sent to a tank where pressure is gradually decreased thus releasing most of the dissolved gases.

These gases also contain traces of methane because of which the gas is purged back into the raw biogas entering the scrubber. The water can also be recycled through a desorption column and can be pumped back into the scrubber (Fig 7).

3.1.2 Organic Solvent Scrubbing

Absorbents based on methanol and polyethylene glycol are used in organic solvent scrubbing which follows the same principle as water scrubbing except that the solvent used here is organic instead of polar.

These solvents, like Selexol, have higher affinity for CO_2 than water as a result of which the size and operational costs of the plant can be scaled down considerably.

Moreover, vapor pressure of Selexol is quite low because of which the chemical losses are minimal with lower volumes of absorbent required for the process.

However, there is an additional step of water removal and gradual heating for desorption of CO_2 at 40 °C.

The main advantage of this method is the anticorrosive organic solvent because of which there is no requirement for stainless steel in the scrubber.

However, precaution may be taken in case of high hydrogen sulfide concentration just the same way as in water scrubbing.

Figure 8 shows a schematic diagram for different steps involved in organic solvent scrubbing process.

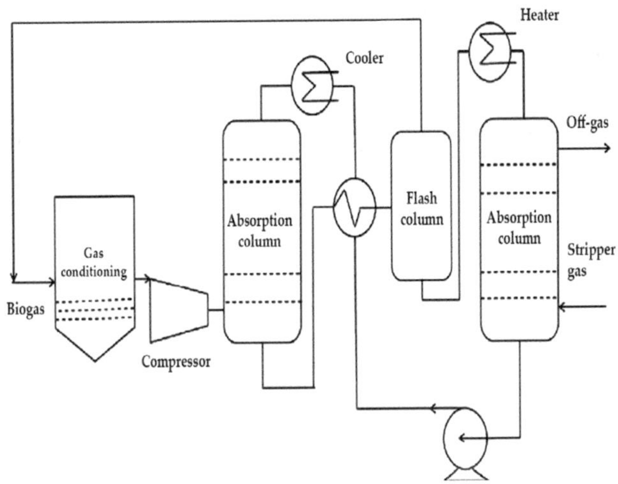

Figure 15: Schematic diagram for organic solvent scrubbing process (source: google images).

3.1.3 Chemical Scrubbing

The gas-liquid mass transfer principle remains the same behind this and water/organic solvent scrubbing except for the fact that chemical reaction occurs between the solvent and the gases.

Gaseous components are physically absorbed and chemically react with the components of the scrubbing liquid (fig 9).

This allows the impurities to strongly bond with the scrubbing liquid with a highly selective reaction thus leading high amounts of methane to remain unreacted with the liquid and be recovered subsequently. These chemicals include absorbents for CO_2 such as alkanol amines and alkalis that can react with it.

- Amine solutions like mono- or di- ethanol amines are used very commonly to absorb CO_2 as they react selectively and result in minimal methane losses (less than 1%) and maximum recovery (~90%).
- The plants operating on amine scrubbing work at a slightly higher pressure of the raw biogas and while high selectivity is favorable in amine scrubbing, it becomes difficult to regenerate the scrubbing solution after the reaction.

The temperature needs to be elevated to 160 °C to release CO_2 for column regeneration.

This is also one reason why hydrogen sulfide needs to be removed prior to the scrubbing as if it gets absorbed during the chemical absorption, temperatures will need to be elevated to a much higher degree for column regeneration.

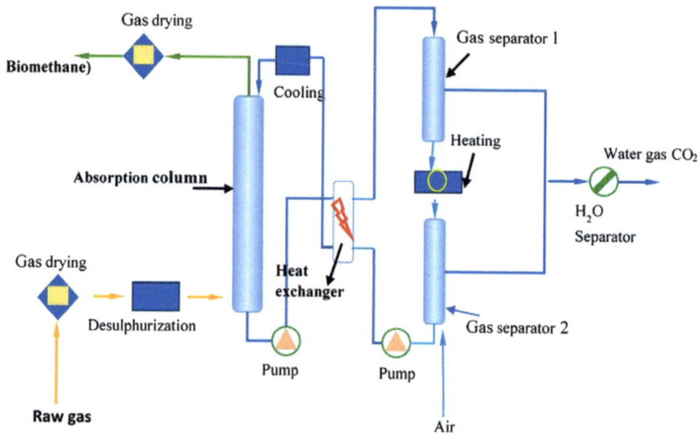

Figure 16: Biogas upgrading by chemical absorption (Amine scrubbing) of CO2 (source: Awe et al., 2017).

The operation is quite similar to water/ organic scrubbing as well with a packed bed unit paired with a desorption unit and a countercurrent flow of the absorbent and the gas (fig 6).

However, one important aspect to consider here is removal of hydrogen sulfide prior to the scrubbing process to avoid amine poisoning of the system (Muñoz et al., 2015).

While chemical scrubbing is highly efficient in CO_2 removal, there are a few problems associated with it including possible precipitation of salts and relatively higher energy consumption in form of heat regeneration (Awe et al., 2017).

3.1.4 Pressure Swing Adsorption (PSA)

Pressure swing, as the name suggests, entails series of adsorption and depressurization steps followed by desorption and pressurization steps within a packed vertical column.

At the high pressure condition, adsorption of the contaminant gases occurs while regeneration of the material occurs at gradual decrease in pressure (fig 10).

The principle behind adsorption based separation of biogas components is different selectivity of the gaseous components on the adsorbent surface under high pressure conditions.

During the pressurization, CO_2 is adsorbed selectively over methane on adsorbents like silica, resins, zeolites and activated carbon in the column (Ryckebosch et al., 2011).

Other gaseous impurities like nitrogen, oxygen and hydrogen sulfide can also be separated from the incoming biogas through selective adsorption over methane gas.

It is also important to let the gas dry down before carrying out adsorption because otherwise it may be adsorbed irreversibly to the adsorbent material which is generally a molecular sieve (Awe et al., 2017).

- Once CO_2 is saturated in one column, biogas is moved to a new column while the CO_2 in the previous column is depressurized to facilitate its release and subsequent purging back into the system to recycle any methane that might be present in the gas mixture.
- During decompression, methane is released earlier at a higher pressure as compared to CO_2 which is desorbed at a lower pressure.

- Pressure is elevated again for the new cycle and the raw biogas is loaded on to the system again.
- Molecular size also plays a role in this kind of separation based on size exclusion in which adsorbents act as molecular sieves with an average pore size of 3.7 Å.

Thus they are able to retain 3.4 Å CO_2 molecules in the pores and excluding 3.8 Å methane molecules (Muñoz et al., 2015).

At an industrial scale, biogas upgrading plants operating at PSA may contain four to nine different vessels connected in series or in parallel in order to enable continuous operation.

Molecular sieves in the PSA need to be non-hazardous and to exhibit linear adsorption isotherm besides having a high selectivity for CO2 adsorption.

The pressure may be elevated up to 4-10 bars during the pressurization phase to facilitate CO2 retention.

Biomethane purity of up to 96-98% may be achieved through PSA which currently cover 21% of the market share of the gas separation technologies currently available.

Other impurities like hydrogen sulfide and siloxanes are also adsorbed within the molecular sieve adsorbents and can be later removed with the help of activated carbon filters (Muñoz et al., 2015).

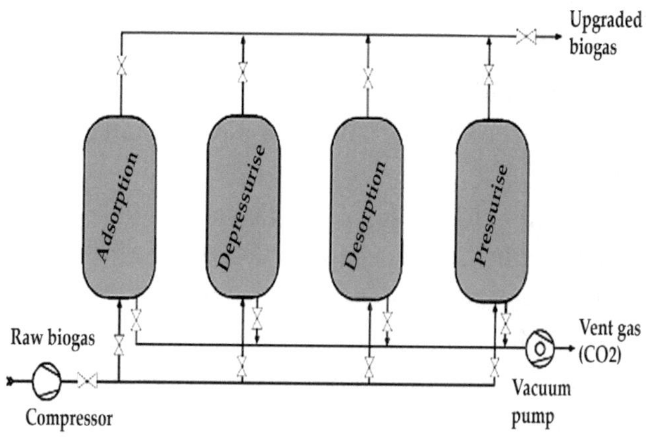

Figure 17: Set-up of (vacuum) pressure swing adsorption (adopted from Ryckebosch et al., 2011).

3.1.5 Cryogenic Separation

Cryogenic separation is based on temperature difference and the different boiling points of the gaseous components of biogas.

This is a relatively new technology and helps separate CO_2 from biogas by cooling it at elevated pressure as the boiling point of CO_2 is -78 °C as opposed to -160 °C for methane (fig 11).

The same principle can be applied for separation of other gaseous impurities in biogas like nitrogen, oxygen, siloxanes and so on by condensation followed by distillation (Awe et al., 2017). The CO_2 gets liquified during the cooling process and can further be used as a solvent for the other impurities.

The gas needs to be dried prior to compression so that it doesn't freeze during the cooling process (Ryckebosch et al., 2011).

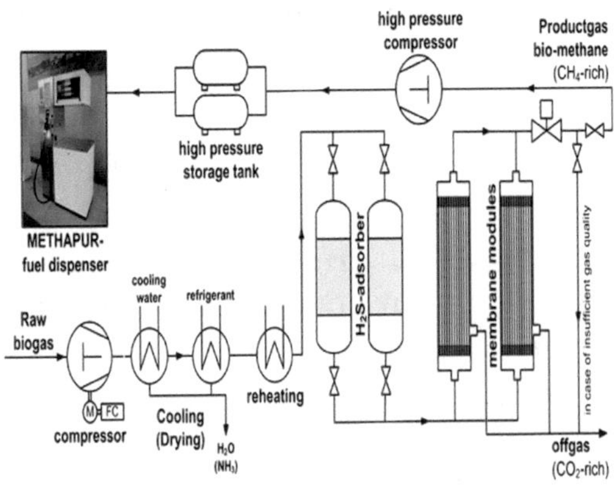

Figure 18: Cryogenic separation for biogas upgradation (source: google images).

During the stepwise decrease in temperature, siloxanes, hydrogen sulfide and halogens are removed at around -25 °C, most of the CO_2 is liquified at -55 °C to be withdrawn from the cooling vessel and eventually till -85 °C at which the leftover CO_2 solidifies.

- High pressure avoids the clogging of delivery pipelines due to instant solidification of CO_2 below -78 °C.
- Incoming biogas is cooled sequentially to facilitate CO_2 liquefaction and then the pressure is reduced to about 8-10 bar at -110 °C so that CO_2 solidifies and the biomethane can be purified.

- Biogas purity of up to 97% can be achieved by cryogenic separation process with less than 2% losses, however, it only shares 0.4% of the market.
- Certain impurities such as water, siloxanes and hydrogen sulfide may be removed before the process as they can cause problems like clogging of pipelines and heat exchangers.
- The final product obtained through this process is called as liquid biomethane (LBM) which is similar to liquid natural gas (LNG).

3.2. CHEMOAUTOTROPHIC BIOGAS UPGRADING

CO_2 can be separated from the biogas by microbial or enzymatic methods into the broth and can be subsequently reduced biologically.

This is the main principle behind biological CO_2 removal from biogas, however the mechanisms can be different and have been discussed in the section below:

In this type of conversion, hydrogenotrophic methanogens use CO_2 as their carbon source and also as an electron acceptor while hydrogen acts as an electron donor.

Chemoautotrophic reactions involve both chemical and biological reactions thus resulting in bioconversion of CO2 to methane by externally administering hydrogen.

This process is advantageous for wastewater treatment and biomass gasification as the CO2 emissions from both these industries can be converted using the hydrogen produced by the processes themselves.

Microorganisms called methanogens such as Methanobacterium, Methanococcus, Methanosaeta and Methanosarcina play an important role in this kind of bioconversion.

These microbes are stable over a pH of 6.5-8, a wide temperature range of mesophilic to thermophilic temperatures and are autotrophic in nature (Muñoz et al., 2015).

The thermophilic methanogens perform the bioconversion faster than the mesophilic methanogens; however, the mesophilic methanogens are capable of achieving higher degree of completion of CO2 conversion.

Hydrogen has poor solubility in water as a consequence of which there is limited mass transfer between the gas-water interface for hydrogen mass transfer thus leading to slower bioconversion of CO2 to methane.

This causes decreased efficiency of methane production and increased biomass formation to tackle which the gas residence time needed to be much longer for at least 90% methane concentration in the biogas upgradation.

Using hydrogen for biogas upgradation involves employment of strict safety procedures for operation as hydrogen turns out be a very flammable gas and there are significant losses in energy that occur with the use of hydrogen.

The only environmentally sustainable methods for hydrogen production these days is water electrolysis from renewable sources of energy. While, hydrogen is a fuel in itself, due to its low density, very large tanks are required for storage.

Moreover, its high inflammability makes it much more difficult to handle compared to methane. Thus, chemoautotrophic biogas upgradation presents itself as an eco-friendly alternative for fuel production from biogas.

3.3 HYDROGEN SULFIDE REMOVAL TECHNOLOGIES

Sulfur is one of the major pollutants in biogas which may be found in hydrogen sulfide, mercaptans and so on and the quantity can range from 100ppm to 10000ppm depending on the type of organic matter, especially if its protein rich there are chances of higher sulfur concentration.

The main reason why presence of sulfur may be detrimental to the upgradation process is the corrosive nature and the potential to corrode metal parts of pipes and engines. Not only that, but sulfur compounds like sulfur dioxide, sulfuric acid and others are also harmful for the environment.

Hydrogen sulfide is also a very potent greenhouse gas and thus, these trace compounds need to be removed before the upgradation process. There are many different available technologies for removal of hydrogen sulfide which are discussed in the sections below.

3.3.1 Sulfide precipitation

Adding different types of metal salts directly to the digester can help in precipitating sulfur in the substrate via the formation of insoluble metal salts like in case of addition of iron salts, there would be formation of iron sulfite. The metal salts can also be added before the AD step and can also be used .for ammonia removal from biogas

This method is also relatively quite inexpensive as compared with the other available methods for desulfurization as it doesn't require any sophisticated installations and very little investment. It can be applied to the existing AD facilities without any trouble and also provides easy enough operation and monitoring.

The only downside of this technique is the lack of control over the extent of desulfurization and also the limited achievable quality of biogas. If the hydrogen sulfide concentration is high otherwise, this technique can be really advantageous.

Moreover, it is a cheap and very reliable method of desulfurization. The precipitated iron sulfide can be removed along with the residual solids in the system. It can also be used as a fertilizer after being oxidized by atmospheric oxygen thus converting into a soluble salt.

The degree of hydrogen sulfide removal is not up to the standards as required by the biomethane to be injected into the gas grid. Reduction rates of up to 200-100 ppm of hydrogen sulfide have been achieved by this method (Al Mamun and Torii, 2015).

3.3.2 Biological Scrubbing

Microorganisms that are chemoautotrophic in nature can also be used to remove hydrogen sulfide from biogas by oxidation. These include species like Thiobacillus where the oxidation requires availability of a certain amount of oxygen within the system.

This oxygen is added either as air or as pure oxygen in small amounts to facilitate biological desulfurization.

To enable this process within the digester, the microorganisms naturally occurring in the digestate need to be immobilized.

Alternatively, an external system can be used through which the biogas passes once it has left the digester and the microorganisms can work on it.

The connected device is shaped as a streaming channel with a stuffed bed interior which contains the microbes which are immobilized.

Biogas is blended with the included oxidizer, meets a counter stream of water containing supplements after entering the trickling filter.

These microorganisms oxidize hydrogen sulfide with atomic oxygen and change over the undesirable gas compound to water and essential sulfur or sulfurous corrosive which is released alongside the wastewater.

The venture needs for this strategy are direct and this innovation is broadly spread along with a very high plant accessibility.

The organic framework is able to expel indeed exceptionally tall sums of hydrogen sulfide from the biogas but its flexibility to fluctuating crude biogas hydrogen sulfide substance is essentially destitute.

Unquestionably, this innovation isn't the most excellent choice in case tall sums of hydrogen sulfide or quick vacillations are anticipated at an anaerobic absorption plant.

3.4 REMOVAL OF TRACE COMPOUNDS

Water vapor is a main contaminant that needs to be removed as biogas is saturated with water vapor as it leaves the reactors. Water vapor can cause corrosion along with the hydrogen sulfide and thus it needs to be removed prior to biogas upgradation.

Reducing the temperature and elevating the pressure can help condense the water vapor and subsequently remove it. It can also be removed by adsorbing it onto silica or activated charcoal columns.

Siloxanes are used in manufacturing products like deodorants and washing soap and so on and can be found in AD plants that operate on municipal waste and landfills. These may pose a serious issue on combustion and can pollute the environment.

Different methods that can be used to remove siloxanes are adsorption, activated silica or absorption.

Commercially, adsorption is the best technology available for removal of siloxanes which is also cost effective and may cost more if the pressure requirement is high. Regeneration of adsorption material can also add to the costs.

3.5 METHANE REMOVAL FROM OFFGAS

Offgas is produced during the upgradation process and there is still some amount of methane that is contained in this off gas which can be recovered by using another separation technology. Moreover, since methane is a harmful greenhouse gas, it is important to remove any methane that may end up in the atmosphere for a sustainable process.

It may also be noted that many countries put a cap on methane emissions during an AD operation. This also means that higher amounts of methane in the offgas would add to the costs and can be economically damaging to the plant operation.

4

Algae for Biogas Production

Navodita Bhatnagar

Algae has been a popular research topic over the past four or five decades on its use as a source of producing biofuels directly or indirectly. Algae holds a high potential for producing biomass over a small area of land required for its growth as compared to the conventional food or energy crops. While it may seem pretty straightforward from here on, considering the higher biomass yield from algae, it comes with its own challenges most important of which is the dilutions in which algae grows in the medium. Commercial processes do not exist which can produce considerable amounts of fuel from algae as the amount of water present is usually quite high in the natural growth conditions of algae. Thus, it is important to be able to harvest algae and concentrate the biomass present while also attaining the energy output that may seem commercially and economically viable.

INTRODUCTION

It is important for bio based energy to replace petroleum based fuels in order to reach renewable energy targets that have been set by different countries in an attempt to reduce greenhouse gas emissions ad fight against global warming and climate change.

Energy use of the world has been increasing by each moment to achieve a higher standard of living by an ever increasing population.

Because of the impacts of energy use and also the carbon dioxide emissions on our environment and the global climate, research focusing on finding new alternatives has directed us towards plant based fuels.

Biofuels or biomass derived fuels are the future of the energy sector.

There is a vast variety of different fuels that can be used to produce biomass derived energy and these fuels include ethanol, biodiesel, biohydrogen and biogas. Although not very popular, but biogas production from algae is one such technology that has huge potential as compared to the energy output that can be achieved by algae to biodiesel or bioethanol production. Another important advantage with AD is that it can be integrated with processes like wastewater treatment or wetlands where the output can be largely scaled up. AD has been growing and expanding in terms of feedstocks and applications throughout the world.

It may, however, be noted that even though plant biomass offers itself as an ideal feedstock for AD, there will always be the debate about food and feed production as there is shortage of food and almost one third of the world suffers from starvation.

Most of the crops that are grown for energy production are also food crops like sugar cane, beet root, maize and so on. Moreover, they are grown on arable land which is yet another disadvantage.

This brings us to the current chapter which is use of algae for AD and biogas production. Algae have many advantages over conventional energy crops most important one being their fast growth rate and requirement for smaller area for cultivation.

Algae get the ability to fix atmospheric carbon dioxide with the help of photolysis by the sunlight to form carbon and they get this ability due to a photosynthetic pigment called chlorophyll. It is very important for sustaining life on this planet because the oxygen released during the process of CO_2 fixation is the oxygen that all living organisms use to respire.

4.1 ALGAE: AN OVERVIEW

Algae seem to be promising for anaerobic digestion and to provide biogas that can be upgraded to biomethane and substitute for natural gas. Biogas purification is necessary to enhance its calorific value and also to reduce the damage that its constituent gases might cause to the equipment.

Algae, however, are quite rich in water and this may become a potential obstacle in the AD process. A significant amount of energy goes into drying the algal biomass and thus it is not very beneficial to incinerate or gasify algal biomass due to very small concentrations.

Biomass based energy can be used to meet global energy demands by providing not only electricity but heat and vehicle fuel as well.

Energy crops like palm, corn, sugarcane and sugar beets have become feedstocks for producing biofuels like bioethanol and biodiesel which may also be referred to as first generation biofuels.

- However, first generation biofuels may not be the most efficient in terms of the input natural resources and the output they generate.
- Most of these require very high consumption of water and there is a big debate on using food crops for energy production as the world also struggles to provide food for all.
- This made way for the second generation biofuels which consisted of lignocellulosic waste materials which were good feedstocks for biofuel generation, but presence of lignin hindered their bioavailability.
- This in turn resulted in compromising with the yields and efficient conversion of the waste into bioenergy.
- The challenges face by the first and second generation biofuels were thus addressed and partly overcome in the third generation biofuels which are the algae based biofuels.
- The major advantage with this category of biofuels is the minimal use of arable land combined with the ability of algae to grow without the need of fertilizers.

Macroalgae or seaweed is also used as human food and also as animal feed. Approximately 29 million tonnes of macroalgae have been produced in 2016 for human consumption and also in the colloid industry.

As compared to energy crops like sugarcane, certain macroalgae could produce almost six times more biomass per unit area annually which again strengthens the research and investment in use of algal biomass as fuel.

Another important advantage of using algal biofuels is the reduction in the emission of greenhouse gas emissions which is a serious environmental concern faced globally. There is almost 40-80% reduction in GHG emissions as compared to use of fossil fuels like natural gas (Maneein et al., 2018).

4.1.1 General Advantages of Algal Bioenergy

Increase in population has also been a reason to look for renewable alternatives to meet the energy demands of an increasing global population. However, it is neither environmentally nor economically sustainable to divert more and more agricultural land and resources to meet these demands.

Currently, agriculture itself is struggling with problems like massive depletion of natural resources, land and water pollution and increasing competition to meet the food and feed requirements.

Non-renewable sources of energy are getting depleted every day especially by the populated countries like India, China and the US which rely on fossil fuels for most of their energy demands.

There was a substantial decrease in their dependence on fossil fuels, however, in the recent years as more and more renewable sources of energy started gaining popularity including the solar, wine, hydro, thermal and biomass based energy sources (Barbot et al., 2016).

Considering the advantages and challenges that lie ahead in diverting energy requirements to be met by algal based fuels, AD offers itself as quite a sustainable technology to overcome the challenges discussed above.

- Firstly, algae provide itself as a cheap and relatively easily available feedstock for AD and secondly, the energy inputs required in the process can be met by the utilizing biogas for energy production.
- Algae are also rich in nutrients which can be recovered at the end of the process and reused in agriculture and other industries.
- The nutrients can also be purged back into the system to enable the algal biomass to utilize it as a source of organic and inorganic minerals.
- Biogas that is generated from this process can be converted directly into heat or may be upgraded to biomethane via various upgradation methods which have been discussed in Chapter 3.
- Biomethane can be injected into the natural gas grid and this is a common practice in many EU nations these days.
- Algae come with added advantages over lignocellulosic biomass in the way that they do not need pretreatment technologies for enhanced biogas production.

4.1.2 Current Scenario

In the present day and time, marine research and development has been at its peak and there is more and more focus on the marine products. This could also mean that the biogas sector will reap the benefits of this development owing to the higher availability of the algal feedstock in future.

Another important fact to remember here is that not all types of algae or seaweed are favorable for the surrounding environment especially for aesthetic reasons in rivers and canals. This means the algae harvested from these areas can be put to a much better use.

Eutrophication is another such example where the load of algae blocking the nutrients for the marine flora and fauna can be used as a feedstock for AD instead.

In this context, AD is not only a source of biofuel production but also a sustainable strategy for waste treatment and management.

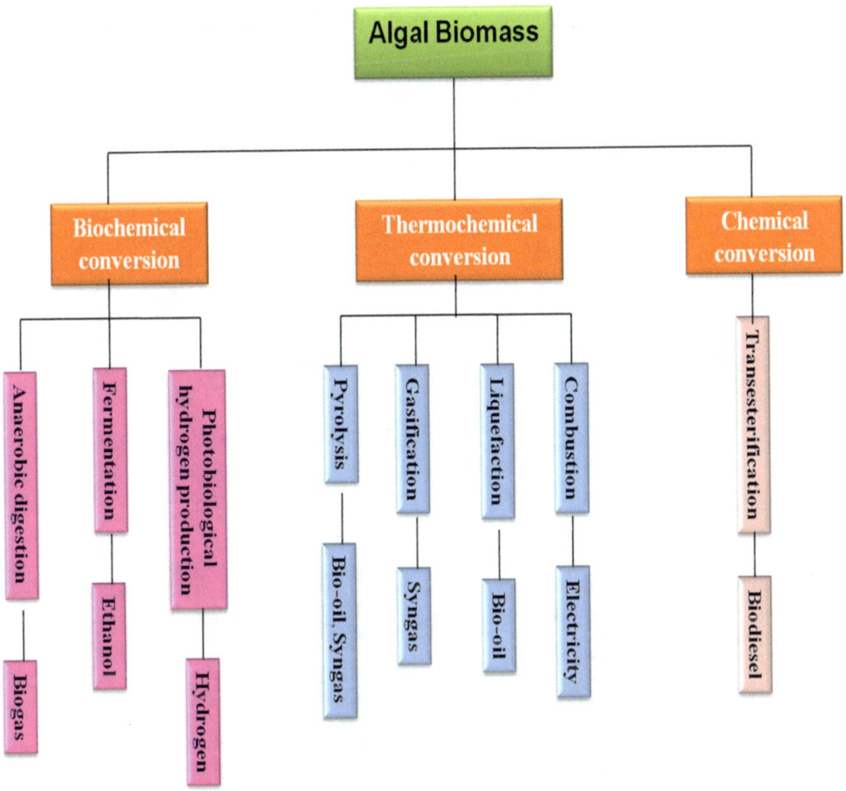

Figure 19: Algal biomass conversion process for biofuel production.

(source: https://www.frontiersin.org/articles/10.3389/fbioe.2014.00090/full)

Figure 19 shows the different pathways through which algae can be converted into different types of biofuels. Unlike biodiesel production from algae, AD is directed towards complete degradation and conversion of the carbon present in algae as compared to the transesterification process (Milledge et al., 2019).

Algae can be found in a many different habitats like marine, freshwater, oceans, rocks and so on. They also have the ability to withstand extreme conditions like deserts, hot springs, ponds and even wastewater from industries. In aquatic ecosystem, algae are also an important primary producer.

The one important condition for algae to thrive is presence of moisture and thus they tend to generally occur in damp places and aquatic environments.

Algae are very important ecologically as they are diverse in nature and are photosynthetic organisms that are aquatic in nature because of which they are at the top of the food chain for other organisms.

Table 3: Biogas yields for algae vs terrestrial plants (source: Behera et al., 2015)

Feedstock	Conditions	Biogas	Reference
ALGAE			
Blue algae	pH-6.8, microcystin (MC) biodegradation	189.89 mL/g of VS	Yuan et al. (2011)
Chlamydomonas reinhardtii	Drying as the pre-treatment, batch fermentation, temp. 38°C	587 mL/g of VS	Mussgnug et al. (2010)
Scenedesmus obliquus		287 mL/g of VS	
Ulva sp.	Batch reactor, Co-digestion with bovine slurry, temp. 35°C	191 mL/g of VS	Vanegas and Bartlett (2013)
Laminaria digitata		246 mL/g of VS	
Saccorhiza polyschides		255 mL/g of VS	
Saccharina latissima		235 mL/g of VS	
TERRESTRIAL PLANTS			
Banana stem	Pre-treatment: 6% NaOH in 55°C for 54 h. 37 ± 1°C for 40 days, batch	357.9 mL/g of VS	Zhang (2013)
Saline creeping wild ryegrass	35°C for 33 days, batch	251 mL/g of VS	Zheng (2009)
Rice straw	Pre-treatment: ammonia conc. 4% and moisture content 70%, temp. 35 ± 2°C, 65 days, 120 rpm, batch	341.35 mL/g of VS	Yuan (2014)
Date palm tree wastes	Pre-treatment: alkaline, particle size 2–5 mm, temp. 40°C	342.2 mL/g of VS	Al-Juhaimi (2014)

Algae have the ability to purify water bodies that have been contaminated by using the minerals present as nutrients for their growth thus opening a huge area of research for wastewater treatment using algae for bioremediation.

Thus, algae are fundamental for the maintenance of carbon dioxide in the carbon cycle in by the process of photosynthesis and algae are also the basis of food chain in aquatic environment.

Majority of the algae that exist are photoautotrophic in nature and have the ability to utilize sunlight to carry out chemical reactions in the form of photosynthesis. Some of the advantages of using algal biofuels are

- The yields are very high as compared to the terrestrial plants (table 3)
- Algae have the ability to grow in extreme conditions and that includes high salt and high mineral content thus providing a path for wastewater treatment
- The biofuels that may be produced using algae would be non-toxic and eco-friendly in nature like biodiesel or biogas
- Besides biofuels, algae also contain pigments that are of great value for various industries
- Since their growth rate is higher than conventional energy crops, this is a plus from a commercial perspective and also it will not have much impact on the food supply globally
- The digestate derived from algal biomass would be of great importance as fertilizer

Cultivation of algae for producing energy has the best potential as compared to the other alternatives like wind, hydro, solar and thermal in the present scenario where renewable energy has become a popular issue globally.

Algae offers a fast growth rate and best efficiency for converting sunlight into carbon as compared with the conventional energy crops.

Algae are an alluring feedstock for the generation of fluid and vaporous biofuels that don't ought to straightforwardly compete with nourishment generation.

Numerous alternatives are accessible with respect to algae sort and strain choice, counting both eukaryotic green growth and prokaryotic cyanobacteria, the source of water for development, development strategy and mode of development, the strategy of collection and so on.

The understanding of organic processes, algal hereditary qualities, carbon capacity digestion system, photosynthesis and algal physiology, have the potential for critical progresses in algal biofuel achievability.

Usually being driven by propels in genomic advances to supply the potential for hereditary and metabolic building, furthermore the improvement of high-throughput strategies for the screening of common strains for appropriate biofuel characteristics.

4.2 COMPOSITION

4.2.1 Macroalgae

Macroalgae or seaweed can be classified into brown, green and red algae based on the pigments present in them which give them their characteristic color.

Table 3 shows the composition of the three different types of seaweed or macroalgae along with their industrial uses and applications.

It can also be seen from the table that the water content for all the different types of seaweeds is quite high almost up to 80% which is much higher than the terrestrial plants.

They are generally rich in carbohydrates which constitute about 60% of the total weight followed by proteins which could vary between 10% to 50% and a very low amount of lipids (up to 3%).

Carbohydrates are mainly composed of polysaccharides like mannitol, starch and mannan which are easily soluble and result in monosaccharide units of glucose, mannose or galactose upon hydrolysis.

AD relies on easy hydrolysis and subsequent fermentation which is favorable in case of algal feedstock due to the presence of the aforementioned carbohydrates.

Complex sugars like cellulose and lignin are generally present in low amounts in algae due to the difference in structural requirements of marine biomass compared to terrestrial biomass.

This is another advantage of algae over conventional lignocellulosic biomass as complex carbohydrates take longer to hydrolyze thus slowing down the entire AD process (Barbot et al., 2016).

Table 4: Composition and uses of different macroalgae (adapted from Barbot et al., 2016)

Compound	Green algae	Red algae	Brown algae
Water content (fresh mass)	70%-85 %	70%-80%	79%-90%
Ash	18%-53%	26%-48%	33%-55%
Total organic	47%-82%	52%-74%	44%-66%

Carbohydrate	25%-50%	30%-60%	30%-50%
Polysaccharide	Alginate Cellulose Mannan Starch Ulvan	Agar Alginate Carraageenan Cellulose Lignin	Agar Alginate Carrageenan Cellulose Fucoidan Laminarin Mannitol
Protein	12%-13%	10%-16%	7%-12%
Lipid	2%-3%	0%-3%	0%-2%
Industrial extracts	Vitamins (Vit C), antiviral and anticoagulants	Vitamins (B and C), mineral nutrients, phycobiliproteins	Fucoidan, polyphenols, nutrients, fucan hydrocolloids (alginate, agar agar etc)
Industrial use	Human food, food supplement, medicinal use	Human and pet food, thickener, emulsifier, gelling agent in labs and industries	Human food, animal feed, alginate, medical fiber, agar in pharma industry, production of organic acids

While in case of the brown seaweed, cellulose is the main component of the cell wall, in green and red seaweeds, polysaccharides like xylan, and mannan are the main cell wall constituent.

- The polysaccharides themselves have different microfibril structures wherein mannans and cellulose have flat ribbon like configuration while xylans are helical.
- They are also associated with polysaccharides int the matrix which may be carboxylated or sulfated in different species.

Sulfated fucans may help in cellulosic backbone interlocking or may also be associated with the proteins.

Phenols also play an important role in structural stability by attaching with alginates to provide rigidity (Maneein et al., 2018).

Selecting a particular method of pretreatment may differ based on the chemical composition of the seaweeds which is quite similar for red and green seaweeds but strikingly different for brown seaweeds especially in terms of the water soluble and insoluble components.

Polysaccharides impart the unique characteristic to each different type of seaweed. Association of sulfate groups to different polysaccharides contributes to the high sulfur content in seaweeds and also varies in its amount in different phylums.

The different macroalgae not only differ in their pigment but also in their structural composition (fig 12 and 13). These structural and biochemical differences may influence the effect of pretreatment on the yields.

Seaweeds are a source of many industrially important carbohydrates like alginate, agar agar, carrageenans and so on.

These components are used commercially as gelling agents, emulsifiers and thickeners. Fucoidan is another sulfated polysaccharide which is used in the pharmaceutical industry and is extracted from the seaweed Fucus vesiculosus.

The by products like glycerol and mannitol are also used in chemical and pharmaceutical industry.

Besides the carbohydrates, there are other products like phenols, potash, phosphorous and so on which can be used for human and animal nutrition.

Seaweed is also consumed as human food in many Asian countries as it is rich in minerals like calcium and magnesium and also provides health benefits over high sodium food.

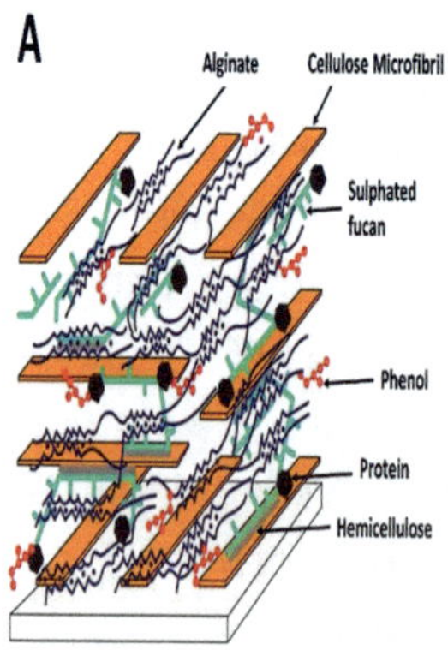

Figure 20: Cell wall model of brown seaweed (adapted from Maneein et al., 2018).

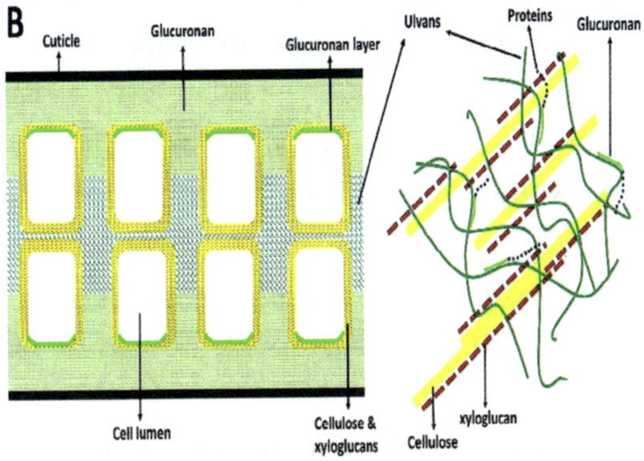

Figure 21: Cell wall polysaccharide distribution of green seaweed (Ulva spp.); far right figure shows closer interactions between polysaccharides (adapted from Maneein et al., 2018).

The composition of seaweed, however, varies with seasonal changes meaning that during warmer weather there is higher production of sugar and volatiles while higher protein and minerals during colder conditions.

This variation is also important in AD as it affects the methane production rates and biogas yield.

4.2.2 Microalgae

Microalgae can be either prokaryotic or eukaryotic microscopic organisms that can fix atmospheric carbon dioxide with the help of sunlight via the process of photosynthesis.

They have shown great potential as a feedstock for production of biodiesel.

The cell structure of microalgae is pretty simple, and they require the presence of sunlight, inorganic nutrients, carbon dioxide, water and minerals for their optimal growth and survival.

Chemically the major components of microalgal cell structure stored within the cell are lipids, proteins and carbohydrates.

High lipid content in microalgae is also a major reason to use it as a feedstock for biodiesel production.

Microalgae could produce up to 300 times higher lipid compared to the oil crops and may contain pigments like phycobiliproteins, carotenoids and chlorophyll.

These pigments are also useful in pharmaceutical, food, chemical and cosmetic industry. Cell wall composition of a few microalgal species can be seen in table 4.

They are flexible in terms of growth requirements as compared to the terrestrial plants, as has also been discussed above to be the case with macroalgae.

Another important advantage for growing microalgae is the growth rate or the doubling time which is about five times shorter than the conventional crops.

Figure 14 shows a few different types of microalgae that are used for commercial purposes.

Similar to any other living organisms, microalgae also need an optimum pH, temperature, light, and nutrients for their growth and also their characteristics. Carbon dioxide is the source of carbon and thus assists directly with cell development.

Table 5: Cell wall composition of microalgae (adapted from Torres et al., 2013)

Microalgae	Cell Wall	Cell Wall composition (%)			References
	(% w/w)	Carbohydrates	Protein	n.d.*	
Chlorella vulgaris (F)	20.0	30.00	2.46	67.54	(Abo-Shady et al. 1993)
Chlorella vulgaris (S)	26.0	35.00	1.73	63.27	(Abo-Shady et al. 1993)
Kirchneriella lunaris	23.0	75.00	3.96	21.04	(Abo-Shady et al. 1993)
Klebsormidium flaccidum	36.7	38.00	22.60	39.40	(Domozych et al. 1980)
Ulothrix belkae	25.0	39.00	24.00	37.00	(Domozych et al. 1980)
Pleurastrum terrestre	41.0	31.50	37.30	31.20	(Domozych et al. 1980)
Pseudendoclonium basiliense	12.8	30.00	20.00	50.00	(Domozych et al. 1980)
Chlorella Saccharophila	-	54.00	1.70	44,30	(Blumreisinger et al. 1983)
Chlorella fusca	-	68.00	11.00	20.00	(Blumreisinger et al. 1983)
Chlorella fusca	-	80.00	7.00	13.00	(Loos & Meindl 1982)

Monoraphidium braunii	-	47.00	16.00	37.00	(Blumreisinger et al. 1983)
Ankistrodesmus densus	-	32.00	14.00	54.00	(Blumreisinger et al. 1983)
Scenedesmus obliquos	-	39.00	15.00	46.00	(Blumreisinger et al. 1983)

The nutrients mainly required are nitrogen and phosphorous which may be derived from wastewater, thus, utilizing microalgae in wastewater treatment strategies. Usually, microalgae are harvested in ponds or photobioreactor based systems.

Lipids could make up to 50% of the total cellular weight. In a stressed environment, microalgae tend to accumulate higher amounts of lipids due to starvation and lack of nutrients (Kumar et al., 2017).

Regardless of the high lipid content, complex structure of the cell wall in microalgae is not favorable for direct oil extraction as the cell wall is covered with glycoproteins and complex carbohydrates. Algal species and growth medium composition directly affect the rigidity of the cell wall along with other physiological factors like temperature, pH, pressure and so on. Biomass concentration and the stage pf growth may also affect this.

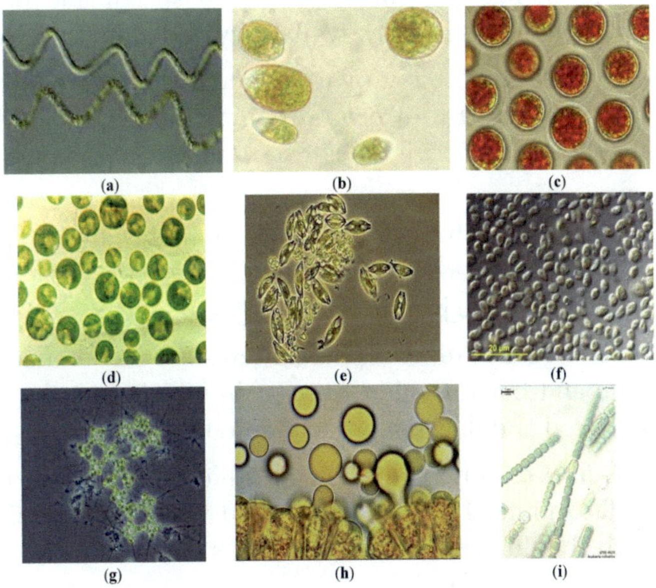

Figure 22: Different types of microalgae commercially cultivated for biofuel production (source: Benemann, 2013) [(a) Arthrospira platensis (Spirulina); (b) Dunaliella salina; (c) Haematococcus pluvialis; (d) Chlorella vulgaris; (e)

Amphora sp. ; (f) Nannochloropsis sp. ; (g) Micractinium sp. ; (h)Botryococcus braunii ; and (i) Anabaena cylindrica.].

Although photobioreactors are more costly in installation compared to ponds, there is better control over the conditions and also lesser area requirement in case of the former.

Lipids that are stored in microalgal cells can be classified as polar or neutral lipids.

Free fatty acids are mostly neutral while glycol- or phospho- lipids are polar lipids due to the presence of polar functional groups.

Polar lipids are generally involved in cell membrane functions while neutral lipids mainly provide energy for the cell.

It is preferred to extract lipids after drying the algae, but it is also quite energy intensive step. High water content may inhibit biodiesel production and also reduce the conversion efficiency. Thus, the challenge lies in the ability to select drying technology which is energy and cost efficient as it is a necessary step if you need to extract biodiesel for microalgae.

Despite these challenges, microalgae are a sustainable feedstock for biodiesel production due to its easy availability and abundance in nature.

Microalgae can either be grown in heterotrophic, phototrophic or mixotrophic cultures. In case of heterotrophic cultures, the carbon source used is organic which may be glucose or sucrose and is the source of energy for the cells.

For phototrophic cultures, however, sunlight along with carbon dioxide are the source of carbon and energy as algae have the ability to fix atmospheric carbon dioxide just the same way as plants.

As the name suggests, mixotrophic cultures are a combination of both the phototrophic and heterotrophic cultures in which way the cell growth rate can be enhanced and thus the production rate of lipids.

In order to select the microalgal specie which can produce high biogas yields, certain factors need to be kept in mind which include the cell wall thickness, which should preferably be as thin as possible.

The larger the cells, the larger the amount of biomass available. Several other qualities like high growth rates and high resilience against contaminants are also favorable.

Methane yields are highly dependent on the composition of microalgae and if the cell wall is too thick or resistant, yields may be compromised.

Algae that contain protein based cell wall instead of cellulosic material are easier to degrade and thus produce higher yields as has also been observed in previous research (Torres et al., 2013).

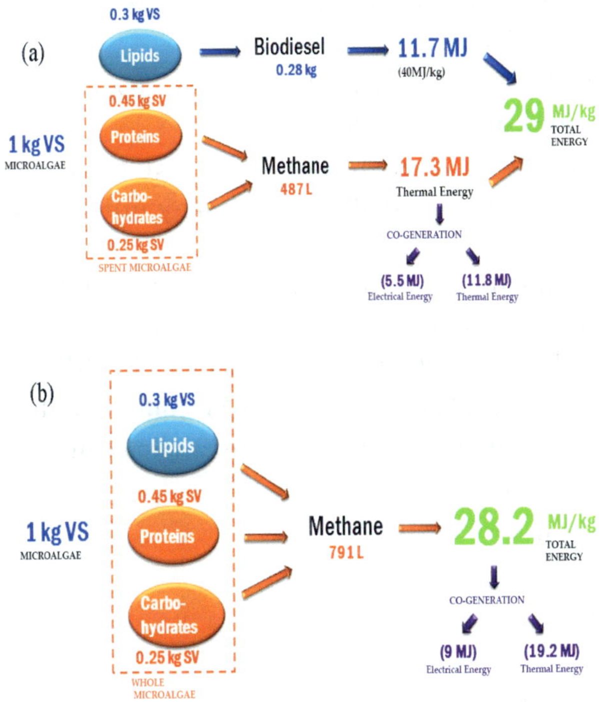

Figure 23: Energy potential of microalgae in scenario a and b (adapted from Torres et al., 2013); a) Biodiesel production from microalgae and AD using the residues to produce biogas; b) AD using all of algal biomass for biogas production.

4.3 PRETREATMENT METHODS FOR OPTIMIZING AD OF ALGAE

Algae have complex cell wall structure which makes them resistant to biodegradability and thus there has been research which focusses towards pretreatment methods to improve yield of biogas by algae AD.

Pretreatment methods are aimed at enhancing the rate of AD and thus to increase biogas and methane yield by increasing substrate bioavailability to the microbes.

The aim of the pretreatment is to assist both in transport and to improve the microbiology of the reactors.

The cellulosic and hemi-cellulosic biomass in algae can be pretreated in order to release sugars which can in turn be utilized during the anerobic fermentation.

Microalgal biomass differs in degradability even if they are closely related to each other at a species level.

This happens because the biogas is formed from the proteins, fats and carbohydrates in the microalgae which may differ in composition for different types.

AD can be limited by the rigid algal cell walls and thus it was realized early on that disruption of the cells was needed to enable efficient hydrolysis followed by the downstream processes(Behera et al., 2015).

Pretreatment helps enhance the rate of AD while solubilizing the organic matter. There are different types of pretreatment that can be done namely:

- Mechanical
- Thermal
- Chemical
- Biological

4.3.1 Mechanical Pretreatment

Mechanical pretreatment involves physical methods like milling, grinding, and so on to disintegrate the algal biomass into smaller sized particles.

This kind of pretreatment method is not highly dependent on the type of algae in question but as compared to the other treatment methods, this method may require higher input energy.

Mechanical pretreatment lets the hydrolyzing agents have better access to the polysaccharides that are important for maintaining the structure of microalgae.

It also increases the surface area of the biomass available for the microbes for AD and also reduces particle size by breaking up the structure of the biomass at a cellular level (Barbot et al., 2016).

This method has several advantages like reduction in the viscosity and also in the energy input required for agitation and mixing. Many feedstocks that are solid undergo mechanical pretreatment prior to the process of AD and may also be combined with other types of pretreatment methods.

One of the shortcomings of this method maybe the presence of grit or other solid materials like sand or pieces of metal which may damage the moving parts like blades and incur additional cost due to these damages.

There are different types of mechanical pretreatment methods available which include size reduction, sonication and beating.

Each of them has their own advantages and disadvantages and research has been done to study effect of these methods directly on the yield and efficiency of AD post the treatment steps.

There is a parameter known as biodegradability index which is the ratio of experimental yield to that of the theoretical yield and it is a great indicator of efficiency of different pretreatment methods.

Macroalgae do not have a lot of lignin and the amount of carbohydrates is quite high which makes them suitable for AD along with low levels of lipids. Mechanical pretreatment also allows easier handling of the feedstock which can improve the digestion and also allows combination with other pretreatment methods.

Grinding and milling are more commonly used for compact biomass using shredders, scissors and knives (Rodriguez et al., 2015).

The degree of polymerization can eb used by ball milling and the particle size can also be reduced thereby allowing the bulk density to increase.

Knife mill is one such equipment which has four to six knives on a steel rotor which spins at a speed of about 500-600 rpm. The feedstock is cut up to the point that it can pass through a screen.

- Size reduction

Size reduction can be achieved by essentially cutting the feedstock into smaller particles to increase the specific surface area for microbial action. This can be done by chopping and milling in an attempt to enhance the rate of hydrolysis of the complex cellulosic and hemi-cellulosic material.

This is the same method used for lignocellulosic material from plants but may not have the same effect on algae as it has on the terrestrial plant material.

The benefits of mechanical treatment are, to a large extent, determined by the structure of algae and the difference in the cell wall structures. The one that have more fibrous cell wall composition will have the maximum benefit from this type of treatment.

However, it has also been noted that VFA accumulation can be one of the outcomes of increased rates of hydrolysis due to mechanical pretreatment .thus leading to a drop in the pH

Disc mill is a milling technique in which the biomass is fed through a small opening in the same axis as the rotor and these rotating discs grind against a fixed disc.

The final effect is generation of pressure and centrifugal force which moves the feedstock to the outer regions.

Hammer mill is another type of milling technique that involves a set of hammers which push the material against a fixed plate. In this process, the material undergoes mixing and shearing simultaneously.

- Beating

Beating is essentially derived from mortar pestle kind of technique to reduce the size and pound the algae against a plate.

This enables the production of a pulp at different settings of the machine leading to different consistencies.

Researchers have found that Hollander beating can be a very effective method of pretreatment for several Laminaria sp. as compared to other methods.

- Washing

Washing is a way of removing solid and inert impurities from the algae such as sand, grit, stones, pieces of metal and so on. It also removes certain salts which can inhibit the process if their concentration exceeds a certain limit.

It has also been recommended by certain researchers to wash seaweed before downstream process involved in biofuel production in order to remove salts and other impurities.

In certain cases, washing led to removal of nitrogenous compounds when compared with the cold washed or unwashed counterparts. A lower ash to volatile solids ratio was observed to produce higher methane yields in case of hot washing which did not correlate with other types of washing.

Washing can also reduce the trace elements like nickel, cobalt and selenium which are very crucial for the microbes in AD and thus may be done with caution.

This can also be a reason for the lack of direct relationship between washing and methane yields and also lower ratios of ash to volatile solids after the washing process.

It can be concluded that washing may not have significant role in improving the yields, however, in the downstream process there could be great benefits and may also help save water in the further treatment steps.

4.3.2 Thermal Pretreatment

Thermal treatment of algae can assist in enhancing the solubilization of the polymeric organic matter which is only partially hydrolysed. Application of heat can be very useful in degradation fo lignin and cellulose thus improving the hydrolysis.

Cellulose crystalline structure consists of hydrogen bonds that can be disrupted by application of heat. Feedstocks that do not have enough water may need some water to be added before they can be thermally pretreated.

Viscosity is another important parameter for biological slurries and sludges which can be reduced via the application of thermal energy. However, methanogens in the AD process are quite sensitive to changes in physiological parameters including temperature and thus application of heat may have an inhibitory effect on them but not for the algae which do not contain bacteria.

While adding certain chemicals may improve the efficiency of the pretreatment, usually, thermal pretreatment does not require the addition of any chemicals. There is yet another benefit of thermal pretreatment which is indirectly sterilizing the feedstock as high temperatures can kill many pathogens.

Thermal pretreatment enables release of sugars and thus help in extraction of polysaccharides from macroalgae which are useful for many other industries besides AD.

It was observed that brown seaweed Nizimuddinia zanardini released almost 84% of the constituent sugars like mannitol and hemicelluloses at a temperature of 121°C and also produce 22% higher methane yield compared to the non-treated counterpart.

Although positive results have been obtained in several studies yet there are certain parts of seaweed which remain unaffected by thermal treatment like the cortex of the seaweed which directly indicated the significance of the seaweed structure.

This also calls for the selection of appropriate method of pretreatment depending upon the type and structure of different algal species.

4.3.3 Chemical Pretreatment

Hydrolysis of seaweed can be improved by addition of strong chemicals like acids or bases and oxidative agents which also enhance the solubilization of the seaweed. These treatments also enhance the biodegradation of the algal biomass while solublising the complex structural polymers.

Most commonly, these acids or alkalis solubilize the lignin and hemicellulose thereby increasing their bioavailability for the microbial enzymes during AD.

Alkalis have been found more effective in solubilizing lignin while acids have been found more effective for hemicellulose (Barbot et al., 2016).

Alkali like sodium, calcium and potassium hydroxides have been most commonly used for pretreatment of lignocellulosic biomass and its effectiveness is dependent on the amount of lignin present in the biomass.

Alkali result in saponification and thus breaking down the lignocellulosic linkages and increasing the surface area. This also results in reduction in the degree of polymerization and also disruption of the crystalline structure of the biomass.

It is not an economically favorable technology but in cases where lignin can otherwise not be degraded at all, alkali treatment may be used.

Acid pretreatment involves use of a strong acid like concentrated sulfuric acid, nitric acid and hydrochloric acid to break down the cellulosic polymers in the algal biomass.

Although its and effective method, it is energy intensive and not very economical. Another disadvantage of using string acids is that they are highly corrosive and thus special parts will be required while constructing the reactor.

Alkalis facilitate subsequent enzyme treatment of the biomass by causing the fibers to swell and the pore sizes to become bigger thereby letting the sugars release from the cell walls.

In case of acids, it was observed that more concentrated acid solutions were more effective in releasing the contents of the cell through the cell wall following hydrolysis.

Another important aspect of the acid pretreatment is production of furfural compounds during bioethanol production.

The differences in the degree of solubilization of seaweed for thermochemical pretreatment has been seen to be dependent on the composition of the seaweed and its biochemical make up. It has been seen the proportion of hydrolysis for different components of the seaweed varies with temperatures if all the other conditions are kept the same.

Moreover, high salt content can also be inhibitory to the fermentation process and thus lead to lower yields of ethanol or biogas. Use of acid and alkali can also up the salinity levels and thus this kind of pretreatment may not be suitable for algae which already have a high salt content (Rodriguez et al., 2015).

Seaweeds also have different affinity towards heavy metals due to different solubilization by acid hydrolysis. Also, ions like aluminum and iron can also affect the concentration of acid required for production of reducing sugars.

Peroxide treatment is another type of chemical pretreatment that uses heat to disrupt the crystalline structure of macroalgae and also cleaves the hydrogen bonds via the hydroxyl radicals. This treatment results in a better conversion rate from cellulose to glucose when compared with alkali or acid pretreatments.

4.3.4 Biological Pretreatment

Biological pretreatment involves the use of microbes and their enzyme to degrade the cellular structure of macro and micro algae. Different types of fungi are used to degrade the algal lignin and hemicellulose.

Laccase and peroxidase are two such enzymes that facilitate the degradation of lignin while. Pre-acidification or the microbial pretreatment is the step where hydrolysis and acidogenesis take place.

It is possible to achieve phase separation which also helps in avoiding the volatile fatty acid accumulation and the inhibitory effects of the same on methanogens.

Enzyme hydrolysis is very commonly used and researched biological pretreatment method and these enzymes occur in AD process already.

These enzymes include hemicellulases, cellulases and carbohydrases which can be a better and more energy efficient alternative to thermal, chemical or mechanical pretreatment methods.

Despite the advantages, this process can be quite slow and may take up to two weeks of residence time.

Direct addition of microbes has been found to be more effective in AD as they are able to degrade the complex carbohydrates as compared to using the enzymes by themselves.

It was observed that decomposed seaweed produced higher biogas amounts when compared with the fresh seaweed which again alludes to the effects of enzymatic hydrolysis.

Improvement in methane yields was observed when a mixture of macroalgae was exposed to *Trametes hirsuta* owing to the lignocellulose degrading enzyme present in it.

Research has been constantly focused towards isolating novel microbes and microbial enzymes which can assist in the hydrolysis of complex polysaccharides in the seaweed.

4.3.5 Microwave Treatment

Microwaves are electromagnetic waves of short wavelength and a frequency range of 0.3-300 GHz. These waves work by increasing the kinetic energy of water and thus making it boil.

During this process, macromolecules are polarized, and this causes disruption of their structure, generation of heat, changes in structure of biomolecules inside the cells thus leading to subsequent cell hydrolysis and release of the cellular components.

Microwaves function involves increase of temperature which is not cause by direct generation of heat but by the alignment of the polarized parts of the macromolecules in a magnetic field leading to breaking of their hydrogen bonds.

Microwave based heat generation is more uniform as compared to the conduction methods. This type of pretreatment can be used for many different types of feedstocks to improve their biogas yields. One important point to note here is that so far, microwave treatment has only been used for microalgae and not for macroalgae.

It is more difficult to lyse the cells in macroalgae due to their structural complexity.

4.4 CHALLENGES IN AD OF ALGAE

Biomass conversion efficiency is greatly dependent on the composition of the feedstock and the rate of hydrolysis which is often the rate limiting step.

Organic material which is easily degradable can be completely degraded as compared to complex polymeric feedstock which are recalcitrant towards degradation.

Certain feedstocks can also release substances that are inhibitory to the microbes required for the process. Macroalgae are rich in carbohydrates like agar, alginate, mannitol and cellulose which release their monomeric sugar units upon saccharification which include glucose, galactose and mannuronic acid molecules which are non-inhibitory for the microbes.

Some if the polysaccharides are comparatively easily hydrolysed by microbial saccharification through enzymatic cleavage of the polymeric chains.

Alginate in algae can increase the viscosity of the system due to its gelling abilities and thus decrease the available surface area for microbial action. It has been found, however, that certain microbes in AD can produce alginate lyase and agarose to hydrolyse alginate and agar respectively.

Research has revealed different factors that can inhibit the microbial conversion of algae via AD. As has also been discussed above, algal cell wall is one the key factors to slow down the rate of hydrolysis.

There are certain antimicrobial substances and toxic compounds that are also produced by algae which may be inhibitory to the AD process.

Another important factor is the imbalance in the carbon to nitrogen ratio considering that algae are rich in carbohydrates and low protein makes the nitrogen concentration quite low. Certain algal species also release polyphenols which have antioxidant activities and may hinder the degradation process.

Besides, there are sulfur compounds that are present in high concentrations, salts and heavy metals which may also affect the AD process.

It is also important to use appropriate microbes for inoculation. Bacterial communities can also be adapted to withstand high salt or heavy metal concentrations in order to ensure the degradation of algal feedstock without inhibition of the microbes.

An important factor to convert organic feedstock into methane was to achieve a positive energy balance which aims at increasing methane yields and helping in reduction of greenhouse gas emissions.

Seaweed can produce up to 200 L methane per gram of volatile solids which is much below the yields that are obtained by commercially used feedstocks like food waste or crop waste.

The yield can, however, depend upon the specie of algae used and on the operation parameters to a large extent and can rage anywhere between 20-80% of their theoretical potential (Barbot et al., 2016).

To summarize, there are different inhibitory factors that can make AD of algae more challenging and these factors include:

- Algal structure and cell wall
- Recalcitrant polysaccharides
- Heavy metals
- Carbon to nitrogen ratio
- Polyphenols, antimicrobial and toxins
- Sulfur

4.4.1 Algal Structure and Cell Wall

Composition of algae varies with season, environmental conditions and also the species which can be challenging in reproducibility of biomethane production from algae. The relative amounts of lipids, proteins and carbohydrates also directly or indirectly affect the methane production as lipids produce more specific methane volumes as compared to proteins or carbohydrates.

Moreover, methanogens and some other microbes in AD are quite sensitive to the feedstock composition.

Seaweed or macroalgae have low levels of lipids which makes them suitable for AD as lipids are more difficult to degrade due to the long fatty acid chains. Microalgae, on the other hand, can have high lipid content which can be increased in conditions of stress.

The problem with lipid digestion is that even though they have a high methane production potential, yet, at high concentrations they can cause blockages sue to recalcitrant fatty acid chains floating and sticking to the surfaces of methanogenic bacteria and hindering their growth.

It was observed that during rumen fermentation, biogas production seemed unaffected by the lipid breakdown pathway and seaweed degradation was found to be unrelated to lipid removal.

Long chain fatty acids break down via beta oxidation and further to acetate and hydrogen while the saturated fatty acids can directly enter the acetate hydrogen pathway.

Protein content may vary in seaweeds and can be quite low as compared to microalgae where protein accounts for more than 50% of their entire biomass.

In macroalgae, the protein concentration is not only low bur also the essential amino acid proportions are not well balanced. In case the protein content is high, carbon to nitrogen ratios will be more balanced which is true for microalgae.

However, it has been observed that higher protein content in microalgal species can also inhibit the AD process due to ammonia toxicity which is not favorable. Algae also have high ash content which can be problematic for disposal due to presence of heavy metals and also due to the quantity.

Cell wall composition of different algae is very different amongst themselves and also as compared to terrestrial plant while there are certain algal species that possess no cell wall at all.

Besides the structural differences between the red, green and brown algae, they also have differences in the cell wall composition of the polysaccharides.

4.4.2 Recalcitrant Polysaccharides

Up to 80% of the total dry weight of macroalgae can be composed of polysaccharides which are different from ones that are found in large plants. Polysaccharides like laminarin, fucoidan, mannitol and alginate are of tremendous industrial importance are found in brown seaweed.

The cellulose microfibrils are intricately intertwined in the complex mixture consisting of hemicellulose, phenols, alginate and fucans along with proteinaceous substances. This complex structure also makes them resistant to enzymatic degradation by the microbes.

Algae also contain wide range of monomeric units of sugars when compared with the terrestrial plants. Cell wall skeleton is primarily composed of cellulose in case of brown algae while for green and red algae it is made of mannan, xylan and cellulose.

Hydrolysis of these polysaccharides in macroalgae often ends up being the rate limiting step in AD. Seaweed contains sulfated polysaccharides which are different from those found in vascular plants.

These include galactans in a linked in a sequence which typically occurs in red algae as agar, in brown algae as fucoidans and alginates and as ulvans in green algae.

4.4.3 Salt and Heavy Metals

Algae that grow in hypersaline conditions results in an increase in their salt content which is much higher than the vascular plants. It is thus important to wash the seaweeds in order to remove as much salt as possible as otherwise it may increase the production costs in further processing.

Low concentrations of salt can actually support microbial growth while higher concentrations can inhibit the process due to increased osmotic pressure and methanogens getting dehydrated in high salt conditions.

High salt is one of the main inhibitory factors in AD of freshwater algae as the presence of sodium and potassium ions at high concentrations can adversely affect the methanogenic populations.

One of the strategies to adapt the anaerobic reactors to high salt conditions is to gradually increase the salt concentration so that the microbes can get acclimatized to high salt.

Osmolytes like trehalose and betaine can also be added to the digester and help the microorganisms withstand the increased osmotic pressures.

It has also been found out that addition of these compounds enables methanogens to flourish even at high slat conditions. However, osmolytes are not inexpensive and thus cost can become an issue in large scale AD.

Salt concentration can also be reduced by simply diluting the algal biomass with other forms of biomass which eliminates the requirement for water addition and also the production costs.

4.4.4 Polyphenols, Antimicrobial and Toxins

Phenols are organic compounds containing a benzene ring attached to a hydroxyl group and their derivatives are called as phenolic compounds. These are found in a variety of green plants and algae.

Algae contain a large amount of phenolics which constitute up to 14% of its dry weight. The phenolics that occur in algae are quite different from those found in vascular plants. Phlorotannins are the most commonly occurring phenols in algae.

Algae are inhibited by the phenolic compounds by their cell permeability being altered by the phenols as a result of which the intracellular components leak out of the cell and the enzymes are deactivated.

This can significantly affect the methane yields during AD as the hydrolytic enzymes would also be affected by the phenolic compounds.

It was observed that phenols did not affect the degradation of simple compounds that could be easily digested but they did inhibit the breakdown of complex polymers in the early stages of AD.

Polyphenols also have high ash content when compared to the volatile solid content and this could have inhibitory effects on methane yields as has also been discussed above.

4.4.5 Sulfur

Algae have a higher organic sulfur content as compared to the vascular plants. High sulfur concentration can result in production of hydrogen sulfide, a very potent greenhouse gas, which is not favorable for the AD process.

Moreover, it also produces foul smell and is corrosive to the metal. Chapter 3 discusses how to remove hydrogen sulfide from biogas.

4.5 CONCLUSIONS

Algal biomass presents itself as a source for producing renewable energy in the form of biomethane by AD. Currently, there is lack of commercial processes that use algae as feedstock as it is not profitable and presents many limitations.

These limitations include the high water content of algae, requirement of pretreatment for recalcitrant species, costs of operation for enhancing the bioconversion efficiency and harvesting costs incurred.

A biorefinery approach may also be a step forward to use algae for food and feed, pharmaceuticals and also for the biofuel production in a sustainable manner.

For realization of the full potential of algal biomass, it is important to identify the most economically and environmentally appropriate method of pretreatment.

Research focusing on the reduction of heavy metals in seaweed would be very beneficial. Algae also have an immense potential in the wastewater treatment systems and it also seems to be a more economically beneficial choice.

5

Impact of Biogas Technology

Navodita Bhatnagar

Biogas based technology is relatively quite cheap and can thus be set up easily with very little investment at a household level. These small scale reactors can eb used in households by operating on kitchen or animal waste and the cost of installation can be covered within a few months of operation. The feedstock that goes in comes free of cost and the biogas generated can be used directly for cooking and also for generating electricity. Farms can also be installed with small scale digesters that can work on crop ad animal waste and provide electricity to operate farm equipment or even to generate heat and light. However, there is always another side to every story and biogas technology is not spared from this rule. This chapter discusses the positive and negative impacts of biogas technology.

5.1 BENEFITS OF BIOGAS

Biogas is a clean source of energy and is generated by AD of organic matter thereby helping in reduce GHG emissions.

- Growing environmental concerns have also become a reason for widespread popularity of biogas technologies in both rural and urban population.
- Biogas has helped divert exploitation of natural resources, especially, coal, petroleum and natural gas, to a renewable and relatively inexpensive source of energy.
- Landfilling of waste had been going on for many years until countries started running out of land for continuing this practice for waste disposal.
- AD has since become an economically and environmentally

sustainable way of waste disposal while also discouraging the practice of landfilling waste.

- This not only saves land and water pollution because of landfill leachate but also avoids the foul smell arising from these landfills.
- It also indirectly helps in improving water quality in the neighboring area along with reduction in pathogen load which in turn helps in reducing incidence of waterborne disease.
- Digestate from biogas is also an enriched in organic matter and nutrients thus making it a perfect alternative or supplement to chemical fertilizers. Commercial fertilizers contain toxic chemicals that can enter the food chain through the food crops. Digestate is an organic fertilizer which doesn't involve such issues.

The advantages of biogas in rural areas are enormous. Usually, cow dung cakes or firewood or coal would be used in rural household to light the kitchen stove which is extremely dangerous to the health of those who inhale the gaseous fumes released when this kind of fuel is burnt.

Millions of people die every year due to air pollution and use of solid fuel for cooking which release toxic fumes. Substituting these materials with anaerobic digestors and producing biogas would prevent these problems ("Benefits of Biogas | PlanET Biogas Solutions," n.d.). Advantages of biogas technology can also be seen in figure 19.

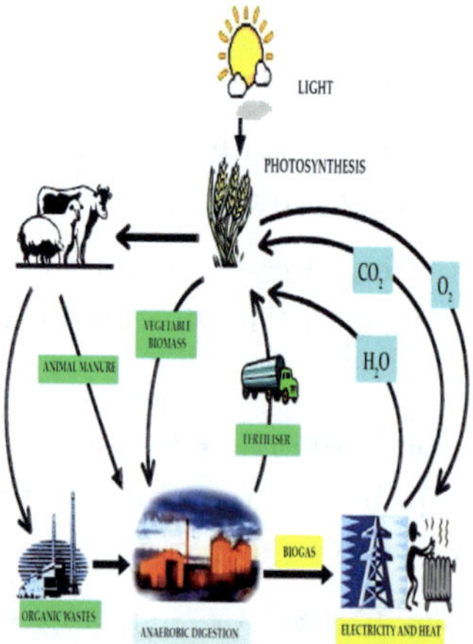

Figure 24: Advantages of biogas (source: "Advantages and Disadvantages of Biogas," n.d.).

Biogas as a renewable source of energy allows the use of agricultural and industrial waste and by products to be utilized with minimal impact on the quality of air compared to direct incineration of such biomass. Moreover, ash from incineration does not have much economic or agronomic value while the by-product of AD is a rich source of nutrients and excellent fertilizer for crops.

AD also controls emission of methane by capturing it from waste which would otherwise be dumped in landfill or left open to release methane into the atmosphere. Methane is a much more potent greenhouse gas than carbon dioxide. Approximately 33% of methane emissions globally are caused due to agricultural practices.

Methane emissions form sources like animal manure can eb used by anaerobically treating them. Not only this but AD also significantly reduces emissions of another very harmful greenhouse gas which is nitrous oxide.

Animal and human waste processed in a close system as in case of anaerobic digesters also improves the hygiene and sanitary conditions of rural communities. Pathogen load is reduced during the process of AD as has also been previously discussed.

As opposed to fossil fuels, biomethane is available as long as there is biomass and thus it is a renewable energy source. It can also be produced in any part of the world unlike petroleum or natural gas which are only produced in certain parts of the world and many countries have to depend on imports for their energy needs.

Biomethane production also doesn't require large scale interventions like drilling holes in the ground or mining for oil and coal respectively thus posing a risk to not only the environment but also to humans who work in these conditions. This way, biogas and biomethane production do not pose a risk to the biodiversity or pollution (Bhardwaj and Das 2017).

Biogas derived energy can also be used in tube well engines in rural areas and also for street lighting. Resource conservation is very important in today's day and age which is achieved in biogas production as no further fuel consumption occurs due to lack of oxygen requirement.

As the technology of biogas consumption is advancing, the applications of biogas are also increasing. Not only this but setting up biogas plants also provides revenue and jobs to the local communities. The capital investment required is minimal while the benefits are enormous.

5.2 NEGATIVE IMPACTS OF BIOGAS COMBUSTION

The positive impacts of biogas outnumber the negative impacts associated with this technology. However, it is still important to discuss the existing or the potential ill effects that biogas technology may have on the environment and human health if at all.

Social opposition is faced very often on installation of biogas plants based on the health issues for humans and the environmental concerns. The resistance depends on many factors including the country in question, their policies and the awareness amongst the communities. Some of the environmental concerns have been discussed in the sections below:

5.2.1 Greenhouse Gas Emissions

Substituting fossil fuels with biogas and thus mitigating greenhouse gas emissions and reducing dependency on fossil fuels have been the main goals of this technology. Anaerobic digestion, however, is itself associated with emission of a few greenhouse gases which include carbon dioxide, methane and hydrogen sulfide.

Thus, it becomes very important to take appropriate measures to first reduce these emissions and second avoid any leakages. These measures include avoiding methane discharge, sealing and covering the tanks, improving efficiency of combined heat and power units and also the utilization of electric power. Biogas has global warming potential that needs to be reduced thus it must be ensured that maximum amount of thermal energy is exploited, and leakages are minimized.

The results too highlighted that solid gauges of greenhouse gas emissions within the case of power generation from biogas can be as it were made on the premise of person observing information, for occasion: diminishment of coordinate methane outflow and spillage, misusing of warm gotten from cogeneration, sum and nature of input fabric, nitrous oxide emanation (e.g. from vitality edit development) and digestate administration.

5.3 BIOGAS IN DEVELOPING COUNTRIES

Circular economy benefits both urban and rural population alike just as it benefits the developed and the developing world. There are many advantages of a circular economy which are more concrete for rural populations more than the urban population.

Anaerobic digestion is a technology which is a classic example of technologies suited for circular economies. Any self-sufficient, environmentally sustainable technology befits perfectly in a circular economy model.

AD is also energy independent and a source of renewable energy which attracts investment from public as well as private sector in many countries.

Biogas as a fuel is becoming more and more popular in developing countries owing to its multifold benefits in terms of environmental sustainability and also creation of revenue and jobs to benefit the locals.

In developed countries, a centralized approach has been more profitable due to government subsidies and large scale applications of AD in agriculture and treatment of household waste. Establishment of farm scale AD reactors has also helped farmers in these countries earn extra income and also produce energy for consumption on the farm itself.

Wastewater treatment plants in conjunction with AD also became quite popular in throughout Europe due to the twofold benefits of waste reduction and energy generation. Wastewater undergoes primary, secondary, tertiary and quaternary treatments to recycle it back into the water supply.

While it is more favorable to use a centralized approach in developed countries, it may not be as well suited for developing countries where the farms are not as large scale, most of the agricultural activities are done in rural areas where there are fewer means of investment in large scale reactors.

AD is the only technology that fully utilizes the potential of organic waste or non-waste to produce energy and nutrients. It is biological and thus also avoid use of any toxic chemicals that may harm the environment. It is also a naturally occurring process in cattle rumen which makes its reproducibility easier.

AD reactors can be installed at a small scale as well and do not always require huge investments as the biogas can be converted into combined heat and power (CHP) instead of upgrading the biogas to biomethane and injecting it into the natural gas grid.

For large scale reactors, it is generally preferred to purify the biogas and inject the upgraded gas to the grid which can be used for both electricity and as transport fuel.

Another important advantage of AD in developing countries is that most of these countries have a tropical to mildly temperate climate. As the process takes place in the mesophilic temperature range, cost of providing external heating can be avoided.

The reasons discussed above makes AD an ideal technology for developing countries and also a relatively low cost technology.

REFERENCES

1. Achinas, S., Achinas, V., Euverink, G.J.W., 2017. A Technological Overview of Biogas production from Biowaste. Engineering 3, 299–307. https://doi.org/10.1016/J.ENG.2017.03.002
2. Advantages and Disadvantages of Biogas [WWW Document], n.d. . Homebiogas Househ. Biogas Dig. Syst. URL https://www.homebiogas.com/Blog/141/Advantages_and_Disadvantages_of_Biogas (accessed 7.12.19).
3. Al Mamun, M.R., Torii, S., 2015. Enhancement of production and upgradation of biogas using different techniques- a review. Int. J. Earth Sci. Eng. 8, 877–892.
4. Amoo, O.M., Fagbenle, R.L., 2013. Renewable municipal solid waste pathways for energy generation and sustainable development in the Nigerian context. Int. J. Energy Environ. Eng. 4, 42. https://doi.org/10.1186/2251-6832-4-42
5. Angelidaki, I., Alves, M., Bolzonella, D., Borzacconi, L., Campos, J.L., Guwy, A.J., Kalyuzhnyi, S., Jenicek, P., van Lier, J.B., 2009. Defining the biomethane potential (BMP) of solid organic wastes and energy crops: a proposed protocol for batch assays. Water Sci. Technol. 59, 927–934. https://doi.org/10.2166/wst.2009.040
6. Arvanitoyannis, I.S., Kassaveti, A., 2008. Fish industry waste: treatments, environmental impacts, current and potential uses. Int. J. Food Sci. Technol. 43, 726–745. https://doi.org/10.1111/j.1365-2621.2006.01513.x
7. Atandi, E., Rahman, S., 2012. Prospect of anaerobic co-digestion of dairy manure: a review. Environ. Technol. Rev. 1, 127–135. https://doi.org/10.1080/09593330.2012.698654
8. Awe, O.W., Zhao, Y., Nzihou, A., Minh, D.P., Lyczko, N., 2017. A Review of Biogas Utilisation, Purification and Upgrading Technologies. Waste Biomass Valorization 8, p.267-283. https://doi.org/10.1007/s12649-016-9826-4
9. Barbot, Y.N., Al-Ghaili, H., Benz, R., 2016. A Review on the Valorization of Macroalgal Wastes for Biomethane Production. Mar. Drugs 14, 120. https://doi.org/10.3390/md14060120
10. Behera, S., Singh, R., Arora, R., Sharma, N.K., Shukla, M., Kumar, S., 2015. Scope of Algae as Third Generation Biofuels. Front. Bioeng. Biotechnol. 2. https://doi.org/10.3389/fbioe.2014.00090

11. Benefits of Biogas | PlanET Biogas Solutions [WWW Document], n.d. URL http://www.planet-biogas.ca/benefits/ (accessed 7.12.19).
12. Benemann, J., 2013. Microalgae for Biofuels and Animal Feeds. Energies 6, 5869–5886. https://doi.org/10.3390/en6115869
13. Caruso, M.C., Braghieri, A., Capece, A., Napolitano, F., Romano, P., Galgano, F., Altieri, G., Genovese, F., 2019. Recent Updates on the Use of Agro-Food Waste for Biogas Production. Appl. Sci. 9, 1217. https://doi.org/10.3390/app9061217
14. Chandra, R., Vijay, V.K., Subbarao, P.M.V., 2012. Vehicular Quality Biomethane Production from Biogas by Using an Automated Water Scrubbing System [WWW Document]. Int. Sch. Res. Not. https://doi.org/10.5402/2012/904167
15. Di Matteo, U., Nastasi, B., Albo, A., Astiaso Garcia, D., 2017. Energy Contribution of OFMSW (Organic Fraction of Municipal Solid Waste) to Energy-Environmental Sustainability in Urban Areas at Small Scale. Energies 10, 229. https://doi.org/10.3390/en10020229
16. Girotto, F., Alibardi, L., Cossu, R., 2015. Food waste generation and industrial uses: A review. Waste Manag., Urban Mining 45, 32–41. https://doi.org/10.1016/j.wasman.2015.06.008
17. Maneein, S., Milledge, J.J., Nielsen, B.V., Harvey, P.J., 2018. A Review of Seaweed Pre-Treatment Methods for Enhanced Biofuel Production by Anaerobic Digestion or Fermentation. Fermentation 4, 100. https://doi.org/10.3390/fermentation4040100
18. Mao, C., Feng, Y., Wang, X., Ren, G., 2015. Review on research achievements of biogas from anaerobic digestion. Renew. Sustain. Energy Rev. 45, 540–555. https://doi.org/10.1016/j.rser.2015.02.032
19. MedCrave, 2017. Cultivation and Chemical Composition of Microalgae Chlorella Vulgaris and its Antibacterial Activity against Human Pathogens. J. Aquac. Mar. Biol. Volume 5. https://doi.org/10.15406/jamb.2017.05.00119
20. Meegoda, J.N., Li, B., Patel, K., Wang, L.B., 2018. A Review of the Processes, Parameters, and Optimization of Anaerobic Digestion. Int. J. Environ. Res. Public. Health 15, 2224. https://doi.org/10.3390/ijerph15102224
21. Mir, M.A., Hussain, A., Verma, C., 2016. Design considerations and operational performance of anaerobic digester: A review. Cogent Eng. 3, 1181696. https://doi.org/10.1080/23311916.2016.1181696

22. Moody, L., Burns, R., Wu-haan, W., Spajic, R., 2009. Use of Biochemical Methane Potential (BMP) Assays for Predicting and Enhancing Anaerobic Digester Performance. Proc. 44th Croat. 4th Int. Symp. Agric. Opatija Croat. 16-20 Febr. 2009.
23. Morales-Polo, C., Cledera-Castro, M.D.M., Moratilla Soria, B.Y., 2018. Reviewing the Anaerobic Digestion of Food Waste: From Waste Generation and Anaerobic Process to Its Perspectives. Appl. Sci. 8, 1804. https://doi.org/10.3390/app8101804
24. Muñoz, R., Meier, L., Diaz, I., Jeison, D., 2015. A review on the state-of-the-art of physical/chemical and biological technologies for biogas upgrading. Rev. Environ. Sci. Biotechnol. 14, 727–759. https://doi.org/10.1007/s11157-015-9379-1
25. Ogbuewu, I.P., Odoemenam, V.U., Omede, A.A., Durunna, C.S., Emenalom, O.O., Uchegbu, M.C., Okoli, I.C., Iloeje, M.U., 2012. Livestock waste and its impact on the environment. Sci. J. Rev. 17.
26. (PDF) A Review: Advantages and Disadvantages of Biogas | IRJET Journal - Academia.edu [WWW Document], n.d. URL https://www.academia.edu/35116037/A_Review_Advantages_and_Disadvantages_of_Biogas (accessed 7.12.19).
27. Pravinkumar, D., Patil, P., P Dehankar, S., Tathe, D., R Hande, V., S Patil, N., 2019. BIOGAS PRODUCTION FROM KITCHEN WASTE: A REVIEW.
28. RedCorn, R., Fatemi, S., Engelberth, A.S., 2018. Comparing End-Use Potential for Industrial Food-Waste Sources. Engineering 4, 371–380. https://doi.org/10.1016/j.eng.2018.05.010
29. Rodriguez, C., Alaswad, A., Mooney, J., Prescott, T., Olabi, A.G., 2015. Pre-treatment techniques used for anaerobic digestion of algae. Fuel Process. Technol. 138, 765–779. https://doi.org/10.1016/j.fuproc.2015.06.027
30. Ryckebosch, E., Drouillon, M., Vervaeren, H., 2011. Techniques for transformation of biogas to biomethane. Biomass Bioenergy 35, 1633–1645. https://doi.org/10.1016/j.biombioe.2011.02.033
31. Sarker, S., Lamb, J.J., Hjelme, D.R., Lien, K.M., 2019. A Review of the Role of Critical Parameters in the Design and Operation of Biogas Production Plants. Appl. Sci. 9, 1915. https://doi.org/10.3390/app9091915

32. Sorathiya, L.M., Fulsoundar, A.B., Tyagi, K.K., Patel, M.D., Singh, R.R., 2014. Eco-friendly and modern methods of livestock waste recycling for enhancing farm profitability. Int. J. Recycl. Org. Waste Agric. 3, 50. https://doi.org/10.1007/s40093-014-0050-6

33. Toma, L., Voicu, G., Ferdeş, M., Dinca, M., 2016. ANIMAL MANURE AS SUBSTRATE FOR BIOGAS PRODUCTION.

34. Torres, Á., Fermoso, F.G., Rincón, B., Bartacek, J., Borja, R., Jeison, D., 2013. Challenges for Cost-Effective Microalgae Anaerobic Digestion. Biodegrad. - Eng. Technol. https://doi.org/10.5772/55975

35. Ye, M., Liu, J., Ma, C., Li, Y.-Y., Zou, L., Qian, G., Xu, Z.P., 2018. Improving the stability and efficiency of anaerobic digestion of food waste using additives: A critical review. J. Clean. Prod. 192, 316–326. https://doi.org/10.1016/j.jclepro.2018.04.244

PART – II
APPLICATIONS OF BIOGAS TECHNOLOGY

6

Recent Updates on Biogas Production - A Review

Ilona Sárvári Horváth[1], Meisam Tabatabaei[2,3], Keikhosro Karimi[4,5], Rajeev Kumar[6]

[1] Swedish Centre for Resource Recovery, University of Borås, 501 90 Borås, Sweden

[2] Microbial Biotechnology Department, Agricultural Biotechnology Research Institute of Iran (ABRII), AREEO, Karaj, Iran

[3] Biofuel Research Team (BRTeam), Karaj, Iran

[4] Department of Chemical Engineering, Isfahan University of Technology, Isfahan 84156-83111, Iran

[5] Microbial Industrial Biotechnology Group, Institute of Biotechnology and Bioengineering, Isfahan University of Technology, Isfahan 84156-83111, Iran

[6] Center for Environmental Research and Technology (CE-CERT), Bourns College of Engineering, University of California, Riverside, California, USA

ABSTRACT

One of the greatest challenges facing the societies now and in the future is the reduction of green house gas emissions and thus preventing the climate change. It is therefore important to replace fossil fuels with renewable sources, such as biogas. Biogas can be produced from various organic waste streams or as a byproduct from industrial processes. Beside energy production,

Citation: Sárvári Horváth, I., Tabatabaei, M., Karimi, K., Kumar, R. (2016). Recent updates on biogas production - a review. Biofuel Research Journal, 3(2), 394-402. doi: 10.18331/BRJ2016.3.2.4

the degradation of organic waste through anaerobic digestion offers other advantages, such as the prevention of odor release and the decrease of pathogens. Moreover, the nutrient rich digested residues can be utilized as fertilizer for recycling the nutrients back to the fields. However, the amount of organic materials currently available for biogas production is limited and new substrates as well as new effective technologies are therefore needed to facilitate the growth of the biogas industry all over the world. Hence, major developments have been made during the last decades regarding the utilization of lignocellulosic biomass, the development of high rate systems, and the application of membrane technologies within the anaerobic digestion process in order to overcome the shortcomings encountered. The degradation of organic material requires a synchronized action of different groups of microorganisms with different metabolic capacities. Recent developments in molecular biology techniques have provided the research community with a valuable tool for improved understanding of this complex microbiological system, which in turn could help optimize and control the process in an effective way in the future.

INTRODUCTION

Biogas production through anaerobic digestion (AD) is an environmental friendly process utilizing the increasing amounts of organic waste produced worldwide. A wide range of waste streams, including industrial and municipal waste waters, agricultural, municipal, and food industrial wastes, as well as plant residues, can be treated with this technology. It offers significant advantages over many other waste treatment processes. The main product of this treatment, i.e., the biogas, is a renewable energy resource, while the byproduct, i.e., the digester residue, can be utilized as fertilizer because of its high nutrient content available to plants (Ward et al., 2008). The performance of the AD process is highly dependent on the characteristics of feedstock as well as on the activity of the microorganisms involved in different degradation steps (Batstone et al., 2002). The conversion of organic matters into biogas can be divided in three stages: hydrolysis, acid formation, and methane production. In these different stages which are however carried out in parallel, different groups of bacteria collaborate by forming an anaerobic food chain where the products of one group will be the substrates of another group. The process proceeds efficiently if the degradation rates of the different stages are in balance (Yong et al., 2015).

This review presents an overview of the biogas industry worldwide and discusses some new technologies aiming at utilizing new substrates and enhancing the efficiency of the process.

BIOGAS, DRIVING FORCES AND THE BIOGAS INDUSTRY

There is an increasing interest in bioenergy production across the world for environmental as well as economic and social reasons. The production of biogas contributes to the production of renewable and sustainable energy since biogas works as a flexible and predictable alternative for fossil fuels. The main political driving forces linked to the biogas system has a countryspecific variation (Huttunen et al., 2014). Within the European Union, well-developed biogas industry can be found in Germany, Denmark, Austria, and Sweden followed by the Netherlands, France, Spain, Italy, the United Kingdom, and Belgium. In these countries, with a strong agro-sector, reduction of nutrient emissions and renewable energy production are equally strong driving forces supporting biogas production. In other countries, like Portugal, Greece, and Ireland, as well as in many of the new East-European member states, the biogas sector is currently under development, due to the identified large potential for biomass utilization there.

The biogas plants in Europe are classified based on the type of digested substrates, the technology applied, or the size of the plant. In this sense, they are usually considered as (1) large scale, joint co-digestion plants or (2) farm scale plants. Nevertheless, there are no major differences between these two categories regarding the technology used.

Joint Co-digestion Plants

Simultaneous digestion of a mixture of two or more substrates is called co-digestion. The coexistence of different types of residues in the same geographic area enables integrated management, offering considerable environmental benefits, like energy savings, recycling of nutrients back to the agricultural land, and reduction of CO_2 emissions (Kacprzak et al., 2010).

Due to the different characteristics of waste streams treated together, co-digestion may enhance the performance of the AD process owing to a positive synergism established in the digestion medium by providing a balanced nutrient supply and sometimes by suitably increasing the moisture content required in the digester (Mata-Alvarez et al., 2000).

Joint biogas plants are referred to large scale plants, with digester capacities ranging from few hundreds m3 up to several thousands m^3 . Different organic waste streams are collected and transported to the plant and co-digested there. The process is running either at thermophilic or mesophilic conditions, using hydraulic retention times (HRT) of 12–25 d. HRT is normally inversely proportional to the process temperature. Generally, the substrates and in particular animal by-products, which are to be sent to the digester, first go through a controlled pre-sanitation phase, to inactivate pathogens and to break their propagation cycles. After the AD process, the digested residue is transferred to storage tanks, which are typically covered with a gas proof membrane for the recovery of the remaining gas and to prevent methane leakage to the atmosphere. The digested residue has a high nutrient content, and therefore, it can be recycled to the fields as fertilizer. The produced biogas is utilized as a renewable energy source.

In Europe, biogas is mainly used for generating heat and electricity. Some of the produced heat is utilized within the biogas plant as process heating and the remaining heat is distributed through districts` heating systems to consumers. The produced power is sold to the grid. In some countries, like Sweden, the produced biogas is upgraded to bio-methane which is utilized as vehicle fuel (Nielsen et al., 2002; Persson et al., 2006). Figure 1 shows the biogas production cycle within an integrated system.

Recently, co-digestion has taken much attention since it is one of the interesting ways of improving the yield of AD. Most of the investigations on co-digestion were carried out in batch operation mode and many researchers have pointed out the influence of synergy, due to a balanced mixture composition, on methane yield (Misi and Forster, 2001; Pagés Díaz et al., 2011; Esposito et al., 2012; Wang et al., 2012; Pagés-Díaz et al., 2014). Pagés-Díaz et al. (2011) reported that it was possible to relate synergetic effects with up to 43% enhancement in methane yield (YCH_4) compared with the expected YCH_4 calculated on basis of methane potentials obtained for the individual substrates. The substrates investigated were four different waste streams, such as slaughterhouse waste, various crop residues, manure, and the organic fractions of municipal solid waste (OMSW).

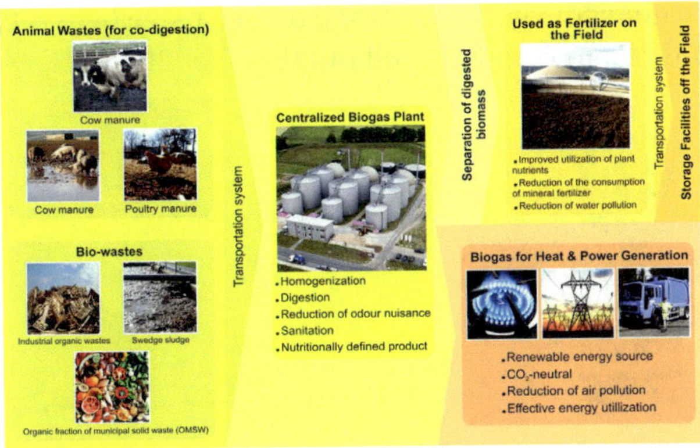

Figure 1: The main streams of the integrated concept of a centralized biogas plant (adapted from Holm-Nielsen et al., 2004).

A successful co-digestion is not simply a digestion of several waste streams treated at the same time. In fact, biogas production and the stability of the process are highly dependent on waste composition, process conditions, and the activity of microbial community in the system. In that sense, for certain mixing ratios, co-digestion may also lead to antagonistic interactions, resulting in methane yields lower than expected (Pagés-Díaz et al., 2014 and 2015).

The Farm Scale Biogas Plants

It has been reported that more than 4,000 farm scale biogas digesters were in operation in Germany; followed by about 350 in Austria, 72 in Switzerland, 65 in the United Kingdom, 35 in Denmark, and 12 in Sweden (Raven and Gregersen, 2007; Wilkinson, 2011). The main substrate fractions, which are utilized in these farm scale biogas plants are animal manure and energy crops. One of the important aspects of biogas production for farmers is to reduce leaching of nutrients from agricultural lands to the aquatic environments (Bojesen et al., 2014). Hence, farm scale plants are usually established at large pig farms, aiming at solving the problems caused by the excessive slurry production. Figure 2 presents the closed cycle of organic waste AD and the main steps involved in the quality management process. The most common and recent digester type that is used in farm scale applications is a vertical tank generally made of concrete and equipped with a flexible membrane and light roof making it possible to be used as digester and gas-storage tank simultaneously. The average digester size here is typically from a couple of hundreds to one thousand m3 (Garcia, 2005).

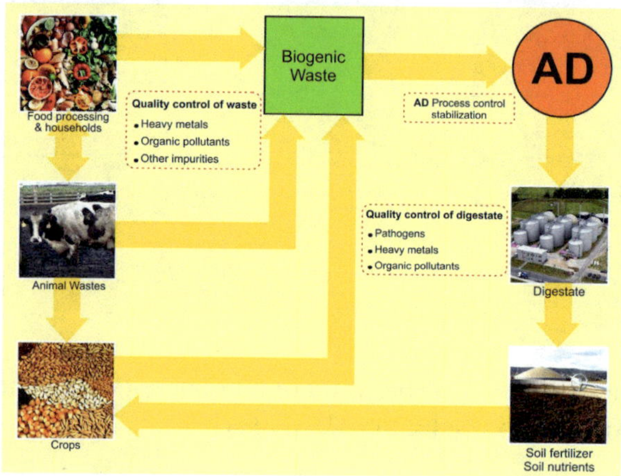

Figure 2: Schematic representation of the closed cycle of anaerobic digestion of organic waste and the main steps involved in the quality management process (adapted from Al Seadi (2002)).

Domestic Biogas Technologies in Developing Countries

Domestic biogas digesters are abundant in developing countries, especially Asian countries, such as Nepal or Vietnam. Prior to thc development of domestic biogas projects, it is important to check the current biogas diffusion in a given country in order to realize the maturity of the sector. The definition of national diffusion targets (i.e., a targeted amount of biogas units that should be built within a specified time frame) by the governments also provides information about the actual diffusion levels. In many countries already promoting domestic biogas production, the governments have implemented national programs aiming at establishing a proper biogas sector. Such programs typically include financing schemes, as well as training campaigns for local workforce, and providing technical support to project developers. These programs involve different players including non-profit organizations cooperating together with the local public institutions and the private sector in order to benefit potential synergies. The German GIZ (Society for International Cooperation, formerly GTZ) and the Dutch SNV are the two main international organizations acting worldwide for domestic biogas advancement, delivering technical service and documentation on this issue.

Some countries like India, Nepal, and China host much more domestic biogas plants than others. It has been reported that about 250,000 domestic plants were installed within the past 20 years in Nepal and 125,000 in Vietnam. Furthermore, 12,500 domestic biogas units are planned to be installed by the end of 2016 in Rwanda, 8,000 in Kenya, and 12,000 plants in Tanzania (Rakotojaona, 2013; TDBP, 2013; Cheng et al., 2014). The domestic biogas plant development is only at an earlier stage in Peru compared with the other Latin American countries. In 2013, the Dutch development organization in cooperation with the Peruvian, planned to set up a national program to construct 10,000 domestic biogas plants within the next 5 years (Rakotojaona, 2013).

CURRENT BIOGAS PROCESS TECHNOLOGIES

The production of biogas through AD offers major advantages over other forms of bioenergy production. In fact, it has been defined as one of the most energy-efficient and environmentally beneficial technology for bioenergy production (Deublein and Steinhauser, 2011). The degradation process can be divided into four phases: hydrolysis, acidogenesis, acetogenesis, and methanogenesis; and in each individual phase, different groups of facultative or obligatory anaerobic microorganisms are involved as shown in Figure 3 (Merlin Christy et al., 2014; Chasnyk et al., 2015; Abdeshahian et al., 2016).

Beside energy production, the degradation of organic waste also offers some other advantages including the reduction of odour release and decreased level of pathogens. Moreover, the nutrient rich digested residue could be used as organic fertilizer for arable land instead of mineral fertilizer, as well as an organic substrate for green house cultivation (De Vries et al., 2012; Abdeshahian et al., 2016). Among the raw substances, organic materials obtained from farm and animal waste streams, as well as from industrial and household activities are pivotal sources for biogas production.

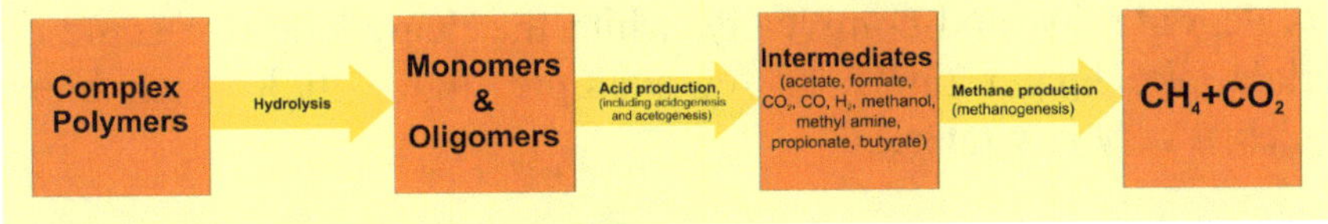

Figure 3: The degradation process taking place during AD, i.e., hydrolysis, acidogenesis, acetogenesis, and methanogenesis.

Substrates Traditionally Used

Through human activities, a huge amount of organic solid waste is generated, which as discussed earlier can be used as feedstock for biogas production. Based on the origin, the different waste streams can be classified as municipal solid waste (MSW), agricultural residues, and wastes from industrial activities. According to a 2012 world bank report, 1.3 billion tons of MSW was generated per year by 3 billion urban residents all over the word, which will increase to 2.2 billion tons by 2025 (Hoornweg and Bhada-Tata, 2012). MSW mainly consists of food waste, paper and paperboard, yard trimmings, wood, plastic, metal, and glass. However, its composition differs depending on regions and countries in which it is collected. To be able to utilize this fraction for biogas production, all the inert material, including plastic, metal, and glass should be removed prior to AD. Moreover, around 15 billion tons of waste, like crops residues and animal manure, is generated worldwide annually from the agricultural sector (Donkin et al., 2013).

Food processing industries also generate waste, however the estimation of its amount is excessively difficult, since it greatly depends on the industry and technology applied. As an example, in the juice producing industry up to 50% of the processed fruit will end up as waste. Moreover, 30% of the weight of a chicken is not suitable for human consumption, and it is therefore removed as waste during slaughtering and other processing steps (Salminen and Rintala, 2002; Forgács et al., 2012).

Although all these different waste fractions are suitable for biogas production, their biogas potential varies significantly. The biogas yield mainly depends on the composition and the biodegradability (under anaerobic conditions) of the waste. Theoretically, the highest biogas yield can be achieved from lipids (1.01 Nm^3 CH_4/ kg VS), followed by proteins (0.50 Nm^3 CH_4/ kg VS), and carbohydrates (0.42 Nm^3 CH_4/ kg VS) (Møller et al., 2004). On the other hand, biodegradability defines how much of a given material is actually utilized during the process. Some compounds like sugars degrade fast and completely, while the degradation of some other materials take longer times, as for example, lignocelluloserich biomass degrades at very low rates.

Pretreatment for Enhanced Biogas Production

The growing global energy demand together with the limited availability of fossil fuels, unstable energy prices, and environmental problems necessitate the use of renewable energies. The currently used feedstocks for AD are limited, and therefore, it is important to explore new substrates for their utilization in AD to reserve the growing needs. The abundance and availability of lignocellulosic biomasses worldwide as well as their high carbohydrate content make these materials an attractive feedstock for biofuel production. Lignocelluloses have been accounted for approximately 50% of the biomass in the world and the production of lignocelluloses can count up to about 200 billion tons per year (Claassen et al., 1999; Zhang, 2008). Currently, the utilization of lignocellulosic residues as feedstock for methane production is not widespread (Lehtomäki, 2006; Seppälä et al., 2007) due to their recalcitrant structure, which is the main challenge (Hendriks and Zeeman, 2009).

During the first step of AD, i.e., in the hydrolysis step, the hydrolytic bacteria convert the insoluble complex organic matters into monomers and soluble oligomers such as fatty acids, amino acids, and sugars (Fig. 3). The enzymes involved in this process are cellulases, hemicellulases, lipases, amylases, and proteases (Taherzadeh and Karimi, 2008). Therefore, in biogas processes, almost all kinds of substrates can be hydrolyzed. However, the rate of the hydrolysis step is highly dependent on the characteristics of a given substrate. Hydrolysis can proceed relatively fast if the necessary enzymes are produced by microorganisms and suitable surface area for physical contact between the enzymes and the substrate is provided (Taherzadeh and Karimi, 2008). Nevertheless, substrates with more recalcitrant structure, like cellulose, need longer period to be degraded, and the degradation is usually not complete (Deublein and Steinhauser, 2011). Hence, the hydrolysis step is often considered as the rate-limiting step when utilizing these kinds of substrates (Vavilin et al., 1996; Taherzadeh and Karimi, 2008).

Therefore, an initial pretreatment step, which converts raw materials to a form that is amenable to microbial and enzymatic degradation is needed (Zhang, 2008). A suitable pretreatment by the disruption of the secondary cell walls structure will reduce biomass recalcitrance and thus facilitate downstream processes. Optimally, a pretreatment should also be costeffective and yield a polysaccharidic-rich substrate with limited amounts of inhibitory by-products.

A numbers of pretreatment methods have been suggested for enhancing biogas production from lignocellulosic biomass, which can be classified as, physical, physicochemical, chemical, and biological pretreatments (Chandra et al., 2007; Taherzadeh and Karimi, 2008; Yang and Wyman, 2008; Hendriks and Zeeman, 2009). Milling, among the physical pretreatments was proven to be effective by shearing, increasing the specific surface area, and reducing the degree of polymerization (DP), thus improving the hydrolysis yield by 5–25%. Degree of such improvement depends on type of biomass, as well as the duration and type of milling (Jin and Chen, 2006; Zeng et al., 2007). Overall, it has been repeatedly shown that smaller particle sizes result in higher yields (Jin and Chen, 2006; Monavari et al., 2009; Lennartsson et al., 2011; Teghammar et al., 2012). That is why the physical pretreatment is often carried out in combination with other pretreatment methods. However, in some cases, the chemical agent used for the pretreatment can act as a potential inhibitor for the microbial community involved in the AD. In a recent study, it was found that the remaining solvent affected the digestion process negatively when forest residues was pretreated with an organic solvent, N-methylmorpholine-N-oxide, even at concentrations as low as 0.008% (Kabir et al., 2013). Besides, the pretreatment process itself might lead to the production of inhibitory products; and despite optimization of pretreatment conditions, some inhibitors will still occur in the pretreated slurry. These may be either degradation products, such as furans through dilute-acid hydrolysis and steam explosion pretreatments, and furfural through alkaline pretreatments, or biomass constituents of varying molecular weights and concentrations (Ahring et al., 1996; Taherzadeh and Karimi, 2008).

Recently, it was shown that using alcohols or weak organic acid for the pretreatment of lignocelluloses seems to be an interesting method. Since they are intermediary products during the anaerobic degradation process, the above-mentioned inhibitory problems can be avoided and moreover, the remaining traces of these solvents after the pretreatment can be consumed for additional methane production. In a recent study, Kabir et al. (2015) applied ethanol, methanol, or acetic acid for the pretreatment of forest residues prior to AD. It was found that although according to the batch experimental results, treatments with ethanol or acetic acid resulted in higher methane yields; the techno-economic calculations showed that treatment with methanol was economically more feasible due to the lower price of methanol and the lower costs for its recovery after the treatment.

Challenges of the Current Processes

In general and as mentioned earlier, the AD of organic material requires combined activity of several different groups of microorganisms with different metabolic capacities (Himmel et al., 1994). To obtain a stable biogas process, all the conversion steps involved in the degradation of organic matters and the microorganisms carrying out these steps must work in a synchronised manner. Methanogens have longer duplication times (of up to 30 d) and are generally considered as the most sensitive group to process disturbances (Griffin et al., 1998). It is therefore important to prevent these groups of microorganisms from being washed out from the system, by decoupling the solid retention time (SRT) and the HRT. Major developments have been therefore made during the last decades with regard to development of high rate systems, lowering the effects of toxic compounds, integrating the biological process with membrane separation techniques, as well as better understanding of anaerobic metabolism, and interactions among different microbial species.

NOVEL ANAEROBIC DIGESTION TECHNOLOGIES

AD systems have undergone several modifications in the last decades to increase the efficiency of the process. In this sense and aiming at overcoming the methanogenesis as the rate-limiting step, efficient retention of the slow-growing methanogenic biomass has been the most important challenge. An important milestone was the development of a new reactor design, i.e., the up flow anaerobic sludge blanket (UASB) reactor, containing a well-settleable methanogenic sludge due to the formation of a dense sludge bed. Another technology making possible to retain active biomass within the system was the application of membrane bioreactors (MBRs). Besides separating cells, the membrane can also be used for the separation of inhibitory compounds, which otherwise would negatively affect the biological process, or for in situ recovery of the product could result in decreased cost of down stream processing. Additionally, the development of molecular biology techniques provided researchers with a valuable tool to understand the complex microbiological system involved in anaerobic degradation of organic matters. By the application of these techniques, it would be possible to regulate and control the process and discover disturbances much earlier then using traditional process parameters for monitoring the process.

High Rate Anaerobic Reactors

The UASB reactor, which was developed by Dr. Gatze Lettinga in the Netherlands during the early 70s, is probably the most popular high-rate reactor system applied for anaerobic biological treatment of "wastewater", as more than 1000 UASB reactors are in operation throughout the World.

This process is attractive because of its compactness, high loading rates, relatively low retention times for anaerobic treatment, low operational cost, low sludge production, and high methane production rates. The granular or flocculated sludge is the main prominent characteristic of this type of reactors as compared with other anaerobic technologies. In an UASB reactor, anaerobic microorganisms can form granules through selfimmobilization of the cells, and the performance of the system is strongly dependent upon the granulation process together with the characteristics of a particular wastewater treated (Schmidt and Ahring, 1996). Thus, changing the waste type will also affect the sludge quality and thereby the efficiency of the process. Moreover, substrates with a high fraction of particulate organic material are not suitable to be treated with this technology. A modified reactor configuration was therefore proposed recently aiming at separating the hydrolysis and acid formation steps from the methanogenesis step when treating MSW using a two-stage process including a continuously stirred tank reactor (CSTR) and an UASB reactor (Aslanzadeh, 2014). Comparing the performance of this two-stage system with that of a traditional one stage digestion, it was found that using this novel technology, organic loading rate (OLR) of 10 gVS/L/d could be achieved while the HRT could be reduced to 3 d.

Anaerobic Membrane Bioreactors (AnMBR)

In membrane bioreactors (MBRs), the membrane forms a selective barrier allowing certain components to pass while retaining others, thereby the biological system can be protected. The application of MBRs provides both increased SRT by avoiding the wash out of the cells and decreasing inhibitor concentrations by the separation of inhibitors (Visvanathan and Abeynayaka, 2012).

Today, there are two different designs for membrane bioreactors applied. The membrane can be placed either in an external loop or submerged within the reactor (Fig. 4).

The submerged system requires less space and energy, since compared with the external loop system, energy input is not required to maintain a continuous flow through the membrane. However, it could be problematic to operate this system at high particulate and/or cell concentrations, due to fouling (Judd, 2010). Membrane technologies developed and applied in waste water treatment processes can also be used for biogas production processes. Different studies on membrane technologies in biogas systems reported yields comparable with those obtained with high rate systems, i.e., UASB systems (Lin et al., 2011; Wijekoon et al., 2011).

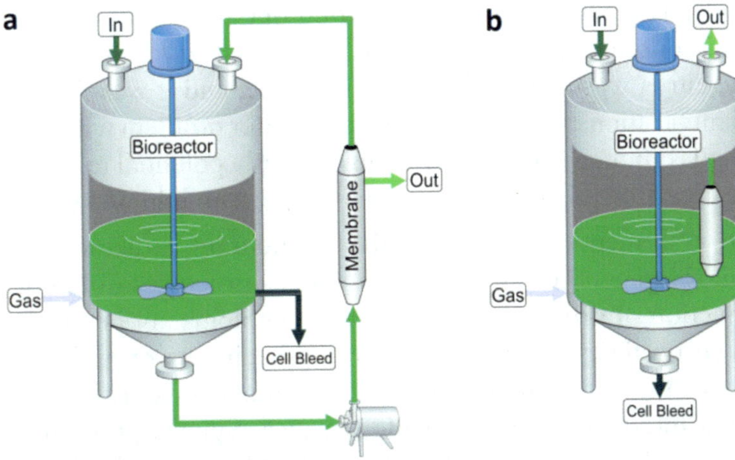

Figure 4: Membrane bioreactor designs; a) external loop, b) submerged (adapted from Ylitervo et al. (2013)).

Encapsulation of methane-producing bacteria was carried out to test the viability of this technique in biogas processes. One-step liquid-droplet forming method was used to form spherical capsules of alginate. Chitosan or Ca^{2+} was used as counter-ions together with the addition of carboxymethyl cellulose. Furthermore, a synthetic Durapore® membrane (hydrophilic polyvinyldifluorid (PVDF)) was also tested by making encapsulating sachets with dimension of 3×3 or 3×6 cm^2 for holding the bacteria. The results indicated that these membranes allowed the penetration of nutrients into the cells while the gas produced could escape out of the capsules by diffusion. Hence, encapsulation can be a promising method, keeping high density of microorganisms in the system (Youngsukkasem et al., 2012). This theory was further investigated by comparing the ability of encapsulated cells with free cells to handle limonene containing synthetic media during AD. Limonene naturally occurs in citrus waste, making the utilization of this waste stream in biogas processes difficult, due to its inhibitory effects on the biogas producing microorganisms. The results showed the protective effect of the PVDF membrane resulting in faster biogas production by the encased bacteria compared to the free cells (Youngsukkasem et al., 2013).

Furthermore, a novel AnMBR configuration was investigated later, where both free cells and encased cells worked simultaneously in a single reactor treating a model substrate, Avicel, with limonene addition (Wikandari et al., 2014). The experiments were carried out at thermophilic conditions under semi-continuous operation at OLR of 1 gVS/L/d and HRT of 30 d. Generally, citrus waste contains 8 g/L limonene, and it was found that this reactor configuration could overcome the inhibitor problem with the addition of up to 5 g/L limonene. Thus, this technique has a potential to be applied for anaerobic digestion of fruit wastes containing certain inhibitory compounds.

As it was mentioned earlier, the recalcitrant structure makes the utilization of lignocellulosic biomass in biogas processes difficult. Besides the introduction of different pretreatment technologies prior to AD with an aim to open up their structure, another approach was recently introduced by processing the lignocellulosic biomass thermochemically instead, aiming at obtaining intermediary gases, called syngas. Syngas primarily contains carbon monoxide (CO) hydrogen (H_2), and carbon dioxide (CO_2). Hence, this gas mixture can be utilized by the anaerobic microorganisms, using the CO and/or CO_2 as carbon source and H2 as energy source, to produce methane. In order to increase the productivity and the efficiency of the conversion, a reverse MBR (RMBR) was applied retaining the cells inside the reactor (Youngsukkasem et al., 2015). Using anaerobic sludge encased in PVDF membranes, the conversion of syngas to methane could be carried out at a retention time of 1 d. Furthermore, co-digestion of syngas with a synthetic organic medium was also successful by allowing the diffusion of both gas and liquid through the surface of the membrane.

Integration of Membranes and High Rate Systems

The combination of anaerobic membrane technology and high rate systems is increasingly being investigated. These integrated systems have several advantages such as improved methane production and less fouling problems and are especially suitable to treat high strength industrial and municipal wastewaters aiming at achieving solids free effluents with a high degree of pathogen removal.

Kraft evaporator condensate was treated at mesophilic conditions with a submerged combined UASB-MBR system achieving a methane yield of 0.35 L CH_4/$gCOD_{removed}$ which was very close to the theoretical yield of 0.397 L CH_4/gCOD at 37°C (Xie et al., 2010). However, seeding the UASB reactor with non-granule sludge required a long start up period (up to 3-4 months) to be able to achieve the formation of granules and hence, a stable biogas production. In that sense, the presence of a membrane in the reactor could eliminate the hydraulic pressure and negatively affect the granular sludge properties (Ozgun et al., 2013). Further investigations are therefore needed to determine the most optimal process configurations, i.e., the reactor type and the way of coupling it with the membrane module.

MICROBIAL COMMUNITY ANALYSIS AND BIOGAS PROCESS CONTROL

As mentioned earlier, AD involves different degradation steps, i.e., hydrolysis, acidogenesis, acetogenesis, and methanogenesis that are facilitated by various groups of microorganisms (Fig. 3). These microorganisms can be divided into three functional groups: hydrolysing and fermenting bacteria, obligate hydrogen-producing acetogenic bacteria, and methanogenic archaea (Ahring, 2003). Hydrolytic acidogenic bacteria (HABs) hydrolyze complex organic polymers into simple compounds during the first step of the degradation. During the acidogenesis process, volatile fatty acids (VFA), alcohols, H_2, and CO_2 are produced. Similarly, acetic acid, H_2, and CO_2 are produced in the acetogenesis step by the obligate H_2-producing acetogens. Syntrophobacter (PUAs: propionateutilizing acetogens) and Syntrophomonas (BUAs: butyrate-utilizing acetogens) represent the major part of acetogens. A key factor in the degradation is that anaerobic oxidation of butyrate and propionate occurs only in syntrophic association with H_2-utilizing methanogens (HUMs), consuming H_2 and CO_2 for methane (CH_4) production, preventing the accumulation of increasing H_2 pressure in the digester. Another way of methane formation is the conversion of acetate to CH_4 and CO_2 by the action of acetate-utilizing methanogens (AUMs) (Climent et al., 2007; Zahedi et al., 2013; Ennouri et al., 2016). In general, the operational parameters as well as substrate characteristics will influence the composition of the anaerobic microbial consortium present in a digester.

Molecular biology techniques provide valuable tools for improved understanding of microbial communities and their function in connection with different aspects of AD, which in turn may help optimize the biogas production process more efficiently. A broad range of studies was published recently on investigations on microbial community structures in biogas reactors. The methodologies applied included analysis of total bacteria and archaeal community by targeting 16S rRNA using 454 next generation sequencing (NGS) technique (Zakrzewski et al., 2012) or terminal restriction fragment length polymorphism (T-RFLP) (Wang et al., 2010); as well as detection and quantification of methanogenic Archaea by quantitative real time polymerase chain reaction (qPCR). qPCR is a commonly used method in microbial community studies to detect and quantify a targeted DNA sequence. The principle of qPCR is very similar to that of conventional PCR. The target gene is amplified over a number of cycles. However, the conventional PCR allows only end point detection, whereas using a fluorescent dye or probe, the concentration of the target gene can be monitored after each cycle in qPCR. The detected change in fluorescence intensity reflects the concentration of the amplified gene in real time (VanGuilder et al., 2008).

Among the first studies aiming at understanding the relationship between biodiversity, operating conditions, and process performance, the prokaryotic community of seven digesters treating sewage sludge was examined by constructing and analyzing a total of 9890 16S rRNA gene clones. The results showed that the bacterial community could be divided in three components: one-third of the phylotypes could be found in most of the digesters, one-third were phylotypes shared among a few digesters, and the rest were specific phylotypes found under certain conditions (Riviere et al., 2009).

Metagenomics Approaches

The traditional molecular biology technologies help with identifying only the most abundant microbial populations present in the reactor. Due to their high sequencing depth, the newly developed sequencing techniques make the determination of both the most abundant and also the minor populations possible. The NGS-based metagenomic approach enables following up changes in the microbial community structure starting from the very initial stage to souring of the digester. Coding gene sequences (mRNA) especially those representing critical steps of specific metabolic pathways can be mapped to assess the functional profiles of microbial communities. The high throughput sequencing-based metagenomic characterization of various microbial communities involved in biomethanation of a range of substrates has been elucidated with the help of 454 pyrosequencing and SOLiD NGS methods (Kovács et al., 2013; Sundberg et al., 2013; Pore et al., 2016). For example, the Ion Torrent PGM technique, which was launched in 2011, provided the highest throughput compared with that of 454 NGS and it was recently used for microbial composition analysis in several studies (Luo et al., 2013; Wang et al., 2013). Investigations on the microbial community in

21 full scale anaerobic digestion plants using 454 pyrosequencing of 16S rRNA gene sequences showed that the bacterial community was always more abundant and more diverse than the archaeal community in all reactors. Moreover, it was found that while acetoclastic methanogens or AUMs were detected in plants digesting sewage sludge, they were absent in co-digestions plants. Hence, methane is generated from acetate mainly via syntrophic acetate oxidation in the co-digestion plants (Sundberg et al., 2013). To date, most studies have strived to investigate the microbial community inside the reactors without taking into account the whole biogas process chain.

Using Ion Torrent PGM technique, investigations on bacterial composition analysis and the presence of bacterial pathogens were performed recently by Luo and Angelidaki (2014) within the whole biogas producing system including the influent, the biogas reactor, and the post-digesters. They found that bacterial community composition of the influent was changed after AD. More specifically, the richness and relative abundance of bacterial pathogens reduced during AD, however, an increase in the relative abundance of pathogens was observed after prolonged post digestion times of 30 d. The authors pointed out that special attention should be therefore paid to the post digestion step aiming at avoiding the re-growth of bacterial pathogens, which otherwise will limit the disposal of the digested residue as bio-fertilizer. Similarly, the denaturing gradient gel electrophoresis (DGGE) technique is still among the promising methods to perform a preliminary analysis of the microbial community profile and to monitor the various experimental stages during the biogas production process. In a recent study, Dias et al. (2016) compared the sequences from DGGE bands with NCBI and RDP databases and reported the significant presence of Proteobacteria (6 from 7 sequences), specifically Gammaproteobacteria in the biogas system from vinasse methanisation.

In another study, the microbial community structure in a solid-state anaerobic digester (SS-AD) treating lignocellulosic residues, i.e., waste from palm oil mill industry or wheat straw was investigated. The samples were analyzed by 16S rRNA gene (rrs) sequence analysis combined with PCRDGGE. The bacterial community in SS-AD was comprised of Ruminococcus sp., Thiomargarita sp., Clostridium sp., Anaerobacter sp., Bacillus sp., Sporobacterium sp., Saccharofermentans sp., Oscillibacter sp., Sporobacter sp., Lachnospiraceae sp., etc. (Heeg et al., 2014; Suksong et al., 2016).

Moreover, the high-throughput Illumina Miseq approach is also widely considered as a promising culture-independent method to perform microbial community analysis of AD systems. By the application of this method, the specific syntrophic relationships between acetogens and methanogens could be better understood, especially in terms of how it can be related to disturbances occurring in the biogas production process. Anaerobic digesters treating lipid-extracted microalgae residue at various inoculum-to-substrate ratios were investigated using Illumina Miseq analysis. Differences in the phylum distribution of the bacterial community were detected in accordance with the changes in inoculum to substrate ratios. The different levels of long chain fatty acids (LCFAs) affected each functional microbial group. Although methanogens were the most sensitive group to LCFA inhibition, the LCFA inhibition factor for hydrolytic bacteria was more highly affected by the inoculum to substrate ratios. Syntrophic acetogens showed a decreased abundance in case of high LCFA concentrations (Ma et al., 2015; Aydin, 2016).

CONCLUDING REMARKS

The increasing demand for renewable energy compels the exploration of new substrates and the development of new technologies for biogas production. Regarding raw materials for AD, it is preferable to utilize waste streams since in this way, the process addresses both waste reduction and energy production. Lignocellulosic residues are readily available; however, further development of novel pretreatment technologies are needed to achieve economically viable processes. Anaerobic degradation of organic material requires a well functioning microbial consortium, and methanogenic microorganisms, responsible for methane production within the final step of the digestion process, are known to be the most sensitive ones to process disturbances. This together with their slow growing rate made it necessary to develop novel process configurations aiming at preventing their wash out from the system. In this sense, the development of UASB reactor was an important milestone. In UASB system the formation of a dense well-settleable granular sludge makes an efficient decoupling of SRT and HRT possible. In better words, a crucial factor for a successful anaerobic high-rate treatment is the retention of all slow-growing microorganisms. Hence, when sludge granulation is hindered or lacking, membranes can be applied for biomass separation and recycling back into the reactor.

Therefore, the interest in using different membrane configurations is driven by the requirement for increasing productivity. However, with high particulate and/or cell concentrations, the operation of these kinds of systems can be problematic due to fouling. Thus, full-scale implementation of the AnMBR technology will be highly dependent on flux levels achieved during long-term operation. Finally, since AD is a complex microbial process, a broad range of studies have recently aimed at understanding the relationship between the microbial community structure, operating conditions, and process performance. By using novel newly-developed molecular biology tools, it would be possible to control and regulate the process in an effective way. To date, these techniques were mainly applied for the digestion step itself, however, it is necessary to pay attention to the whole biogas production system, including storage and feeding together with the post digestion step in the future as well.

REFERENCES

1. Abdeshahian, P., Lim, J.S., Ho, W.S., Hashim, H., Lee, C.T., 2016. Potential of biogas production from farm animal waste in Malaysia. Renew. Sust. Energy Rev. 60, 714-723.
2. Ahring, B.K., Jensen, K., Nielsen, P., Bjerre, A.B., Schmidt, A.S., 1996. Pretreatment of wheat straw and conversion of xylose and xylan to ethanol by thermophilic anaerobic bacteria. Bioresour. Technol. 58(2), 107-113.
3. Ahring, B.K., 2003. Perspectives for anaerobic digestion, in: In Biomethanation I. Advances in biochemical engineering/biotechnology. 81, 1-30.
4. Al Seadi, T., 2002. Quality management of AD residues from biogas production. Proc. IEA Bioenergy, Task 24 - Energy from Biological Conversion of Organic Waste.
5. Aslanzadeh, S., 2014. Pretreatement of cellulosic waste and high rate biogas production. University of Borås, Borås, Sweden. Aydin, S., 2016. Enhancement of microbial diversity and methane yield by bacterial bioaugmentation through the anaerobic digestion of Haematococcus pluvialis. Appl. Microbiol. Biotechnol.100, 5631-5637.
6. Batstone, D.J., Keller, J., Angelidaki, I., Kalyuzhnyi, S.V., Pavlostathis, S.G., Rozzi, A., Sanders, W.T.M., Siegrist, H., Vavilin, V.A., 2002. The IWA Anaerobic Digestion Model No 1 (ADM 1). Water Sci. Technol. 45(10), 65-73.
7. TDBP, 2013. Tanzania Domestic Biogas Programme.
8. Chandra, R.P., Bura, R., Mabee, W.E., Berlin, A., Pan, X., Saddler, J.N., 2007. Substrate pretreatment: The key to effective enzymatic hydrolysis of lignocellulosics?, In: Olsson L.(Ed.), Biofuels. Springer Berlin Heidelberg, pp. 67-93.
9. Chasnyk, O., Sołowski, G., Shkarupa, O., 2015. Historical, technical and economic aspects of biogas development: Case of Poland and Ukraine. Renew. Sust. Energy Rev. 52, 227-239.
10. Cheng, S., Li, Z., Mang, H.P., Huba, E.M., Gao, R., Wang, X., 2014. Development and application of prefabricated biogas digesters in developing countries. Renew. Sust. Energy Rev. 34, 387-400.
11. Claassen, P.A.M., Van Lier, J.B., Contreras, A.M.L., Van Niel, E.W.J., Sijtsma, L et al., 1999. Utilisation of biomass for the supply of energy carriers. Appl. Microbiol. Biotechnol. 52(6), 741-755.

12. Climent, M., Ferrer, I., Baeza, M.d.M., Artola, A., Vázquez, F., Font, X., 2007. Effects of thermal and mechanical pretreatments of secondary sludge on biogas production under thermophilic conditions. Chem. Eng. J. 133(1), 335-342.

13. Deublein, D., Steinhauser, A., 2011. Biogas from waste and renewable resources: an introduction. Mörlenbach, Germany: Wiley-VCH Verlag GmbH & Co. KGaA.

14. de Vries, J.W., Vinken, T.M.W.J., Hamelin, L., De Boer, I.J.M., 2012. Comparing environmental consequences of anaerobic mono- and codigestion of pig manure to produce bio-energy - A life cycle perspective. Bioresour. Technol. 125, 239-248.

15. Dias, M.F., Colturato, L.F., de Oliveira, J.P., Leite, L.R., Oliveira, G., Chernicharo, C.A., de Araújo, J.C., 2016. Metagenomic analysis of a desulphurisation system used to treat biogas from vinasse methanisation. Bioresour. Technol. 205, 58-66.

16. Donkin, S.S., Doane, P.H., Cecava, M.J., 2013. Expanding the role of crop residues and biofuel co-products as ruminant feedstuffs. Anim. Front. 3(2), 54-60.

17. Ennouri, H., Miladi, B., Diaz, S.Z., Güelfo, L.A.F., Solera, R., Hamdi, M., Bouallagui, H., 2016. Effect of thermal pretreatment on the biogas production and microbial communities balance during anaerobic digestion of urban and industrial waste activated sludge. Bioresour. Technol. 214, 184-191.

18. Esposito, G., Frunzo, L., Panico, A., Pirozzi, F., 2012. Enhanced biomethane production from co-digestion of different organic wastes. Environ. Technol. 33(24), 2733-2740.

19. Forgács, G., Pourbafrani, M., Niklasson, C., Taherzadeh, M.J, Sárvári Horváth, I., 2012. Methane production from citrus wastes: Process development and cost estimation. J. Chem. Technol. Biotechnol. 87(2), 250-255.

20. Garcia, S.G., 2005. Farm scale anaerobic digestion integrated in an organic farming system. JTI (Institutet för Jordbruks-och Miljöteknik). Report, 34.

21. Griffin, M.E., McMahon, K.D., Mackie, R.I., Raskin, L., 1998. Methanogenic population dynamics during start-up of anaerobic digesters treating municipal solid waste and biosolids. Biotechnol. Bioeng. 57(3), 342-355.

22. Heeg, K., Pohl, M., Sontag, M., Mumme, J., Klocke, M., Nettmann, E., 2014. Microbial communities involved in biogas production from wheat straw as the sole substrate within a two-phase solid-state anaerobic digestion. Syst. Appl. Microbiol. 37(8), 590-600.
23. Hendriks, A.T.W.M., Zeeman, G., 2009. Pretreatments to enhance the digestibility of lignocellulosic biomass. Bioresour. Technol. 100(1), 10- 18.
24. Himmel, M.E., Baker, J.O., Overend, R.P., 1994. Enzymatic Conversion of Biomass for Fuels Production. American Chemical Society, Washington DC.
25. Holm-Nielsen, J.B., Al-Seadi, T., Lens, P., Hamelers, B., Hoitink, H., Bidlingmaier, W., 2004. Manure-based biogas systems-Danish experience, in: Lens, P., Hamelers, B., Hoitink, H., Bidlingmaier, W. (Eds.), Resource recovery and reuse in organic solid waste management. IWA Publishing, London, UK, pp. 377-394.
26. Hoornweg, D., Bhada-Tata, P., 2012. What a waste: a global review of solid waste management. World Bank.
27. Huttunen, S., Kivimaa, P., Virkamäki, V., 2014. The need for policy coherence to trigger a transition to biogas production. Environ. Innovat. Soc. Transit. 12, 14-30.
28. Jin, S., Chen, H., 2006. Superfine grinding of steam-exploded rice straw and its enzymatic hydrolysis. Biochem. Eng. J. 30(3), 225-230.
29. Judd, S., 2010. The MBR book, principles and applications of membrane bioreactors for water and wastewater treatment. Elsevier, Oxford.
30. Kabir, M.M., del Pilar Castillo, M., Taherzadeh, M.J., Sárvári Horváth, I., 2013. Enhanced methane production from forest residues by N-methylmorpholine-N-oxide (NMMO) pretreatment. BioResources. 8(4), 5409-5423.
31. Kabir, M.M., Rajendran, K., Taherzadeh, M.J., Sarvari Horvath, I., 2015. Experimental and economical evaluation of bioconversion of forest residues to biogas using organosolv pretreatment. Bioresour. Technol. 178, 201-208.
32. Kacprzak, A., Krzystek, L., Ledakowicz, S., 2010. Co-digestion of agricultural and industrial wastes. Chem. Pap. 64(2), 127-131.
33. Kovács, E., Wirth, R., Maróti, G., Bagi, Z., Rákhely, G., Kovács, K.L., 2013. Biogas production from protein-rich biomass: fed-batch

anaerobic fermentation of casein and of pig blood and associated Changes in microbial community composition. PLoS One. 8(10), e77265.

34. Lehtomäki, A., 2006. Biogas production from energy crops and crop residues. University of Jyväskylä.
35. Lennartsson, P.R., Niklasson, C., Taherzadeh, M.J., 2011. A pilot study on lignocelluloses to ethanol and fish feed using NMMO pretreatment and cultivation with zygomycetes in an air-lift reactor. Bioresour. Technol. 102(6), 4425-4432.
36. Lin, H., Liao, B-Q., Chen, J., Gao, W., Wang, L., et al., 2011. New insights into membrane fouling in a submerged anaerobic membrane bioreactor based on characterization of cake sludge and bulk sludge. Bioresour. Technol. 102(3), 2373-2379.
37. Luo, G., Angelidaki, I., 2014. Analysis of bacterial communities and bacterial pathogens in a biogas plant by the combination of ethidium monoazide, PCR and Ion Torrent sequencing. Water Res. 60, 156- 163.
38. Luo, G., Wang, W., Angelidaki, I., 2013. Anaerobic Digestion for Simultaneous Sewage Sludge Treatment and CO Biomethanation: Process Performance and Microbial Ecology. Environ. Sci. Technol. 47(18), 10685-10693.
39. Ma, J., Zhao, Q-B., Laurens, L.L.M., Jarvis, E.E., Nagle, N.J., et al., 2015. Mechanism, kinetics and microbiology of inhibition caused by long-chain fatty acids in anaerobic digestion of algal biomass. Biotechnol. Biofuels. 8(1), 1-12.
40. Mata-Alvarez, J., Mace, S., Llabres, P., 2000. Anaerobic digestion of organic solid wastes. An overview of research achievements and perspectives. Bioresour. Technol. 74(1), 3-16.
41. Merlin Christy, P., Gopinath, L.R., Divya, D., 2014. A review on anaerobic decomposition and enhancement of biogas production through enzymes and microorganisms. Renew. Sust. Energy Rev. 34, 167-173.
42. Misi, S.N., Forster, C.F., 2001. Batch co-digestion of multicomponent agro-wastes. Bioresour. Technol. 80(1), 19-28.
43. Møller, H.B., Sommer, S.G., Ahring, B.K., 2004. Methane productivity of manure, straw and solid fractions of manure. Biomass Bioenergy. 26(5), 485-495.
44. Monavari, S., Galbe, M., Zacchi, G., 2009. Impact of impregnation

time and chip size on sugar yield in pretreatment of softwood for ethanol production. Bioresour. Technol. 100(24), 6312-6316.

45. Nielsen, L.H., Hjort-Gregersen, K., Thygesen, P., Christensen, J., 2002. Samfundsøkonomiske analyser af biogasfllesanlg. Fødevareøkonomisk Institut, Rapport, 136.
46. Ozgun, H., Dereli, R.K., Ersahin, M.E., Kinaci, C., Spanjers, H., van Lier J.B., 2013. A review of anaerobic membrane bioreactors for municipal wastewater treatment: Integration options, limitations and expectations. Sep. Purif. Technol. 118, 89-104.
47. Pagés Díaz, J., Pereda Reyes, I., Lundin, M., Sárvári Horváth, I., 2011. Co-digestion of different waste mixtures from agro-industrial activities: Kinetic evaluation and synergetic effects. Bioresour. Technol. 102(23), 10834-10840.
48. Pagés-Díaz, J., Pereda-Reyes, I., Taherzadeh, M.J., Sárvári-Horváth, I., Lundin, M., 2014. Anaerobic co-digestion of solid slaughterhouse wastes with agro-residues: Synergistic and antagonistic interactions determined in batch digestion assays. Chem. Eng. J. 245, 89-98.
49. Pagés-Díaz, J., Westman, J., Taherzadeh, M.J., Pereda-Reyes, I., Sárvári Horváth, I., 2015. Semi-continuous co-digestion of solid cattle slaughterhouse wastes with other waste streams: Interactions within the mixtures and methanogenic community structure. Chem. Eng. J. 273, 28-36.
50. Rakotojaona, L., 2013. Domestic biogas development in developing countries ENEA Consulting.
51. Persson, M., Jönsson, O., Wellinger, A., 2006. Biogas upgrading to vehicle fuel standards and grid injection. IEA Bioenergy, Task, 37.
52. Pore, S.D., Shetty, D., Arora, P., Maheshwari, S., Dhakephalkar, P.K., 2016. Metagenome changes in the biogas producing community during anaerobic digestion of rice straw. Bioresour. Technol. doi:10.1016/j.biortech.2016.03.045
53. Raven, R., Gregersen, K.H., 2007. Biogas plants in Denmark: successes and setbacks. Renew. Sust. Energy Rev. 11(1), 116-132.
54. Riviere, D., Desvignes, V., Pelletier, E., Chaussonnerie, S., Guermazi, S., et al., 2009. Towards the definition of a core of microorganisms involved in anaerobic digestion of sludge. ISME J. 3(6), 700-714.
55. Salminen, E., Rintala, J., 2002. Anaerobic digestion of organic solid poultry slaughterhouse waste-a review. Bioresour. Technol. 83(1), 13-

26.

56. Seppälä, M., Paavola, T., Rintala, J., 2007. Methane yields of different grass species on the second and third harvest in boreal conditions. Proc. 11th IWA World Congress on Anaerobic Digestion, Brisbane, Australia. pp. 23-27.

57. Schmidt, J.E., Ahring, B.K., 1996. Granular sludge formation in upflow anaerobic sludge blanket (UASB) reactors. Biotechnol. Bioeng. 49(3), 229-246.

58. Bojesen, M., Boerboom, L., Skov-Petersen, H., 2014. Towards a sustainable capacity expansion of the Danish biogas sector. Land Use Policy 42, 264-277.

59. Suksong, W., Kongjan, P., Prasertsan, P., Imai, T., O-Thong, S., 2016. Optimization and microbial community analysis for production of biogas from solid waste residues of palm oil mill industry by solid-state anaerobic digestion. Bioresour. Technol. 214, 166-174.

60. Taherzadeh, M., Karimi, K., 2008. Pretreatment of lignocellulosic wastes to improve ethanol and biogas production: a review. Int. J. Mol. Sci. 9(9), 1621-1651.

61. Teghammar, A., Karimi, K., Sárvári Horváth, I., Taherzadeh, M.J., 2012. Enhanced biogas production from rice straw, triticale straw and softwood spruce by NMMO pretreatment. Biomass Bioenergy. 36, 116- 120.

62. VanGuilder, H.D., Vrana, K.E., Freeman, W.M., 2008. Twenty-five years of quantitative PCR for gene expression analysis. Biotechniques. 44(5), 619- 626.

63. Visvanathan, C., Abeynayaka, A., 2012. Developments and future potentials of anaerobic membrane bioreactors (AnMBRs). Membr. Water. Treat. 3, 1-23.

64. Vavilin, V.A., Rytov, S.V., Lokshina, L.Y., 1996. A description of hydrolysis kinetics in anaerobic degradation of particulate organic matter. Bioresour. Technol. 56(2), 229-237.

65. Wang, H., Vuorela, M., Keränen, A-L., Lehtinen, T.M., Lensu, A., Lehtomäki, A., Rintala, J., 2010. Development of microbial populations in the anaerobic hydrolysis of grass silage for methane production. FEMS Microbiol. Ecol. 72(3), 496-506.

66. Wang, L.H., Wang, Q., Cai, W., Sun, X., 2012. Influence of mixing proportion on the solid-state anaerobic co-digestion of distiller's grains and food waste. Biosyst. Eng. 112(2), 130-137.
67. Wang, W., Xie, L., Luo, G., Zhou, Q., Angelidaki, I., 2013. Performance and microbial community analysis of the anaerobic reactor with coke oven gas biomethanation and in situ biogas upgrading. Bioresour. Technol. 146, 234-239.
68. Ward, A.J., Hobbs, P.J., Holliman, P.J., Jones, D.L., 2008. Optimisation of the anaerobic digestion of agricultural resources. Bioresour. Technol. 99(17), 7928-7940.
69. Wijekoon, K.C., Visvanathan, C., Abeynayaka, A., 2011. Effect of organic loading rate on VFA production, organic matter removal and microbial activity of a two-stage thermophilic anaerobic membrane bioreactor. Bioresour. Technol. 102(9), 5353-5360.
70. Wikandari, R., Youngsukkasem, S., Millati, R., Taherzadeh, M.J., 2014. Performance of semi-continuous membrane bioreactor in biogas production from toxic feedstock containing D-Limonene. Bioresour. Technol. 170, 350-355.
71. Wilkinson, K.G., 2011. A comparison of the drivers influencing adoption of on-farm anaerobic digestion in Germany and Australia. Biomass Bioenergy. 35(5), 1613-1622.
72. Xie, K., Lin, H.J., Mahendran, B., Bagley, D.M., Leung, K.T., Liss, S.N., Liao, B.Q., 2010. Performance and fouling characteristics of a submerged anaerobic membrane bioreactor for kraft evaporator condensate treatment. Environ. Technol. 31(5), 511-521.
73. Yang, B., Wyman, C.E., 2008. Pretreatment: the key to unlocking low-cost cellulosic ethanol. Biofuels, Bioprod. Biorefin. 2(1), 26- 40.
74. Ylitervo, P., Akinbomia, J., Taherzadeha, M.J., 2013. Membrane bioreactors' potential for ethanol and biogas production: a review. Environ. Technol. 34(13-14), 1711-1723.
75. Yong, Z., Dong, Y., Zhang, X., Tan, T., 2015. Anaerobic codigestion of food waste and straw for biogas production. Renew. Energ. 78, 527-530.
76. Youngsukkasem, S., Akinbomi, J., Rakshit, S.K., Taherzadeh, M.J., 2013. Biogas production by encased bacteria in synthetic membranes: protective effects in toxic media and high loading rates. Environ. Technol. 34(13-16), 2077-2084.

77. Youngsukkasem, S., Chandolias, K., Taherzadeh, M.J., 2015. Rapid bio-methanation of syngas in a reverse membrane bioreactor: membrane encased microorganisms. Bioresour. Technol. 178, 334- 340.

78. Youngsukkasem, S., Rakshit, S.K., Taherzadeh, M.J., 2012. Biogas production by encapsulated methane-producing bacteria. BioResources. 7(1), 56-65.

79. Zahedi, S., Sales, D., Romero, L.I., Solera, R., 2013. Optimisation of single-phase dry-thermophilic anaerobic digestion under high organic loading rates of industrial municipal solid waste: population dynamics. Bioresour. Technol. 146, 109-117.

80. Zakrzewski, M., Goesmann, A., Jaenicke, S., Jünemann, S., Eikmeyer, F., Szczepanowski, R., Al-Soud, W.A., Sørensen, S., Pühler, A., Schlüter, A., 2012. Profiling of the metabolically active community from a production-scale biogas plant by means of highthroughput metatranscriptome sequencing. J. Biotechnol. 158(4), 248-258.

81. Zeng, M., Mosier, N.S., Huang, C.P., Sherman, D.M., Ladisch, M.R., 2007. Microscopic examination of changes of plant cell structure in corn stover due to hot water pretreatment and enzymatic hydrolysis. Biotechnol. Bioeng. 97(2), 265-278.

82. Zhang, Y.H.P., 2008. Reviving the carbohydrate economy via multiproduct lignocellulose biorefineries. J. Ind. Microbiol. Biotechnol. 35(5), 367-375.

7

Enhancement and Optimization Mechanisms of Biogas Production for Rural Household Energy in Developing Countries: A review

Yitayal Addis Alemayehu

Department of Environmental science; Faculty of natural and computational sciences; Kotebe University College, Addis Ababa, Ethiopia

ABSTRACT

Anaerobic digestion is common but vital process used for biogas and fertilizer production as well as one method for waste treatment. The process is currently used in developing countries primarily for biogas production in the household level of rural people. The aim of this review is to indicate possible ways of including rural households who own less than four heads of cattle for the biogas programs in developing countries. The review provides different research out puts on using biogas substrates other than cow dung or its mix through different enhancement and optimization mechanisms. Many biodegradable materials have been studied for alternative methane production. Therefore, these substrates could be used for production by

Citation: Y. Alemayehu, "Enhancement and Optimization Mechanisms of Biogas Production for Rural Household Energy in Developing Countries: A review," International Journal of Renewable Energy Development, 4(3), 189-196, Oct. 2015.

addressing the optimum conditions for each factor and each processes for enhanced and optimized biogas production.

Keywords: Biogas, digestion, factor, process, substrate

INTRODUCTION

One requirement for sustainable development is the availability of adequate energy services for satisfying basic needs, improving social welfare, and achieving economic development (Rogner et al. 2004). Consequently, the challenge of energy for sustainable development will require uninterrupted effort on the part of international organizations, national governments, the energy community, civil society, the private sector and individuals (Green et al. 2004).

The energy consumption patterns of the people in developing countries tend to add to their misery and aggravate their poverty. Therefore, a direct improvement in energy services would allow these people to enjoy both short-term and long-term advances in living standards through advancing their energy strategies based on increasing the use of energy carriers other than biomass, or on using biomass in modern ways (Reddy et al. 2004). This is to mean that turning solid biomass to liquid and gas by efficient and greener technologies is essential in order to minimize environmental pollution, land degradation and social welfare.

In many developing countries women and children, who do most of the domestic labor in many cultures, spend more and more hours searching for wood for fuel. In some places, it takes eight hours, or more, just to walk and back with a load of twigs and branches that will only last a few days. For people who live in cities, the opportunity to scavenge fire wood is generally non-existent and fuel must be bought from the market which is too expensive. In Addis Ababa, Ethiopia, for example, 25% of households' income is spent on wood for cooking (Cunningham et al. 2003).

In addition, people often dry and burn animal manure as an additional source of household energy. When cow dung is burned in open fires, more than 90 percent of the potential heat and most of the nutrients are lost compared to the efficiency of using dung to produce methane gas which is an excellent fuel (Cunningham et al. 2003). Use of these traditional fuels in cooking stoves with efficiency as low as 10 to 12 percent exacerbate the exposure of people to indoor pollution (Chipman and Dizioubinski 1999).

To prepare for a transition to more sustainable sources of energy, viable alternatives for conservation, supplementation, and replacement must be explored. One of these is bio-fuel production from agricultural, municipal, and industrial wastes which is efficiently accomplished through conversion to biogas, a mixture of mostly methane (CH_4) and carbon dioxide (CO_2), via anaerobic digestion. In practice, microbial anaerobic conversion to methane is a process for effective waste treatment, biological fertilizer and sustainable energy production. It has the potential for reducing the use of traditional biomass, the demand for fossil fuels like coal, oil, and natural gas which continued exploitation will significantly impact our environment and affect the global climate (Wilkie 2008).

Accordingly, developing countries in collaboration with the Netherland Development Organization (SNV) have embarked on biogas programs in developing countries so as to address the rural energy crisis and indoor pollution caused by the burning of traditional biomass (Anonymous 2011). However, the production potential is restricted as different factors have yet not been critically considered, and focus has been given only for a single substrate, manure which has lower energy content than green shrubs since animals that produce it have already digested the substrate (Gannon 2005; Marshal 2010). In addition, the program has an exclusive nature as it only considers those who possess more than four heads of cattle (Getachew et al. 2006). These together make little attainment of objective of biogas technology dissemination programs as there could be minimum substitution of solid biomass to gaseous one so that the health and environmental impacts would not be progressed as intended and expected.

The purpose of this review is therefore to assess potential substrates that could be used for biogas production so that non-biogas users would be included in the program, and users would produce relatively more biogas using additional substrates together with cow dung. In addition, this paper intended to review and indicate favorable conditions for optimum gas production from mixed substrates. Moreover, it would forward enhancement mechanisms for more gas production.

RATIONALE OF USING BIOGAS THAN SOLID BIOMASS

Physical form and contaminant content are the two characteristics of fuels that most affect their pollutant emissions when burned. It is generally difficult to pre-mix solid fuels sufficiently with air to assure good combustion in simple small-scale devices such as household stoves. Even though most biomass fuels contain few noxious contaminants, they are usually burned incompletely in household stoves and so produce a wide range of health-damaging pollutants (Mckinney and Schoch 2003).

Biomass burning is a major contributor to air pollution and the green house gas effect when done unsustainably (i.e. when replacements are not planted for all the materials burned). It can emit carbon monoxide, nitrogen oxides (NO_X), and particulate matter (such as ash and soot) into the air (Botkin and Keller 1995). Burning municipal solid waste can also be extremely dangerous from this perspective, potentially releasing known carcinogens and heavy metals into the environment. However, such emissions from biofuels are less than those produced from comparable fossil fuels (Mckinney and Schoch 2003). For example, wood burning contributes less to acid precipitation than coal. Because wood has little sulfur, it produces few sulfur gases and burns at lower temperatures than coal; thus it produces fewer sulfur oxides (Cunningham et al. 2003). Two billion people, about 40% of the total world population, depend on fire wood and charcoal as their primary energy source. Of these people, three-quarters (1.5 billion) do not have an adequate and affordable supply. Most of them are in the less developed countries where they face a daily struggle to find enough fuel to warm their homes and cook their food. The problem is intensifying because rapidly growing populations in many developing countries create increasing demands for fire wood and charcoal from a diminishing supply (Cunningham et al. 2003). Each person in Ethiopia, Tanzania and Gambia (Africa) and in Thailand (south eastern Asia) uses more than a ton of wood each year and that nearly everyone in these countries is completely dependent on wood for energy (Karen A. 1994; Ethiopia's Energy Sector 2010).

All in all, the efficiency and the environmental concerns of energy use can be progressed in one of the strategies forwarded by UNDP (2004) i.e. the 'energy ladder'. It is a framework for examining trends and impacts of household fuel use and ranks these fuels along a spectrum running from simple biomass fuels (dung, crop residues, wood) through fossil fuels (kerosene and gas) to the most modern form (electricity). The energy assessment further elaborated that the fuel-stove combinations that represent rungs in the ladder tend to become cleaner, more efficient, more storable, and more controllable in moving up the ladder. Biogas, produced through anaerobic digestion, provides both fuel and fertilizer, while options on biomass uses without anaerobic digestion (burning, applying on a field, applying to a field and ploughing under, and composting) provide either one or the other, but not both (Wilkie 2008). Nitrogen can be lost during digestion only by reduction of nitrates to nitrogen gas and volatilization of ammonia into biogas. Since organic matter is degraded during digestion to produce biogas, the percentage of nitrogen in the slurry rises, compared with solid content. Nitrogen is conserved during anaerobic digestion i.e. a reduction in total solids concentration is accompanied by a corresponding increase in the nitrogen content of the remaining solids which is not observed in an aerobic use of biomass (Marchaim 1992).

In addition to creating clean-burning energy, anaerobic digestion also eliminates harmful pathogens and odors, and most importantly, the biogas process produces high-quality, nitrogen-rich fertilizer that can be used to replace chemical fertilizers made with fossil fuels. Unlike compost, biogas slurry is a liquid and can be applied on a commercial scale with existing farm equipment. The fertilizer replacement value of biogas slurry far exceeds any potential revenue from the gas itself (Weisman 2009). For example, a study on the effect of biogas slurry in crop yield shows increment yield of 13.6 % for wheat, 25 % for vegetables (Karki et al. 1995 cited in Gurung 1997), and 18 % for maize and cotton (Gurung 1997).

OPTIMIZATION LEVEL

Marshal (2004), Jemmett (2006) and (Rahmat et al. 2014) reported that depending on the digestion process, the methane content of biogas is generally between 55-80 %. The remaining composition is primarily CO_2 with trace quantities (0-15000 ppm) of corrosive H_2S and H_2O. In general, it would be best and clean if the methane content is above 75 % by absorbing the CO_2 with lime and H_2S by lead acetate (see equations 7 and 8). Together with ignition test, the gas characteristics should be tested, evaluated and optimized according to Table 1 after production.

Table 1: Biogas Characteristics

Characteristic	Value/Behavior
Density	1.2 kg/m^3
Calorific Value	4-7.5 kwh/m^3
Ignition temperature	700 °C
Ignition concentration gas content	6-12 %
Smell	Rotten eggs

Source: EEMBPM 2002; Sidhu 2006

ENHANCING THE BIOCHEMICAL PROCESSES IN ANAEROBIC DIGESTION

In anaerobic digestion there are four stages: hydrolysis, acidification, acetogenesis and methane production (Sagagi et al. 2009; Rai 2004). Different microorganisms play a significant role in each stage of the processes. Therefore, intervention and follow up in all the stages is essential for enhanced and optimum methane production.

Enzymatic Hydrolysis

Enzymatic hydrolysis is the process where the fats, starches and proteins contained in cellulosic biomass are broken down into simple compounds (Rai 2004). Polymers are transformed into soluble monomers through enzymatic hydrolysis.

$$(C_6H_{10}O_5) + nH_2O \xrightarrow{\text{Hydrolysis}} n(C_6H_{12}O_6) \qquad (1)$$

These monomers become substrates for the microorganisms in the second stage where they are converted in to organic acids by a group of bacteria (Sagagi et al. 2009). This process is natural, but could be enhanced by external application of enzymes for better methane production. For example, the hydrolysis of cellulose and hemi-cellulose is considerably enhanced by the application of hydrolytic enzymes of fungal origin on selected substrates, and increases methane yield (Quinones 2010).

Acid Formation (Acidogenesis)

It is a process where microorganisms of facultative and anaerobic group, collectively called as acid formers, hydrolyze and ferment compounds into acids and volatile solids. As a result, complex organic compounds are broken down to short chemical simple organic acids. In some cases, these acids may be produced in such large quantities that the pH may be lowered to a level where all biological activity is arrested. This initial acid phase of digestion may last about two weeks and during this period a large amount of carbon dioxide is given off (Rai 2004). In this stage, there should be intervention in buffering the contents of the digester as the micro-organisms may be affected irreversibly. Generally, if pH is below 4.5 in the acidogenesis phase, there is a need of adding adjusting substances like lime, ash or ammonia as the gas producing bacteria couldn't ferment the acid and restore balance as reported by Saxon (1998).

Acetogenesis

Simple molecules created through the acidogenesis phase are further digested by acetogens to produce largely acetic acid as well as carbon dioxide and hydrogen (Buswell and Sollo 1948). Acetogen are the vital link between hydrolysis, acidogenesis and the methanogenesis in anaerobic digestion. Acetogenesis provides the two main substrates for the last step in the methanogenic conversion of organic material, namely hydrogen and acetate. Both the acidogenesis and acetogenesis produce the methanogenic substrates, acetate, H_2 and CO_2 (Chynoweth and Isaacson 1987).

$$n(C_6H_{12}O_6) + nH_2O \xrightarrow{\text{Acetogensis/Acidogenesis}} n(CH_3COOH) \quad (2)$$

Methane Formation (Methanogenesis)

Organic acids as formed above are then converted into methane (CH_4) and CO_2 by the bacteria which are strictly anaerobes, called methane fermenters (Rai 2004;). In this step, methanogenic bacteria generate methane by two routes, by fermenting acetic acid to methane (CH_4) and CO_2 and by reducing CO_2 via hydrogen gas or by other bacterial species.

$$CH_3COOH \xrightarrow{\text{Methane for min g bacteria}} CH_4 + CO_2 \quad (3)$$

$$CO_2 + 4H_2 \xrightarrow{\text{Reduction}} CH_4 + 2H_2O \quad (4)$$

Similarly, CO_2 can be hydrolyzed to carbonic acid and to methane as in equations 5 and 6

$$CO_2 + H_2O \xrightarrow{\text{Hydrolysis}} H_2CO_3 \quad (5)$$

$$4H_2 + H_2CO_3 \xrightarrow{\text{Reduction}} CH_4 + 3H_2O \quad (6)$$

The carbon dioxide and hydrogen sulfide in the biogas are undesirable. They are removed for optimum performance of biogas as fuel. Carbon dioxide is removed by passing the gas into lime water which turns milky due to formation of calcium carbonate.

$$Ca(OH)_2(aq) + CO_2(g) \rightarrow CaCO_3 + H_2O \quad (7)$$

H_2S is removed by passing the gas through a lead acetate solution (Sagagi et al. 2009).

$$(CH_3COO)_2Pb(aq) + H_2S(g) \rightarrow CH_3COOH(aq) + PbS(s) \quad (8)$$

For efficient digestion, these acid formers and methane fermenters must remain in a state of dynamic equilibrium. This equilibrium is a very critical factor which decides the efficiency and rate of generation (Rai 2004).

ENHANCEMENT OF BIO-DIGESTION FOR OPTIMUM PRODUCTION

Enhancing in each step of the different biochemical processes is essential for optimum biogas production. This could be done through selecting and adjusting the favorable conditions for the microorganisms that are used for the conversion in each process. The different factors that should be adjusted are reviewed as follows:

Temperature

Both fixed temperature at which bio-digestion take place and temperature fluctuation affect the quantity and quality of biogas production. For the former case, methane bacteria (methanogenes) work best at a temperature of 35-38 °C and fall in gas production starts at 20 °C and stops at a temperature of 10 to 13 °C (Velsen et al. (1979); Dahlman and Forst (2001); Rai 2004)). In this connection, there are two significant temperature zones in anaerobic digestion and two types of microorganisms that should be considered, mesophilic and thermophilic. The optimum mesophilic temperature lies at about 35 °C, while the thermophilic temperature is around 55 °C. Most of the sewage digestion tanks are heated at 35 °C so as to reduce the time required for digestion and minimize the capacity of the tanks (Velsen et al., 1979; Rai 2004). Others reported that more biogas is produced at mesophilic temperature zones than at room temperature as shown in Table 2.

Table 2: Rate of biogas production from manures

Temperature	Manure Type	Rate (mL/hr/litre of digester)
	Cow	3
Room	Chicken	9
	Pig	1
	Cow	4
Mesophilic	Chicken	13
	Pig	2
Thermophilic	Cow	3
	Pig	1

Source: Prasad 2012

Furthermore, since methanogenesis is sensitive to temperature fluctuations, effective insulation should be done during the digestion process. According to NRCS (2005), the daily fluctuations of digester temperature should be limited to less than 1 °C. Bilhat (2008) reported that uninsulated bottles (biogas digesters) showed a change of temperature from 20.5 °C to 47 °C i.e. a 26.5 °C rise. But, sand jacketed bottle temperature went from 20.5 °C to 23 °C, a 2.5 °C rise at the same time when both were immersed in a 50 °C water bath and were stayed for 30 minutes and above. Therefore, insulating biogas digesters with concrete or bury it underground should be taken as a pr-requisite before the beginning of biogas production.

pH or Hydrogen Ion Concentration

pH of slurry changes at various stages of digestion. In the initial acid formation stage, in the fermentation process, the pH is around 6 and much of CO_2 is given off. In the latter 2-3 weeks time, the pH increases as the volatile acid and N_2 compounds are digested and CH_4 is produced. To maintain a constant supply of gas, it is necessary to maintain a suitable pH range in the digester. The digester is usually buffered to maintain the pH at 6.5 to 7.5. In this pH range, the micro-organisms will be very active and bio-digestion will be very efficient (Rai 2004).

The introduction of too much raw material can cause excess acidity and the gas-producing bacteria will not be able to digest the acids quickly enough. The addition of a little ammonia can raise the pH value very fast. If the pH grows too high (not enough acid), fermentation will slow until the digestive process forms enough acidic carbon dioxide to restore balance (Saxon 1998).

Carbon to Nitrogen Ratio of the Input Material

Besides carbon the quantity of nitrogen present in the input material is a crucial factor in the production of biogas. The elements of carbon (in the form of carbohydrates) and nitrogen (as protein, ammonium nitrates) are the main food of anaerobic bacteria. Carbon is used for energy and nitrogen for building the cell structure. The bacteria use up carbon about 30 times faster than they use up nitrogen. So, carbon and nitrogen should be present in the proper proportion i.e. C: N is 30:1 (Rai 2004). If the ratio is higher, the nitrogen will be exhausted while there is still a supply of carbon left. This causes some bacteria to die, releasing the nitrogen in their cells and eventually restoring the equilibrium. Digestion proceeds slowly as this occurs. On the other hand, if there is too much nitrogen, fermentation (which will stop when the carbon is exhausted) will be incomplete and the "leftover" nitrogen will not be digested. This lowers the fertilizing value of the slurry (Saxon 1998). Hills (1979) reported that the greatest methane production per unit occurred when the C: N ratio of the feed was 25:1.

Such 25:1to 35:1 ratio could be maintained only when cow dung is supplemented and mixed with plant materials depending on the substrate used. According to Rai (2004), substituting a portion of vegetable waste instead of dung will enable to get more gas for the same amount of substrate. Moreover, most vegetable matter has a much higher C/N ratio than dung has. As a result, the former produces about eight times as much biogas as manure, so some nitrogen producers (preferably organic) must generally be added to the vegetable matter (EEMBPM 2002). This could lead to a conclusion that plant materials are preferable substrates for biogas production to cow dung. Consequently, biogas programs are recommended to consider alternative feed stocks than entire dependence on a single one, cow dung.

Total Solid (TS) and Volatile Solid (VS) Content

The cow dung is mixed usually in the proportion of 1:1(by weight) in order to bring the total solid content to 8-10 %. The raw cow dung contains 80- 82 % of moisture. The balance 18-20 % is termed as total solids. The adjustment of total solid content helps in bio-digesting the material at the faster rate, and also in deciding the mixing of the various crop residues, weeds, and plants etc. as feed stocks in biogas digester (Rai 2004).

Elias (2010) studied that the TS and moisture contents of fresh cow dung and VS (as percentage of TS) and ash content were 16 %, 84 %, 79.11 % and 20.89 %, respectively. For cow dung, the TS was in the range of 15-20 %, and the TS and moisture contents (MC) of Cladodes were 14 % and 86 %, respectively. Nallathambi et al. (1990) studied the biogas production potential of Parthenium in batch digester. They observed that the maximum gas production was 35 liters per kg fresh plant at a TS concentration of 5 % and the methane content of the biogas was 75 %. Similarly, Steffen et al. (2000) reported that the composition of garden wastes, fruit wastes, food remain and typical animal's wastes consists of 50-70 % of TS and VS is 70- 90 % of TS. For cow dung, the VS as percent of TS are in the range 75-80 %. Tesfaye (2007) studied that the TS and VS (as percentage of TS) of chat waste and found them to be 29 % and 90 %, respectively. This is also evident that more biogas is produced from green biodegradable species and co-digestion with cow dung than cow dung alone.

Retention Time or Rate of Feeding

The period of retention of the material for biogas generation depends on the type of feed stocks and the temperature (Rai 2004). Normal retention period is between 30 and 45 days and in some cases 60 days (Rai 2004). Depending on the waste material and operating temperature, a batch digester starts producing biogas after two to four weeks, slowly increasing in production then droping off after three or four months (Jemmett 2006). Here it should be noticed that gas may be produced before these periods. However, the gas produced could be other than methane majorly carbon dioxide. This can be checked by ignition test (Yitayal et al. 2011).

According to Gannon (2005), the amount of time the substrates spend in the digesters is one of the critical factors in methane production. Too short retention time means an inefficient extraction of methane, so full revenue is not realized. Too long retention time means too much is spent on surplus capacity or not enough substrate is being added to maximize revenue.

Feed Stocks and Co-digestion

According to Wilkie (2008), the feed stocks for biogas generation can be composed of carbohydrates, lignocelluloses, proteins, fats or mixtures of these components.

Table 3: Theoretical Methane yield

Component	Max. Methane Yield (m^3/t VS)
Carbohydrate	370
Proteins (leucine)	998
Fats (lauric acid)	585
Plants	470

Source: Wilkie (2008); EEMBPM (2002)

All plant and animal wastes or leaf of plants prepared for this purpose could be used as the feed materials for a digester. When feedstock is woody or contains more of lignin, then bio-digestion becomes difficult. Cow and buffalo dung, human excreta, poultry droppings, pig dung, waste materials of plants, cobs, etc can all be used as feed stocks. To obtain an efficient bio-digestion, these feed stocks are combined in proportions (Rai 2004).

The feed stocks for anaerobic digestion vary considerably in composition, homogeneity, fluid dynamics and biodegradability. In intensive animal farming, pig and cow slurries are reported to contain dry matter contents in the range of 3 to 12 %. Chicken manure contains 10 to 30 % TS. Some agro-industrial wastes may contain less than 1 % TS, while others contain high TS contents of more than 20 %. This results in some substrates being able to be fermented only when mixed with other substrate or diluted (Braun and Wellinger 2002).

Consequently, the need for different substrates such as manure, organic wastes and green plant materials for biogas production is increasing from time to time. For example, the substrates used in Germany are composed of 48 % animal excrement, 26 % organic waste and 26 % renewable green raw materials (Fachagentur 2010). In line with this, studies on the substrates for biogas production such as Cladodes of Opuntia ficus-inica (Elias 2010), food waste (Bilihat 2008), Chat waste (Tesfaye 2007) and Justicia shimperiana (Yitayal et al. 2011) showed that these materials could be used as a substrate either to substitute or supplement cow dung. In addition, maize (Zea mays L.), herbage (Poacae), clover grass (Trifolium), Sudan grass (Sorghum sudanense), fodder beet (Beta vulgaris) and others that serve as energy crops have been studied (Vindis et al. 2009).

It has been reported that the performance of digesters could be considerably improved by means of co-substrate addition and hence increase degradation efficiency and biogas production (Kaparaju et al. 2001). Shivappa et al. (1980) studied different plant wastes like soybean wastes, sunflower wastes, and banana trashes. Up to the fourth month, the total gas production from 1:1 and 1:3 proportions of soybean and cattle dung was more than that of cattle dung alone. Similarly, Rai (2004) recommended that the weight of dung in a dung vegetable mixture should be maintained above 50 %. Similarly, Anand et al. (1991) carried out an experiment on anaerobic digestion of leaf biomass and water hyacinth for the production of biogas and reported that the decomposition of the leaf biomass and water hyacinth substrates used was rapid, taking 45 and 30 days for the production of 250 and 235 L biogas per kg of TS, respectively.

In addition, Rajasekaran et al. (1989) studied the anaerobic digestion of Euphorbia tirucally L. They stated that the slurry consisting of 375 g fresh cow dung, 375 g of one cm bits of Euphorbia tirucally and 750 g of water produced 19.2 liters of biogas in nine weeks compared to 16.5 liters produced by cow dung alone. The carbon dioxide content of the biogas was

38 percent from Euphorbia tirucally L. plus cow dung and 35 % from cow dung alone.

Adding human urine is another enhancement mechanism for optimum biogas production. For example, Sau et al. (2014) reported that the rate of biogas production is enhanced at a volume of 150 mL urine. Thus, a combination of two or more substrates, co-fermentation, will enhance and optimize the degradation properties of the feed stocks and hence increase the methane yield.

Pre-treatment of Feed Stocks

Pretreatment provides three benefits for digestion: makes the substrate more accessible to enzymes, minimize lose of organic matter and limiting the formation of inhibitors for production (Liqian et al. 2011; Karaalp 2014).

Plant biomass mainly consists of cellulose, hemicelluloses and lignin which is poorly degraded in anaerobic conditions, and the rate and extent of lignocelluloses utilization is severely limited due to the intense cross-linking of cellulose with hemicelluloses and lignin as these materials form a scum and can easily clog the system. So pre-treatment of the substrate in order to break the polymer chain so that increasing surface area and reduce lignin content is highly important (Fan et al. 1981; Sagagi et al. 2009).

Treatments may be physical, biological or chemical and the most important physical pretreatment of crop biomass is particle size reduction leading to increase in available surface area and release of intracellular components (Palmowski and Muller 1991). According to Badger et al. (1979), mechanical chopping and grinding could provide greater volumes of gas from carbonaceous residues.

Four pre-treatment methods (ultrasonic, chemical, thermal, and thermo-chemical) are studied and suggested for enhancement of biogas production (Almukhtar et al. 2012). It is reported that ultrasonic pretreatment increase methane yield of sugar beet leaf and maize by 43 and 41%, respectively, and chemical pretreatment by 68 and 102%, respectively, and the thermal one increases digestion of straw by 54 % (Liqian et al. 2011). Therefore, before feeding biogas digesters with a substrate pretreatment is needed for enhanced and optimized production. For example, if the substrate is a plant material (leaf), chopping could be needed prior to digestion for the household level.

Uniform Feeding and Dilution

One of the prerequisites of good digestion is uniform feeding of the digesters so that the microorganisms are kept in a relatively constant organic solid concentration at all times. Therefore, the digester must be fed at the same time every day with a balanced feed of the quality and quantity.

All waste materials fed to a biogas plant consist of solid substance, volatile organic matter and nonvolatile matter (fixed solids) and water. During anaerobic fermentation process, volatile solids undergo digestion and non-volatile solids remain unaffected (Almukhtar et al. 2012). For optimum gas yield through anaerobic fermentation, normally, 8-10 % TS in feed is required (Rai 2004). This is achieved by making slurry of fresh cattle dung in water in the ratio of 1:1. However, if the dung is in dry form, the quantity of water has to be increased accordingly to keep the desired amount and consistency of the input (i.e., ratio could vary from 1:1.25 to even 1:2) (Rai 2004). If the dung is too diluted, the solid particles will settle down into the digester and if it is too thick, the particles impede the flow of the gas formed at the lower part of the digester. In both cases, gas production will be less than optimum (Anonymous 1987).

Mixing of the Contents of the Digester

According to Rai (2004), since bacteria in the digester have very limited reach to their food, it is necessary that the slurry is properly mixed and bacteria get their food supply. He reported that slight mixing improves the fermentation; however a violent slurry agitation retards the digestion.

Some method of stirring the slurry in a digester is always advantageous. If not stirred, the slurry will tend to settle out and form a hard scum on the surface, which will prevent release of the biogas. This problem is much greater with vegetable waste than with manure, which will tend to remain in suspension and have better contact with the bacteria as a result (Saxon 1998).

Inoculation

The use of a source high in anaerobic microbes (digester effluent for example) to start up an anaerobic system is called inoculation. According to Wilkie (2008), the quality and quantity of inoculums are critical to the performance, time required, and stability of biomethanogenesis during commissioning (startup) or restart of an anaerobic digester. In manures and some wastes the microbes needed for digestion may be already present in the waste in small numbers, albeit sufficient to act as inoculums, and will develop into a fully functional bacterial population if the right conditions are provided (Wilkie 2008).

CONCLUSION

This work collects the findings reported in different literatures to be applied in the biogas programs of developing countries, especially for inclusion of the poor people in the rural areas. The programs launched in developing countries could include more households by using different biodegradable plant and waste materials through application of enhancement and optimization mechanisms than depending on cow dung alone as a substrate. Moreover, fulfilling the criteria for the different factors of biogas production would optimize gas yield, minimize time of production, and reduce substrate need and cost.

ACKNOWLEDGEMENT

My special thanks go to Kotebe University College and environmental science, and environmental management staffs for their support and encouragement.

REFERENCES

1. Almukhtar R.S., Asawer A., Alwasiti A.A., Mohammed T., Naser M.T. (2012) Enhancement of Biogas production and organic reduction of sludge by different pre-treatment processes. Iraqi Journal of Chemical and Petroleum Engineering, 13(1): 19- 31
2. Anand V., Chanakya H.N., Rajan M.G. (1991) Solid phase fermentation of leaf biomass to biogas. Resources Conservation and Recycling 6:23-33
3. Anonymous (1987) Fixed Dome Biogas Plants. A Design, Construction and Operation Manual, TERI, New Delhi, pp 3
4. Anonymous (2011) Domestic Biogas Newsletter. SNV. Issue 4, http://www.snvworld.org Accessed on 24 February 2013
5. Badger C.M., Bogue M.J., Stewart D.J. (1979) Biogas production from crops and organic Wastes, New Zealand Journal of Science 22: 11-20
6. Bilihat L. (2007) Contribution and potential of food left over from campus students' cafeteria in biogas energy production, a Master Thesis, Addis Ababa University, pp 34, 66
7. Botkin D., Keller E. (1995) Environmental science. Earth as a living planet, John Wiley and Sons, New York, pp 345-347
8. Braun R., Wellinger A. (2002) Potential of co-digestion. IEA Bio-energy, Task 37
9. Buswell A.M, Sollo F.W. (1948) The mechanism of Methane Fermentation. American Chemical Society 70: 1778-1780
10. Chipman R., Dzioubinski O. (1999) Trends in Consumption and Production. Household Energy Consumption, DESA discussion paper number 6, pp 3
11. Chynoweth D.P., Owens J.M., Legrand R. (2001) Renewable methane from anaerobic digestion of biomass. Renewable Energy 22: 1-8
12. Cunningham M.A., Cunningham W.P., Saigo B.W. (2003) Environmental science. A global concern, 7th edn, McGraw-Hill higher education, USA, pp515-517
13. Dahlman J., Forst C. (2001) Technologies demonstrated at echo. Floating drum biogas digester; echo, 17391 Durance Rd, North Ft. Myers FL 33917, USA
14. Elias J. (2010) Study on Renewable Biogas Energy Production from Cladodes of Opuntia ficus-indica, Msc thesis, Addis Ababa University,

pp 26-28, 34

15. Ethiopia's Energy Sector (EES) (2010) Italian development cooperation. http://www.itacaddis.org/italy/index.cfm?fuseaction=basic Accessed on 02 November 2014
16. European Energy Manager Biogas Preparation Material (EEMBPM) (2006), pp2-5
17. Fachagentur N. (2010) Responding to the world's growing demand for affordable energy and a Cleaner environment Sustainability for Anaerobic Digestion. http://www.bioenergie-portal.info Accessed on 02 May 2015
18. Fan L.T, Gharpuray M.M, Lee Y.H. (1981) Evaluation of pretreatments for enzymatic Conversion of agricultural residues. Biotechnology and Bio-energy 11: 29–45
19. Gannon B. (2005) Biogas Energy Systems. Key Elements of Biogas Energy Systems, Anaerobic digesters. Bio-cycle Conference paper, Germany, pp 3
20. Getachew E., Sonder K., Heegde F.t. (2006) Report on the feasibility study of a national program for domestic biogas in Ethiopia. SNV – Ethiopia, pp 52
21. Green D., Nagata E., Slotnick J. (2009) World energy assessment. Energy and the Challenge of Sustainability: energy, the environment and health chapter 3, pp 63, 65
22. Gurung J.B. (1997) Review of Literature on effects of slurry use on crop production. The biogas Support program final report. Kathmandu, Nepal pp 32, 36
23. Hills DJ. (1979) Effect of C: N ratio on anaerobic digestion of dairy manure. Agricultural Wastes 1: 247-320
24. Jemmett R. (2006) Methane-biogas production guide; version 1.0, UK, pp3-4
25. Kaparaju P., Luostarinen S., Kalmari E., Kalmari J., Rintala J. (2001) Co-digestion of Energy crops and industrial confectionery by-products with cow manure. Batch scale and Farm-scale Evaluation in Anaerobic Digestion. Anaerobic Conversion for Sustainability. 9th World Congress, Antwerpen – Belgium, September 2-6
26. Karaalp D., Arslan K., Azbar N. (2014) Enhancement of Biogas

Production from Laying Hen Manure via Sonolysis as Pretreatment. J Bioprocess Biotech 4 (6): 179 doi: 10.4172/2155-9821.1000179

27. Karen A. (1994) Environmental Science, 2nd edn., pp208
28. Karki, Amrit B.And Krishana M., Guatam (1995) Effect of slurry from anaerobic digestion of organic wastes on crops and vegetables and its residua. Effect on soil. Kathmandu
29. Liqian W., Mattsson M., Rundstedt J., Karlsson N. (2011) Different Pretreatments to Enhance Biogas Production -A comparison of thermal, chemical and ultrasonic methods. Master Thesis, Halmstad University
30. Marchaim U. (1992) Biogas processes for sustainable development. Biomass uses following anaerobic digestion. Bulletin FAO Agricultural Services, Rome, 95: 165-193
31. Marshal A. (2010) Stages of anaerobic Digestion, November. http://www.enwikipedia.org/wiki/file Accessed on 02 November 2014
32. Mckinney M.L, Schoch R.M. (2003) Environmental Science. Systems and Solutions, Jones and Bartlett: Boston, Massachusetts, London and Singapore, pp 227-232
33. Nallathambi G.V., Lakshmanaperumalsamy P. (1990) Biogas production potential of parthenium, Biological Wastes 33:311-314
34. NRCS (2005) Conservation Practice Standard. Anaerobic digester in controlled temperature. Number code 366. IA, pp 4-6
35. Palmowski L. and Muller J. (1999) Influence of the size reduction of organic waste on their anaerobic digestion. In: Mata-Alvarez, J., Cecchi, F. & Tilche, A. (eds). Proc. 2nd International Symposium on Anaerobic Digestion of Solid Waste: 137–144. IWA Publishing, London
36. Prasad R.D. (2012) Empirical Study on Factors Affecting Biogas Production. International Scholarly Research Network. Volume 2012, 1-7. ID 136959, doi:10.5402/2012/136959
37. Quinones T.S., Matthias Plochl M., Budde J., Heiermann M. (2010) Results of batch anaerobic digestion test: effect of enzyme addition. Pp 1-19
38. Rahmat, B., Priyadi, R., Kuswarini, P. (2014) Effectiveness Of Anaerobic Digestion On Reducing Municipal Waste, International Journal of Scientific & Technology Research, 3 (3): 98-101
39. Rai G.D. (2004) Non-conventional energy resources, 2nd edn., Khpu Khanna, India. Pp331-337,369

40. Rajasekaran P., Swaminthan K., Jayapragasm M. (1989) Biogas production potential of Euphorbia tirucally along with cattle manure. Biological Wastes, 30: 75-77
41. Reddy N., Amulya K., Wendy A., Kornelis B., David B., Brenda B., Anton E., Jamuna R., Quentin W., Anita K., Mehdi Z. (2004) World Energy Assessment. Energy and Social issues, part I, Chapter 2, pp 40
42. Rogner H., Fritz B., Maritess C., Andre F., Marc G., David H., Vladimir K., Serguei K., Thierry L., Roberto M., Notstaller R. , Peter O., Martin T. (2004) World energy Assessment. Energy and the Challenge of Sustainability. Energy resources, Chapter 5, pp 136
43. Sagagi B.S., Garba B., Usman N.S. (2009) Studies on biogas production from fruits and Vegetable waste. Bayero Journal of Pure and Applied Sciences 2(1): 115 – 118
44. Sau S.K., Manna T.K., Apurba G.A., Nandi P.K. (2014) Enhancement of Biogas Production from Predigested Substrate by Human Urine under Different Thermal Condition. International Journal of Engineering and Innovative Technology (IJEIT) 4 (3):71-77
45. Saxon J. (1998) Troubled Times. Renewable Energy 2: 519-520
46. Shivappa S.K., Naharaju H.C., Bhagyaraj D.J., Patil R.B. (1980) Studies on the use of agricultural and animal wastes for biogas production proceedings. RRAI symposium, Punjab Agricultural University, Ludhiana, pp 333-340
47. Sidhu K.S. (2006) Non-conventional Energy Resources, PEC campus, Chandigarh, pp-6-7
48. Steffen R., Szolar O., Braun R. (2000) Feedstock for anaerobic digestion. Anaerobic Digestion, making energy and solving modern waste problem, AD-Nett report
49. Tesfaye N. (2007) Study on anaerobic digestion of chat wastes. MSc Thesis, AddisAbaba University pp-28
50. UNDP (2004) Over view of World Energy Assessment. Policies and Actions to Promote Energy for Sustainable Development, part VI, pp-68
51. Velsen V., Lettinga G., and Ottelander D. (1979) Anaerobic digestion of Piggery Waste. Influence of temperature. Netherlands Journal of Agricultural Sciences 27:255-267

52. Vindis P., Mursec B., Rozman C., Janzekovic M., Cus F. (2009) Mini digester and biogas Production from plant biomass. Journal of Achievements in materials and Manufacturing Engineering 35:191-196
53. Weisman W. (2009) Biogas wrongly ignored as an alternative source of energy: http://special.registerguard.com/csp/cms/sites/web/opinion/22324802-47/story.csp. Accessed on 02 December 2014
54. Wilkie AC. (2008) Bioenergy: Biomethane from biomass, Biowaste and Biofuel, J.Well et al. Washington DC, pp195-199
55. Yitayal A., Mekibib D., Araya A. (2011) Study on biogas production potential of Justicia shimperiana. MSc Thesis, AddisAbaba University pp-28

8

Methanogens: Biochemical Background and Biotechnological Applications

Franziska Enzmann[1] , Florian Mayer[1] , Michael Rother[2] and Dirk Holtmann[1]

[1] DECHEMA Research Institute, Industrial Biotechnology, TheodorHeussAllee 25, 60486 Frankfurt am Main, Germany
[2] Technische Universität Dresden, Institut für Mikrobiologie, Zellescher Weg 20b, 01217 Dresden, Germany

ABSTRACT

Since fossil sources for fuel and platform chemicals will become limited in the near future, it is important to develop new concepts for energy supply and production of basic reagents for chemical industry. One alternative to crude oil and fossil natural gas could be the biological conversion of CO_2 or small organic molecules to methane via methanogenic archaea. This process has been known from biogas plants, but recently, new insights into the methanogenic metabolism, technical optimizations and new technology combinations were gained, which would allow moving beyond the mere

Citation: Enzmann, F., Mayer, F., Rother, M., & Holtmann, D. (2018). Methanogens: biochemical background and biotechnological applications. AMB Express, 8(1), 1. doi:10.1186/s13568-017-0531-x conversion of biomass. In biogas plants, steps have been undertaken to increase yield and purity of the biogas, such as addition of hydrogen or metal granulate. Furthermore, the integration of electrodes led to the development of microbial electrosynthesis (MES). The idea behind this technique is to use CO2 and electrical power to generate methane via the microbial metabolism. This review summarizes the biochemical and metabolic background of methanogenesis as well as the latest technical applications of methanogens. As a result, it shall give a sufficient overview over the topic to both, biologists and engineers handling biological or bioelectrochemical methanogenesis.

INTRODUCTION

Methanogens are biocatalysts, which have the potential to contribute to a solution for uture energy problems by producing methane as storable energy carrier. The very diverse archaeal group of methanogens is characterized by the ability of methane production (Balch et al. 1979). The flammable gas methane is considered to be a suitable future replacement for fossil oil, which is about to be depleted during the next decades (Ren et al. 2008). Methane can be used as a storable energy carrier, as fuel for vehicles, for the production of electricity, or as base chemical for synthesis and many countries do already have well developed natural gas grids (Ren et al. 2008). In terms of the necessary transition from chemical to biological processes, methanotrophic bacteria can use methane as a carbon and energy source to produce biomass, enzymes, PHB or methanol (Strong et al. 2015; Ge et al. 2014). The biological methanation is the main industrial process involving methanogens. These archaea use CO_2 and H_2 and/or small organic molecules, such as acetate, formate, and methylamine and convert it to methane. Although the electrochemical production of methane is still more energy efficient than the biological production [below 0.3 kWh/cubic meter of methane (0.16 MPa, Bär et al. 2015)], the biological conversion may be advantageous due to its higher tolerance against impurities (H_2S and NH_3) within the educt streams, especially if CO_2 rich waste gas streams shall be used (Bär et al. 2015). Apart from that, research is going on to increase the energy efficiency of the biological process, so that it might be the preferred way of methane production in the future (Bär et al. 2015). Biological methanation occurs naturally in swamps, digestive systems of animals, oil fields and other environments (Garcia et al. 2000) and is already commonly used in sewage water plants and biogas plants. New applications for methanogens such as electromethanogenesis are on the rise, and yet, there is still a lot of basic research, such as strain characterization and development of basic genetic tools, going on about the very diverse, unique group of methanogens (Blasco-Gómez et al. 2017). This review will summarize important facts about the biological properties and possibilities of genetic modification of methanogenic organisms as well as the latest technical applications. It shall therefore give an overview over the applicability of methanogens and serve as a start-up point for new technical developments.

BIOCHEMICAL AND MICROBIAL BACKGROUND

Methanogens are the only group of microorganisms on earth producing significant amounts of methane. They are unique in terms of metabolism and energy conservation, are widespread in different habitats and show a high diversity in morphology and physiological parameters.

Phylogeny and Habitats of Methanogens

For decades known methanogenic archaea belonged exclusively to the phylum ***Euryarchaeota***. There, methanogens were classified first into five orders, namely ***Methanococcales***, ***Methanobacteriales***, ***Methanosarcinales***, ***Methanomicrobiales*** and ***Methanopyrales*** (Balch et al. 1979; Stadtman and Barker 1951; Kurr et al. 1991). Between the years 2008 and 2012 another two orders of methanogens, namely ***Methanocellales*** (Sakai et al. 2008) and ***Methanomassiliicoccales*** (Dridi et al. 2012; Iino et al. 2013), were added to the phylum ***Euryarchaeota***. Hydrogenotrophic methanogenesis from H_2 and CO_2 is found in almost all methanogenic orders with the exception of the ***Methanomassiliicoccales***. Due to its broad distribution it is postulated that this type of methanogenesis is the ancestral form of methane production (Bapteste et al. 2005). Methane formation from acetate, called aceticlastic methanogenesis, can be found only in the order ***Methanosarcinales***. In contrast to that, methylotrophic methanogenesis, which is the methane formation from different methylated compounds such as methanol, methylamines or methylated thiols, is found in the orders ***Methanomassiliicoccales***, ***Methanobacteriales*** and ***Methanosarcinales***. Extensive recent metagenomic analyses suggested that methanogens may no longer restricted to the ***Euryarchaeota***. Two new phyla, namely the ***Bathyarchaeota*** (Evans et al. 2015) and the ***Verstraetearchaeota*** (Vanwonterghem et al. 2016) were postulated. Genome sequences from both phyla indicate a methylotrophic methane metabolism in these -as of yet uncultivated- potential methanogens.

Methanogens are a relative diverse group of archaea and can be found in various anoxic habitats (Garcia et al. 2000). For example, they can be cultured from extreme environments such as hydrothermal vents or saline lakes. ***Methanocaldococcus jannaschii*** was isolated from a white smoker chimney of the East Pacific Rise at a depth of 2600 m (Jones et al. 1983) and ***Methanopyrus kandleri*** from a black smoker chimney from the Gulf of California in a depth of 2000 m (Kurr et al. 1991). From a saline lake in Egypt the halophilic methanogen ***Methanohalophilus zhilinae*** was cultured (Mathrani et al. 1988). But methanogens also colonize non-extreme environments. They can be isolated from anoxic soil sediments such as rice fields, peat bogs, marshland or wet lands. For example, ***Methanoregula boonei*** was obtained from an acidic peat bog (Bräuer et al. 2006, 2011) and several strains of ***Methanobacterium*** as well as ***Methanosarcina mazei*** TMA and ***Methanobrevibacter arboriphilus*** were isolated from rice fields (Asakawa et al. 1995).

Some methanogens can also associate with plants, animals and could be found in the human body. ***Methanobacterium arbophilicum*** could be isolated from a tree wetwood tissue and uses the H_2 resulting from pectin and cellulose degradation by ***Clostridium butyricum*** for methanogenesis (Schink et al. 1981; Zeikus and Henning 1975). From the feces of cattle, horse, sheep and goosc the methanogens ***Methanobrevibacter thaueri***, ***Methanobrevibacter gottschalkii***, ***Methanobrevibacter wolinii*** and ***Methanobrevibacter woesei*** have been isolated, respectively (Miller and Lin 2002). In addition, different ***Methanobrevibacter*** species could be found in the intestinal tract of insects such as termites (Leadbetter and Breznak 1996). Beside the intestinal tract of herbivorous mammals also the rumen contains methanogens. One of the major species here is ***Methanobrevibacter ruminantium*** (Hook et al. 2010). Methanogenic archaea are also present in the human body. ***Methanobrevibacter smithii*** and ***Methanosphaera stadtmanae*** as well as ***Methanomassiliicoccus luminyensis*** could be detected in human feces (Dridi et al. 2009, 2012; Miller et al. 1982). Further ***Methanosarcina*** sp., ***Methanosphaera*** sp. and ***Methanobrevibacter oralis*** were discovered in human dental plaque (Belay et al. 1988; Ferrari et al. 1994; Robichaux et al. 2003).

Methanogens can be also found in non-natural habitats such as landfills, digesters or biogas plants. There, the microbial community varies with the substrate. In biogas plants, due to hydrolysis of complex polymers to sugars and amino acids, followed by fermentation and acetogenesis, acetate, H_2 and CO_2 is produced as substrates for methanogenesis. Therefore, hydrogenotrophic and aceticlastic methanogens are prevalent in mesophilic biogas plants, often dominated by species of ***Methanosarcina*** (***Methanothrix*** at low acetate concentrations) or ***Methanoculleus*** (Kern et al. 2016b; Karakashev et al. 2005; Lucas et al. 2015; Sundberg et al. 2013). However, under certain conditions syntrophic acetate oxidation may be the dominant path towards methane (Schnürer and Nordberg 2008; Westerholm et al. 2016).

Diversity of Methanogens in Morphology and Physiological Parameters

Methanogens show not only a wide diversity in regard to their habitats but are also highly diverse in terms of morphology, temperature optimum, pH and osmolarity. The shapes of methanogens (only some typical methanogens are mentioned here) can be coccoid as for ***Methanococcus***, ***Methanosphaera*** or ***Methanococcoides***, long or short rods as for ***Methanobacterium*** or ***Methanobrevibacter***, or rods in chains as for ***Methanopyrus*** (Kurr et al. 1991). ***Methanoplanus*** (Ollivier 1997) has a plate-shaped morphology and ***Methanospirillium*** (Zeikus and Bowen 1975), as the name says, a spirally shape. ***Methanosarcina*** (Balch et al. 1979; Bryant and Boone 1987; Kern et al. 2016a; Mah 1980) are irregularly shaped cocci, most often arranged to sarcina cell packages. In addition long filaments formed with rods were observed by species of ***Methanothrix*** [formerly designated ***Methanosaeta*** (Kamagata et al. 1992)]. The formation of multicellular aggregates irrespective of the individual cell shape can also occur, like for species of ***Methanolobus***(Mochimaru et al. 2009), ***Methanosarcina*** (Kern et al. 2016a), or ***Methanobacterium*** (Kern et al. 2015).

The diversity of methanogens is also reflected in the different growth conditions. Many methanogens have a mesophilic temperature spectrum, as, e.g. ***Methanosarcina***, ***Methanobacterium***, or most ***Methanococcus***. However, thermophilic and even hyperthermophilic methanogens are known, like ***Methanothermobacter thermautotrophicus*** or ***M. jannaschii*** which grow at temperatures of up to 75 and 86 °C, respectively. Even growth up to 110 °C is possible in hot environments as shown for the hyperthermophilic strain ***M. kandleri***(Kurr et al. 1991). In contrast, also cold-loving methanogenic strains could be isolated. One example is the methanol-converting archaeon ***Methanolobus psychrophilus***, which grows optimally at 18 °C and shows still metabolic activity at 0 °C (Zhang et al. 2008).

Beside the temperature, salt concentration may also be an important physiological parameter for methanogens. A few methanogens have colonized niches such as saline lakes, which are extreme environments for microorganisms because of their high salinity. Microorganisms living under such salty conditions have to protect themselves from losing water and "salting-out". Due to the fact that biological membranes are permeable to water, a higher solute concentration outside the cell, as in the case of environments with a high salinity, would drag water out of the cell and would lead to cell death. To prevent the loss of water, and as a countermeasure, microbes increase the cytoplasmatic osmolarity to survive in such salty environments. This can be done in two ways. The first is the synthesis and accumulation of osmoprotectants, also known as compatible solutes, which have a small molecular mass and a high solubility. This has been shown for example for ***M. mazei***. At a NaCl concentration of 400 mM the methanogen synthesizes glutamate in response to hypersalinity. At higher salt concentration (800 mM NaCl) *N*-acetyl-β-lysine is synthesized in addition to glutamate (Pflüger et al. 2005, 2003). But *N*-acetyl-β-lysine is not essential for growth and can be also substituted by glutamate and alanine at high salinity (Saum et al. 2009). Moreover it has been also shown that ***M. mazei*** can take up the osmoprotectant glycine betaine from its environment (Roeßler et al. 2002). The second way to protect the cell from loosing water, and to balance the cytoplasm osmotically with the high salinity of the environment, is an influx of potassium and chloride into the cytoplasm (Oren 2008). This as "high-salt-in strategy" known way may be also used by the recently discovered "Methanonatronarchaeia" (Sorokin et al. 2017). They appear to be extremely halophilic, methyl-reducing methanogens related to the haloarchaea.

Although most (by far) methanogens grow optimally around neutral pH, some, which are halophilic or halotolerant, show also an adaptation to alkaline pH. ***Methanocalculus alkaliphilus*** grows alkaliphilically with an optimum at pH 9.5 and a moderate salinity up to 2 M of total Na^+, whereas ***Methanosalsum natronophilum*** can even tolerate higher salinities, up to 3.5 M of total Na^+, at the same alkaline pH (Sorokin et al. 2015). Moderately acidic environments can also be inhabited by methanogens as, for example,

Methanoregula booneii, which was isolated from an acidic peat bog and has an pH optimum for growth of 5.1 (Bräuer et al. 2006, 2011).

Substrates and Metabolism of Methanogens

Methanogens use the substrate CO_2 and the electron donor H_2 during hydrogenotrophic methanogenesis. In the first step, CO_2 is reduced and activated to formyl-methanofuran (Wagner et al. 2016) in which reduced ferredoxin (Fd_{red}) is the electron donor for this reaction (Fig. 1).

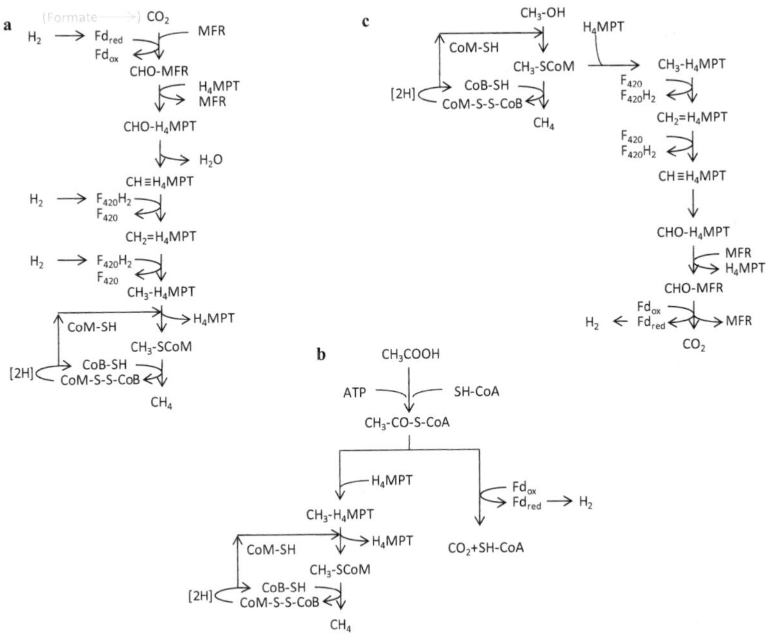

Figure 1: Schematic overview of hydrogenotrophic (a), aceticlastic (b) and methylotrophic (c) methanogenesis. Hydrogenotrophic methanogenesis for Ech-containing methanogens is shown. The methylotrophic methanogenesis from methanol is displayed. Abbreviations are mentioned in the text. (Adapted from (Thauer et al. 2008; Welte and Deppenmeier 2014; Welander and Metcalf, 2005)).

Inthesecondstepthe formylgroupistransferredtotetrahydromethanopterin (H_4MTP) obtaining formyl-H_4MTP. Then the formyl group is dehydrated and reduced to methylene-H_4MTP and subsequently to methyl-H_4MTP with reduced F_{420} ($F_{420}H_2$) as electron donor. The methyl group is then transferred to coenzyme M (HS-CoM). Finally, methyl-CoM is reduced to methane with coenzyme B (HS-CoB) as electron donor. The resulting heterodisulfide (CoM-S-S-CoB) is reduced with H_2 to recycle the coenzymes (Liu and Whitman 2008; Thauer et al. 2008). It is also important to note that several methanogens can use formate instead of H_2 as electron source for CO_2 reduction. There, four formate molecules are first oxidized to CO_2 by formate dehydrogenase (Fdh) followed by the reduction of one molecule of CO_2 to methane (Liu and Whitman 2008). Instead of H_2, a few methanogens can also use alcohols like ethanol or 2-propanol as electron donors (Frimmer and Widdel 1989; Widdel 1986).

Some methanogens can also use carbon monoxide (CO) for methanogenesis. In ***Methanosarcina barkeri***and ***M. thermautotrophicus*** four molecules of CO are oxidized to CO_2 by CO dehydrogenase (CODH) followed by the reduction of one molecule of CO_2 to methane with H_2 as electron donor (Daniels et al. 1977; O'Brien et al. 1984). Thus, growth on H_2 and CO_2 is still possible with both methanogens. In contrast, CO metabolism of ***Methanosarcina acetivorans*** seems to be different. It can also use CO, but is unable to grow on H_2 and CO_2 due to the lack of a functioning hydrogenase system. Further, the organism produces high amounts of acetate and formate from CO during methanogenesis (Rother and Metcalf 2004). The genera ***Methanosarcina*** and ***Methanotrix*** can use acetate for methane production. In this aceticlastic methanogenesis acetate has to be activated first. It is converted with ATP and coenzyme A (CoA) to acetyl-CoA, which is then split by the CODH/acetyl-CoA synthase complex. The methyl group is transferred to H_4MTP [which is tetrahydrosarcinapterin (H_4SPT) in ***Methanosarcina***] and further converted to methane like in the CO_2 reduction pathway. The carbonyl group is oxidized to CO_2, thus providing the electrons for the methyl group reduction (Welte and Deppenmeier 2014).

The third way of biological methanation is methylotrophic methanogenesis in which methylated substrates as methanol, methylamines or methylated sulfur compounds like methanethiol or dimetyl sulfide, are utilized. Most methylotrophic methanogens belong to the ***Methanosarcinales***. In the first step the methyl-group from the methylated substrate is transferred to a corrinoid protein by a substrate-specific methyltransferase (MT1) and subsequently to HS-CoM by another methyltransferase (MT2), thus forming methyl-CoM (Burke and Krzycki 1997). One methyl-CoM is oxidized to CO_2 (via the hydrogenotrophic pathway in reverse) generating the reducing equivalents to reduce three methyl-CoM to methane and also generating a proton motive force (Timmers et al. 2017; Welte and Deppenmeier 2014).

Energy Conservation in Methanogens

In general, methanogens can be divided into two groups according to their mode of energy conservation: methanogens without and with cytochromes (Mayer and Müller 2014; Thauer et al. 2008). Most of the methanogenic archaea do not contain cytochromes. They have a methyl-H_4MPT:coenzyme M methyltransferase (Mtr) which couples the methyl group transfer to a primary, electrochemical Na^+gradient over the membrane (Becher et al. 1992; Gottschalk and Thauer 2001). Furthermore, the H_2-dependent reduction of CoM-S-S-CoB in cytochrome-free methanogens is catalyzed by a complex consisting of a (methyl viologen-reducing) hydrogenase and heterodisulfide reductase (Mvh-Hdr), which also couples this exergonic process to the concomittant endergonic reduction of oxidized ferredoxin (Fd_{ox}) via flavin-based electron bifurcation (Buckel and Thauer 2013). Due to the existence of a Na^+ binding motif in the ***c*** subunits of A_1A_O ATP synthases of almost all non-cytochrome containing methanogens (one exception is ***Methanosalsum zhilinae***), the established Na^+ gradient can be used for ATP synthesis (Mayer and Müller 2014; Grüber et al. 2014).

Cytochrome-containing methanogens such as ***M. mazei*** or ***M. barkeri***, also employ Mtr, thus, generating a Na^+ gradient over the membrane. However, reduction of CoM-S-S-CoB is catalyzed by a membrane-bound heterodisulfide reductase (HdrED), which obtains electrons from reduced methanophenazine ($MPhH_2$, functionally analogous to quinoles) via its cytochrome ***b*** subunit, which is coupled to the generation of a proton motive force. During hydrogenotrophic methanogenesis, a membrane-bound (F_{420} non-reducing) hydrogenase (Vho) oxidizes H_2 and transfers electrons via cytochrome ***b*** to oxidized methanophenazine (MPh), again generating a proton motive force. Further, another membranous energy converting hydrogenase, Ech (which is similar to complex I) couples the endergonic reduction of Fd_{ox} with H_2 to the intrusion of H^+, i.e., uses the proton motive force (Mayer and Müller 2014; Thauer et al. 2008; Welte and Deppenmeier 2014). Under environmental conditions, e.g. as in a biogas plants, cytochrome-containing ***Methanosarcina*** are outcompeted by "true" hydrogentrophic methanogens, which produce methane from CO_2 and H_2 exclusively.

Methanosarcina acetivorans lacks both Vho and Ech. Instead it employs an Rnf complex which is thought to establish a Na^+ gradient over the membrane by transferring electrons from Fd_{red} (accrued from, e.g., oxidation of CO or oxidation of the carbonyl-group from acetyl-CoA) to

MPh. Subsequent electron transport from $MPhH_2$ to HdrED again generates a H^+ gradient (Mayer and Müller 2014; Schlegel et al. 2012b; Welte and Deppenmeier 2014).

The fact that methanogenesis in cytochrome-containing methanogens is coupled to the generation of both a H^+ and a Na^+ gradient (Schlegel and Müller 2013) may be also reflected by the ion dependence of their A_1A_O ATP synthases. It has been shown that the A_1A_O ATP synthase from ***M.*** .(***acetivorans*** can use both ion gradients (Schlegel et al. 2012a

During methylotrophic growth of cytochrome-containing methanogens oxidation to CO_2 involves reduction of cofactor F_{420}, which is a 5-deazaflavin derivative. $F_{420}H_2$ is re-oxidized by $F_{420}H_2$dehydrogenase (Fpo), which is a membrane-bound complex (similar to Nuo of ***E. coli***) and transfers electrons to MPh, thereby establishing a H^+ gradient over the membrane in addition to the H^+ gradients at Hdr and Ech, and the use of the Na^+ gradient at Mtr (Welte and Deppenmeier 2014).

Analyses of genomes from ***Bathyarchaeota*** (Evans et al. 2015) and ***Verstraetearchaeota*** (Vanwonterghem et al. 2016) suggest a methylotrophic methane metabolism for members of these two new phyla. Reduction of the CoM-S-S-CoB in the ***Verstraetearchaeota*** might be accomplished by the Mvh-Hdr complex which might be coupled to re-oxidation of Fd_{red} by an Ehb or and Fpo-like complex. However, what type of ion gradient (H^+ and/or Na^+) might be established over the membrane, is unclear, although H^+ are predicted to be the coupling ion of the respective A_1A_O ATP synthase (Vanwonterghem et al. 2016). It is obvious that pure culture isolation of ***Verstraetearchaeota*** is required in order to address the physiology and energy conservation in these potential methanogens.

In the ***Bathyarchaeota*** energy conservation is even more of a mystery. Two available metagenomes, BA1 and BA2 (proposed to be 91.6 and 93.8% complete, respectively), are missing most of the genes encoding for methanogenic energy conservation. Mtr is incomplete, Fpo as well as an energy-converting hydrogenase (like EhaB, establishing a H^+ or Na^+ gradient over the membrane), are missing. In the genome of BA1 only an Ech hydrogenase is encoded. Also, genes encoding for an A_1A_O ATP synthase are absent, which would restrict the organism to ATP synthesis by substrate level phosphorylation (SLP) (Evans et al. 2015).

ELECTROACTIVITY OF METHANOGENS

Electron Transfer

When electrodes are inserted into a reactor with methanogens, these electrodes can eventually be used by the organisms to produce methane. An external potential leads to the electrolysis of water at the anode; oxygen and protons are produced, electrons are transferred to the anode. Otherwise, excess electrons out of metabolic reactions can be transferred to the anode, like it would happen in a microbial fuel cell. The electrons migrate to the cathode through an external circuit. At the cathode surface, the electrons are transferred to the methanogens, which can use them to produce methane. The complete mechanism is not yet elucidated, but mainly, three possibilities are suggested (Fig. 2) (Sydow et al. 2014; Geppert et al. 2016). Probably, more than one of these mechanisms contributes to the electron transfer (Zhen et al. 2015).

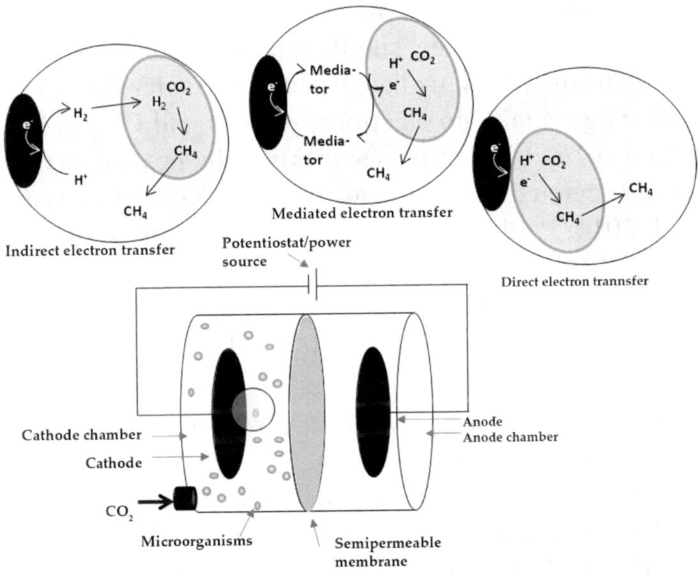

Figure 2: Extracellular electron transfer. Means of electron transfer within a separated, electromethanogenic system at the cathode: indirect electron transfer (IET), mediated electron transfer (MET) and direct electron transfer (DET).

One possible way would be the transfer of electrons from the cathode to protons, which have been produced at the anode and migrated through the membrane between anodic and cathodic chamber. Thereby, hydrogen is produced at the cathode, which is then consumed by the methanogens. This indirect electron transfer (IET) would allow the production of methane out of hydrogen and CO_2 (Villano et al. 2010). As an example, IET was observed in ***M. thermautotrophicus*** (Hara et al. 2013). It has also been shown that some ***Methanococcus maripaludis*** secrete hydrogenases and probably formate dehydrogenases which catalyze the formation of hydrogen and formate directly at the electrode surface; the produced hydrogen and formate is then metabolized by the cells (Deutzmann et al. 2015). This has to be seen as an indirect electron transfer, since the cells were not directly attached to the electrode; from the experimental results, it may be mistaken for a direct electron transport, since the abiotically (without catalyzing hydrogenases) produced amounts of hydrogen and formate cannot explain the amount of methane produced (Deutzmann et al. 2015).

Another possibility suggests that mediator molecules could accept the electrons at the cathode surface, shuttle it to the methanogens and donate it to the microorganisms. This mediated electron transfer (MET) would imply that the methanogens take up electrons, protons and CO_2 to form methane (Choi and Sang 2016). Flavins, phenazines or quinones can serve as mediator, either naturally secreted by the organisms or added to the reaction medium (Sydow et al. 2014; Patil et al. 2012). A natural secretion of mediators with a redox potential suitable for microbial electrosynthesis (should be < -0.4 V vs. SHE) has not been observed yet (Sydow et al. 2014). In methanogens, MET could be performed by using neutral red as an electron shuttle (Park et al. 1999).

The third option would be the direct electron transfer (DET) from the cathode surface to the methanogens, e.g. via surface proteins or conductive filaments (so-called nanowires). To generate methane, the microorganisms would use electrons, protons and CO_2 (Cheng et al. 2009). Several studies suggest that direct electron transfer indeed occurs in methanogens (Zhen et al. 2016; Lohner et al. 2014). For a hydrogenase-deficient strain of ***M. maripaludis*** hydrogenase-independent electron uptake was demonstrated (Lohner et al. 2014), ruling out IET.

In a mixed microbial consortium, direct interspecies electron transfer (DIET) is another possible way of electron transfer. There, one microbial strain takes up electrons at the cathode surface and transfers it to another strain. This may happen, e.g. via conductive filaments (Gorby et al. 2006). It has been reported that this (syntrophic) electron transfer can be very specific between two species, e.g. based on conductive filaments between ***M. thermautotrophicus*** and ***Pelotomaculum thermopropionicum*** (Gorby et al. 2006) or between ***M. barkeri*** and ***Geobacter metallireducens*** (Rotaru et al. 2014). Apart from this direct interspecies electron transfer, an interspecies hydrogen transfer can occur. Here, one organisms takes up electrons, produces hydrogen as an intermediate and transfers them to a second organism that forms another product. An example is the defined co-culture between the iron-corroding, sulfate-reducing bacterium 'Desulfopila corrodens' IS4 (former name: ***Desulfobacterium corrodens***) for electron uptake and ***M. maripaludis*** for methane production (Deutzmann and Spormann 2017).

Electroactive Methanogens

Up to date, most investigations on electromethanogenesis have been carried out with mixed cultures, e.g. from wastewater treatment plants, biogas plants or microbial fuel cells. In technical applications, mixed cultures might be more resistant against environmental stress (Babanova et al. 2017), but it is hard to conclude how many and which methanogenic strains are electroactive by themselves. From analysis of the mixed cultures studied, it can be concluded which methanogens are enriched and are therefore likely to be electroactive, although mixed culture experiments cannot replace pure culture studies to prove electroactivity. These are for example ***Methanobacterium palustre*** (Cheng et al. 2009; Batlle-Vilanova et al. 2015; Jiang et al. 2014), ***Methanosarcina thermophila*** (Sasaki et al. 2013), ***M. thermautotrophicus***(Sasaki et al. 2013; Fu et al. 2015), ***Methanoculleus thermophilus*** (Sasaki et al. 2013), ***Methanobacterium formicicum*** (Sasaki et al. 2013), ***M. maripaludis*** (Deutzmann and Spormann 2017), ***Methanococcus aeolicus*** (Feng et al. 2015), ***M. mazei*** (Feng et al. 2015), ***M. arboriphilus*** (Jiang et al. 2014), ***Methanocorpusculum parvum*** (Jiang et al. 2014) and ***Methanocorpusculum bavaricum*** (Kobayashi et al. 2013). In other studies, the dominant methanogenic organism has not been defined exactly or not explicitly mentioned (Batlle-Vilanova et al. 2015; Bo et al. 2014; Zhen et al. 2015). Only few studies have been carried out with pure cultures instead of mixed cultures, so these methanogens are the only ones that are certainly electroactive. To mention are ***M. thermautotrophicus*** (Hara et al. 2013), and a ***Methanobacterium***-like strain IM1 (Beese-Vasbender et al. 2015).

Yet, just a minority of methanogenic strains has been tested for electroactivity, mostly under similar growth conditions. Unfortunately, no specific marker for electroactivity has been found yet (Koch and Harnisch 2016). It is therefore possible that more electroactive methanogens, active even under more extreme conditions, exist.

GENETIC TOOLS FOR METHANOGENS

Many properties of a (model) organisms are unraveled by biochemical and physiological analysis; however where neither of the two lead to satisfactory insight, genetic analysis is often desirable. Furthermore, the accessibility of an organism relevant for applied purposes to genetic manipulating opens the possibility for targeted engineering by removal of -or amendment with- metabolic or regulatory functions. The principal requirements for such a system are sufficiently efficient methods to (a) isolate clonal populations (e.g., via plating on solid media), to (b) transfer genetic material (i.e., protocols for transformation, transduction, or conjugation), and to (c) link the transfer of the genetic material to an identifiable (i.e., screenable or selectable) phenotype (e.g., conferred by marker genes).

The biochemistry of the methanogenic pathway, the trace elements required, as well as the nature and structure of unusual (C1-carrying) cofactors involved has been elucidated using various ***Methanobacterium***strains (some of them now reclassified as ***Methanothermobacter***). Therefore, it was a logical next step to develop genetic systems for these models. Plating of ***Methanothermobacter*** on solid media could be achieved which allowed isolation (and consequently characterization) of randomly induced mutations (Harris and Pinn 1985; Hummel and Böck 1985). However, this species could not be developed into model organisms for genetic analysis because the transfer of genetic material is too inefficient (Worrell et al. 1988). Furthermore, the use of selectable phenotypes was (and still is) restricted because antibiotics (in conjuncture with the respective genes conferring resistance) commonly used in bacterial genetics are ineffective in archaea due to the differences in the target structures (e.g., cell wall, ribosomes). Therefore, the establishing of an antibiotic selectable marker (the ***pac*** gene from ***Streptomyces alboniger*** conferring resistance to puromycin) in ***Methanococcus voltae*** (Gernhardt et al. 1990) was key to the development of gene exchange systems in methanogens. Another feature of ***Methanococcus***, which facilitated method development, is absence of pseudomurein from its cell wall; instead, the organism is surrounded by a proteinacous surface

(S-) layer that can be removed with polyethylene glycol (PEG), resulting in protoplasts, which apparently can take up DNA. Combined with its comparably robust and fast growth on $H_2 + CO_2$ ***Methanococcus*** species, most prominently ***M. maripaludis***, prevailed as the genetic model for hydrogenotrophic methanogens, for which many useful genetic tools have been developed (Table 1, and see Sarmiento et al. 2011 for a review).

Table 1: Genetic elements used for manipulation of methanogens

Element(s)	**Type**	**Use/phenotype**	**Organism**[a]	**References**
(Counter-) selectable markers				
pac	Resistance gene	Puromycin resistance	***Methanococcus, Methanosarcina***	Gernhardt et al. (1990) and Metcalf et al. (1997)
APH3'II	Resistance gene	Neomycin resistance	***M. maripaludis***	Argyle et al. (1996)
apH-2b	Resistance gene	Neomycin resistance	***M. mazei*** [b]	Mondorf et al. (2012)
ileS3	Resistance gene	Pseudomonic acid resistance	***M. acetivorans***	Boccazzi et al. (2000)
serC	Biosynthesis gene	Auxotrophic marker	***M. barkeri***	Metcalf et al. (1996)
proC	Biosynthesis gene	Auxotrophic marker	***M. acetivorans***	Pritchett et al. (2004)
hpt	Purine salvage gene	Counter-selectable marker (8-aza-2,6-diaminopurine)	***Methanosarcina***	Pritchett et al. (2004)
upt	Purine salvage gene	Counter-selectable marker (6-aza-uracil)	***M. maripaludis***	Moore and Leigh (2005)
Controlling gene expression				
PhmvA	Promoter	Constitutive gene expression	***Methanococcus***	Beneke et al. (1995)
Psl	Promoter	Constitutive gene expression	***Methanococcus***	Sun and Klein (2004)
PmcrB/Tmcr	Promoter/ terminator	Constitutive gene expression	***Methanococcus, Methanosarcina***	Gernhardt et al. (1990) and Mūller et al. (1985)
PmcrB(tetO)/ TetR	Promoter/repressor	Tetracycline-dependent gene de-repression	***Methanosarcina***	Guss et al. (2008)
PmtaCB1	Promoter	Methanol-dependent gene induction	***M. acetivorans***	Rother et al. (2005)

PmtbC	Promoter	Methylamine-dependent gene induction	***M. mazei***	Mondorf et al. (2012)
***Pnif*/NrpR**	Promoter/repressor	N-dependent gene repression	***M. maripaludis***	Cohen-Kupiec et al. (1997 and Lie and Leigh (2003)
tc-RS4	Riboswitch	Tetracycline-dependent gene repression	***M. acetivorans***	Demolli et al. (2014)
Random mutagenesis				
***Himar1*/Tnp**	Insect transposon/ transposase	Random transposon mutagenesis	***M. acetivorans*, *M. maripaludis***	Zhang et al. (2000) and Sattler et al. (2013)
***Tn5*/Tnp**	Transposon/ transposase	Random transposon mutagenesis	***M. maripaludis***	Porat and Whitman (2009)
Reporter genes				
lacZ	Reporter gene	Quantification of gene expression/promoter function	***M. maripaludis***	Gardner and Whitman (1999)
uidA	Reporter gene	Quantification of gene expression/promoter function	***M. voltae*, *Methanosarcina***	Beneke et al. (1995) and Pritchett et al. (2004)
bla	Reporter gene	Quantification of gene expression/promoter function	***M. acetivorans***	Demolli et al. (2014)
Facilitating recombination/mutagenesis				
ΦC31 ***int*** ***attB/P***	Phage recombination system	Site specific chromosomal integration	***Methanosarcina***	Guss et al. (2008)
***flp*/FRT**	Yeast recombination system	Marker rescue	***Methanosarcina*, *Methanococcus***	Welander and Metcalf (2008) and Hohn et al. (2011)
Cas9	Nuclease	Targeted genome editing	***M. acetivorans***	Nayak and Metcalf (2017)

[a]If genus is given, the element is effective in various model species

[b]The construct used does not confer resistance to neomycin in ***M. acetivorans*** (Matschiavelli and Rother, unpublished)

For methylotrophic methanogens containing cytochromes (***Methanosarcina*** species) genetic methodology was initially developed on existing tools. PEG-mediated transformation was reported to be ineffective [but later shown to require only modest modifications of the existing protocol (Oelgeschläger and Rother 2009)], but cationic liposomes could be used to transform ***Methanosarcina*** species (Metcalf et al. 1997). Like in ***Methanococcus***, presence of an S-layer and availability of an autonomously replicating cryptic plasmid [pC2A in ***Methanosarcina*** (Metcalf et al. 1997) and pURB500 in ***Methanococcus*** (Tumbula et al. 1997)], which could be engineered into shuttle vectors also replicating in ***E. coli***, made (heterologous) gene expression comparably easy. Markerless insertion and/or deletion of genes was achieved by establishing counter-selective markers, which are used to remove "unwanted" DNA from the chromosome (Pritchett et al. 2004; Moore and Leigh 2005).

Chromosomal integration and deletion of DNA in methanogens, which can be rather inefficient, mostly relies on homologous recombination requiring sequences of substantial length (500–1000 bp) to be cloned. Thus, establishing site-specific recombination by engineering a ***Streptomyces*** phage recombination system (ΦC31) to integrate DNA into (Guss et al. 2008) -and the yeast Flp/FRT system to remove DNA from- the chromosome (Welander and Metcalf 2008), was a major progress for the genetic manipulation of ***Methanosarcina***. The recent successful -and highly efficient- application of the CRISPR/Cas9-system from ***Streptococcus pyogenes*** (Doudna and Charpentier 2014) for gene deletion and insertion in ***M. acetivorans***(Nayak and Metcalf 2017) holds the promise of an even easier way to genetically manipulate these important organisms. Most tools (Table 1) developed for one methanoarchaeal model organism can usually be adapted for use in another, as exemplified by exploiting the insect transposable element ***Himar1*** together with its transposase for random mutagenesis in ***Methanosarcina*** (Zhang et al. 2000) and, later, in ***Methanococcus*** (Sattler et al. 2013). Thus, any progress made will likely be useful for all other model systems.

Although it might not be possible to use genetically modified methanogens in the established methanogenic processes like biogasproduction or wastewater treatment, new genetic tools are necessary to guarantee the progress in methanogenic research. It will get clear in the next sections that modified methanogens can be used for bioproduction.

APPLICATIONS OF METHANOGENS

Methanogenic archaea are a very diverse group and some strains can grow under extreme conditions, like extremely high or low temperatures, high osmolarities or pH values. Therefore, the development and optimization of industrially applicable processes making use of methanogens is desirable. This is not only true in terms of methane production as a technical relevant fuel (Ravichandran et al. 2015), but also for other products and applications.

Hydrogen Production

It has been observed that several methanogenic strains can also produce hydrogen (Valentine et al. 2000; Goyal et al. 2016). This can happen if the amount of available hydrogen is limited (sub-nanomolar), so that the methanogens seem to start metabolic hydrogen production instead of hydrogen consumption; it has turned out that not methane, but formate and possibly other metabolites can be the source of H_2; this cannot be seen as reverse methanogenesis (Valentine et al. 2000; Lupa et al. 2008). The hydrogen production observed by Valentine et al. reached 0.25 μmol/mg cell dry mass for ***Methanothermobacter marburgensis***, 0.23 μmol/mg cell dry mass for ***Methanosaeta thermophila*** strain CALS-1 and 0.21 μmol/mg cell dry mass for ***M. barkeri*** strain 227 (Valentine et al. 2000). Several strains of ***M. maripaludis*** produced 1.4 μmol/mg of hydrogen per milligram of cell dry mass, out of formate (Lupa et al. 2008). This application is still restricted to the lab scale (Valentine et al. 2000) and to create a reasonable process, genetic engineering would have to be done to increase the hydrogen yield (Goyal et al. 2016). It is assumed that the hydrogenases present in methanogens are the enzymes catalyzing the hydrogen production (Valentine et al. 2000). A possible way to increase the hydrogen yield could therefore be the detection of the relevant hydrogenase and afterwards overexpressing it.

Biotechnological Production by Genetically Modified Methanogens

During recent years, genetic tools for methanogens have been improved, opening a new field of research on these important microorganisms. As a first step, the product spectra of methanogens could be increased. For example, it has been possible to modify ***M. maripaludis*** to produce geraniol instead of methane from $CO_2 + H_2$ or from formate (Lyu et al. 2016). Apart from allowing different products, it has also been possible to broaden the substrate range. As an example, the introduction of a bacterial esterase allowed ***M. acetivorans*** to grow on methyl-esters (like methyl acetate and methyl propionate, Lessner et al. 2010). In wild type methanogens, "trace methane oxidation" (i.e., "reverse methanogensis") has been reported to occur during net methane production (Timmers et al. 2017). It has been possible to use this effect for acetate production: Heterologous expression in ***M. acetivorans*** of genes encoding methyl-CoM reductase from anaerobic methanotrophic archaea (ANME-1) resulted in a strain that converted methane to acetate three times faster than the parental strain (Soo et al. 2016). Also, additional expression of the gene encoding 3-hydroxybutyryl-CoA dehydrogenase (Hbd) from ***Clostridium acetobutylicum*** resulted in formation of l-lactate (0.59 g/g methane) from methane with acetate as intermediate, possibly by Hbd exhibiting lactate dehydrogenase activity in the heterologous host (McAnulty et al. 2017). Thus, the principal possibility might exist to engineer ***M. acetivorans*** for industrial production. However, as both conversion rates and product yields were low and for neither case the conversion stoichiometries reported, the applicability of such a system remains in question. The same holds true for the production of other high value products like amino acids or vitamins with methanogens, and due to their slow growth, a technical application is not yet developed (Schiraldi et al. 2002). But since there is continuous progress in the development of genetic tools for methanogens, as described above, it is thinkable that new processes with heterologeous methanogens will emerge during the next years.

Methane from Oil and Coal Beds

Nearly two-thirds of the fossil oil remains within the oil fields if using conventional production methods (Gieg et al. 2008). It was observed that the residual oil can be converted to natural gas by a methanogenic consortium, which was added to the oil field (Gieg et al. 2008). The consortium used was gained from subsurface sediments and could be enriched with crude oil. ***Methanosaeta*** spec. was the dominant archaeon in the enrichment, which also contained syntrophic sulfate-reducing bacteria, ***Clostridiales***, ***Bacteroidetes***and ***Chloroflexi***. The consortium was added to samples of petroliferous cores from different oilfields, with residual oil saturation of the sandstone grains of approximately 30–40%. Methane could be produced with yields of up to 3.14 mmol/g crude oil (Gieg et al. 2008). Apart from oil fields, also oil sands tailing ponds or other oil–water emulsions could be treated that way (Voordouw 2011). But since costs for natural gas remain relatively low, whilst those for crude oil are significantly higher, this approach remains experimental due to lack of benefit (Voordouw 2011). A natural source of methane is coal bed methane. It has been discovered that about 40% of this methane are produced by microbial consortia containing methanogens (e.g. ***Methanosarcinales***); the substrates for this production are methoxylated aromatic compounds within the coal beds (Mayumi et al. 2016). It was recently discovered that pure cultures of ***Methermicoccus shengliensis*** can produce up to 10.8 μM/(g coal) methane (Mayumi et al. 2016). Coal bed methane is already industrially used; it might be possible to use ***M. shengliensis*** for methane production from other sedimentary organic material (Mayumi et al. 2016).

Biogas Production from Organic Matter

The main technical application of methanogens is the production of biogas by digestion of organic substrates. It is estimated that up to 25% of the bioenergy used in Europe could be produced using the biogas process until 2020 (Holm-Nielsen et al. 2009). Digestion of organic matter can be seen as a four-stage process. During the first step (hydrolysis), complex organic matter (proteins, polysaccharides, lipids) is hydrolyzed by exo-enzymes to oligo- and monomers (amino acids, sugars, long chain fatty acids), which can be taken up by microorganisms (Vavilin et al. 2008). The second step, fermentation or acidogenesis, leads to an oxidation of the compounds formed during hydrolysis to typical fermentation products like butyrate, propionate, acetate, formate, ethanol, H_2 and CO_2. Acetogenesis represent the third step, where the fermentation products are oxidized, mostly to acetate and CO_2 with the concomitant formation of H_2(Batstone et al. 2002). However, this process is only sufficiently exergonic for the organisms if the H_2partial pressure is kept very low (McInerney et al. 2008). This requires the fourth step, methanogenesis, where acetate (and methylated compounds) and CO_2 and H_2 is converted to methane by the methanogens. This implicates that a syntrophic consortium of microorganisms is always needed, whereas the exact composition of this consortium can not only change over time, but also vary between different reactors (Solli et al. 2014). Depending on the microbial community and the type of methanogens within, this process can be carried out in psychrophilic, mesophilic or thermophilic temperature range (Vanegas and Bartlett 2013). For stable biogas production, hydrolysis, acidogenesis, acetogenesis and methanogenesis have to run within the digester in balanced reaction rates to prevent the overacidification of the reactor by surplus protons. However, the microorganisms responsible for these different steps often have different optimal growth conditions, so it is crucial that conditions are maintained, which favor all steps (Niu et al. 2015). Therefore, careful control of process parameters like temperature

(Vanegas and Bartlett 2013), hydraulic retention time (Rincón et al. 2008), pH (Lay et al. 1997) and ammonia concentration (Karakashev et al. 2005) are necessary. Apart from that, the biogas yield and the process operation and conditions strongly depend on the type of substrate used (Niu et al. 2015). It was for example observed that the methanogenic consortium, which is strongly depending on the substrate type, is usually dominated by ***Methanosaetaceae*** in digesters with sludge as substrate, while solid waste digesters operated with manure explained in the following section usually host a majority of ***Methanosarcinaceae*** (Karakashev et al. 2005). In both cases, methanogens that can metabolize acetate (see also "Substrates and metabolism of methanogens" section) are preferred in biogas systems, compared to those feeding on hydrogen and CO_2. Apart from the substrate type itself, a differentiation is made between wet and dry fermentation, whereby the more common wet fermentation includes up to 10% of solids in the substrate, and the dry fermentation between 15 and 35% (Stolze et al. 2015).

Treatment of Sewage Water

The treatment of sewage water by anaerobic digestion does not only lead to biogas production but also to clean water. Using a methanogenic process to convert the organic matter within wastewater to biogas reduces the amount of sludge to be disposed, lowers its pathogenic potential and usually needs less additional energy than aerobic processes, since biogas as energy fuel is produced and no energy intense aeration is necessary (Martin et al. 2011). Apart from that, the greenhouse gas emission of the anerobic process is lower when treating high strength waste waters, although no greenhouse gas savings could be detected for low strength sewage water (Cakir and Stenstrom 2005).

A commonly used system for the anaerobic treatment of wastewater is the upflow anaerobic sludge blanket (UASB) reactor; wastewater enters the reactor from the bottom and flows to an outlet in the upper part of the reactor. Sludge particles out of the waste water agglomerate and form a sludge blanket, which has to be passed by the incoming wastewater. In this zone, the methanogenic consortium digests organic material and produces biogas, which leaves the reactor at its top. Since the solubility of methane in water is low compared to that of CO_2, the holdup of methane within the water is negligible (Sander 2015). The contact between organic material and microorganisms is sufficient for efficient methane production due to the sludge blanket, thus allowing higher loading rates than in other reactor types. The system only requires a low energy input, but needs a long start-up phase of several months, until the sludge blanket has fully established (Rajeshwari et al. 2000). To overcome long start up phases, continuously stirred tank reactors can be used, but here, organic loading rates are about tenfold lower than in the UASB reactor (Rajeshwari et al. 2000). It is important to consider the type of sewage water (e.g. from breweries, paper mills, oil mills, dairy production or other) when estimating the biogas yield of digestion. Different organic loads or different substrate composition lead not only to fluctuating amounts of biogas, but also to changes of the biogas composition (reviewed by Tabatabaei et al. 2010). Instead of treating sewage water itself via anaerobic digestion, it is also possible to purify the water by aerobic processes and anaerobically digest the remaining sewage sludge (Van Lier et al. 2008). Usually, a pretreatment of the sludge can increase the biogas yield. This can for example, but not only, be an alkaline pretreatment, ozonation, ultrasonic pretreatment or electric pulses to increase the biodegradability of the sludge (Wonglertarak and Wichitsathian 2014; Bougrier et al. 2007; Rittmann et al. 2008; for a recent review see: Neumann et al. 2016).

Treatment of Solids

The largest amount of biodegradable waste for biogas production can be obtained from the agricultural sector. This includes animal manure and slurry from the production of pig, poultry, fish and cattle (Holm-Nielsen et al. 2009). The treatment of agricultural wastes like animal manure with methanogenic consortia is not only beneficial in terms of the biogas produced. It also reduces odors and pathogens and is therefore increasing the fertilizer qualities of the manure (Sahlström 2003). The process of biogas formation does not necessarily have to be coupled to waste treatment. Biogas plants can also be operated with energy crops cultured for the biogas production, like sugar beet or maize silage (Demirel and Scherer 2008; Lebuhn et al. 2008). Another possibility is the anaerobic digestion of microalgae, which lowers the necessary cultivation area (Mussgnug et al. 2010). Especially if energy crops without addition of manure are digested, it can be necessary to add micronutrients to ensure optimum growth conditions (Choong et al. 2016). It is also important to consider that lignocellulosic materials are not fully convertible without pretreatment, which leads to lower methane yields (Zheng et al. 2014). Table 2 shows production yields for different solid substrates.

Table 2: Biogas production from organic wastes

Substrate	**Biogas (ml/gVS)**	**Methane (ml/gVS)**	**Methane content (%)**	**References**
Food waste	784	518	66.1	Liu et al. (2009)
Green waste	631	357	56.5	Liu et al. (2009)
Bovine manure	150	40	46.5	Fantozzi and Buratti (2009)
Chicken manure	220	110	66.6	Fantozzi and Buratti (2009)
Pig manure	412	216	52	Amon et al. (2006)
Sugar beet	730	387	53	Weiland (2010)
Grass	211	150	71	Yu et al. (2002)
Maize	560	291	52	Weiland (2010)
Microalgae (***Chlamydomonas reinhardtii***)	784	518	66.1	Mussgnug et al. (2010)
Microalgae (***Arthrospira platensis***)	631	357	56.5	Mussgnug et al. (2010)

VS volatile solids

A crucial aspect of the biogas process is the design of the anaerobic digester (Nizami and Murphy 2010). There are several digester types for the anaerobic digestion of wastewater. For the digestion of solids, biogas plants are usually designed as continuously stirred tank reactors (CSTRs). Even though this might be the easiest and cheapest way of biogas production, it turned out that the efficiency can be increased by using a serial system. Here, two CSTRs are used; biogas yield was increased by a longer overall retention time (Boe and Angelidaki 2009; Kaparaju et al. 2009). Instead of CSTRs, plug-flow systems have been invented by different companies to perform continuous processes (Fig. 3); in a serial digestion, they would usually be taken for the first stage (Weiland 2010). Another possibility is the use of a batch process, especially for substrates with low water contents, for example in a garage type fermenter (Li et al. 2011; Nizami and Murphy 2010).

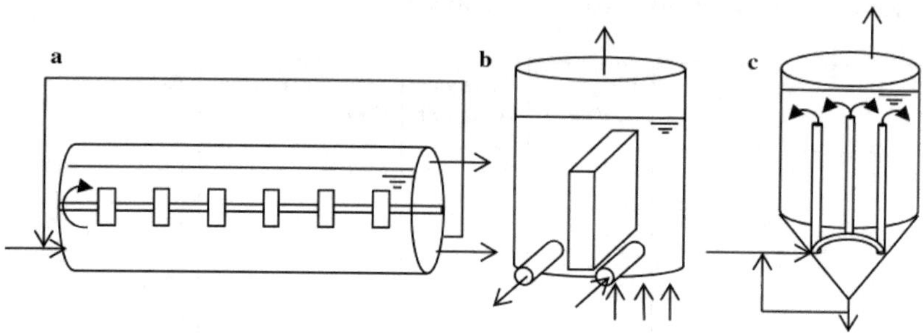

Figure 3: Plug flow digesters for biogas production. a "Kompogas" reactor. Horizontal plug flow reactor. Additional mixing by axial mixer. Increased process condition stability by partial effluent recycling. Gas outlet on top of the outlet side. 23–28% total solids. b Valorga reactor. Substrate entry at the bottom; plug flow over a vertical barrier to the outlet. Additional mixing by biogas injection at the bottom. 25–35% total solid content. c Dranco reactor. Substrate entry wit partial effluent recycling at the bottom, upward flow through substrate pipes. Downward plug flow to outlet. 30–40% total solids (Li et al. 2011; Nizami and Murphy 2010).

Micro Biogas Systems

An interesting application of the biogas process is the use of micro biogas plants in developing countries. These plants of up to 10 m^3 can be operated using domestic organic waste or feces, while the produced gas can be used directly for heating and cooking. There are also attempts to convert the biogas out of those digesters with volumes of up to 10 m^3 to electricity, which might be valuable in rural areas (Plöchl and Heiermann 2006). These reactors are particularly popular in China and India and programs to equip households with biogas energy are supported by the government (Bond and Templeton 2011). Domestic biogas plants are especially beneficial in warm regions (e.g. Africa around the equator, South-East Asia) with sufficient water available. In general, 3 types of digesters are used, which are the fixed dome, the floating cover digester, which was further developed to the ARTI biogas system, and the plug flow (or tube) system (Fig. 4).

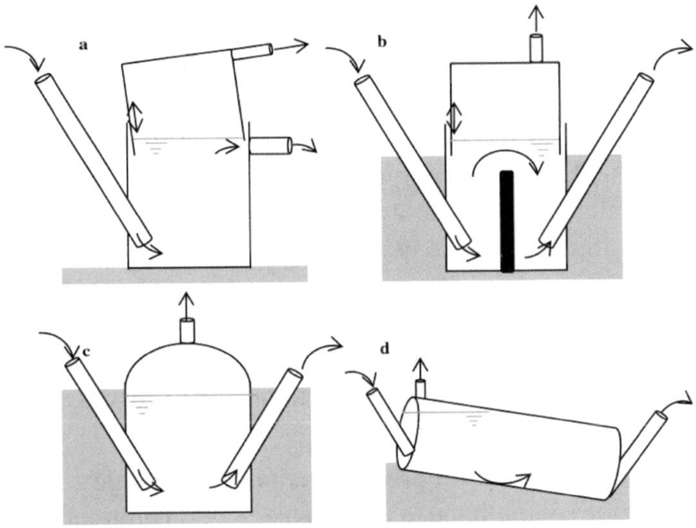

Figure 4: Micro biogas systems. a Arti biogas (India). Material two plastic water tanks (working volume of 1 m^3). Substrate mainly kitchen waste. Disadvantage of gas losses of up to 20% (Voegeli et al. 2009). b Floating cover (India). Material bricks and metal cover. Top rises when gas is produced. Substrate mainly pig and cow manure (Bond and Templeton 2011). c Fixed dome (China). Material bricks and clay. Substrate mainly pig and cow manure (Plöchl and Heiermann 2006). d Plug flow. Material affordable plastic foils (Bond and Templeton 2011).

Although micro biogas systems might not solve the energy problems in developing countries, and the investment costs may not be covered without governmental subsidy, some positive impacts of this technology can be observed. The deforestation in rural areas decreases since wood is not needed for heating, at the same time risks caused by open fire in closed buildings are minimized by the use of a biogas driven stove. The amount of pathogens in the substrate (waste and feces) is decreased, so that it can be reused as fertilizer (Bond and Templeton 2011). Therefore, microbiogas systems are an important contribution to the development of third world countries and a use of the biogasprocess not standing in conflict to food-production, since organic waste is the main substrate.

Biogas Composition and Process Optimizations

The composition of biogas does not only include methane, but also up to 40% CO_2, water, hydrogen sulfide and other trace gases. Biogas is usually flammable due to the high yield of methane (40–75%), but for the use in engines or for injection into the natural gas grid it has to be purified and upgraded in methane content. This leads to higher calorific values of the biogas and avoids the presence of corrosive gases like hydrogen sulfide, which could cause damages to engines and pipes if remaining in the biogas (Ryckebosch et al. 2011). There are several upgrading techniques, which take place after digestion (extensively reviewed by Bauer et al. 2013). Process optimization can influence the biogas composition already during the process, lowering the costs of after-process purification. Numerous investigations on improvement of the biogas process have been undertaken, either to increase the overall amount of biogas, or to increase the methane content of the biogas.

It has turned out that careful pretreatment of the organic substrates leads to higher percentages of methane in the biogas. Several pretreatment methods such as chopping, alkali treatment and thermal treatment are reviewed in Andriani et al. (2014). From a biotechnological point of view, biological pretreatment of substrate is especially interesting. Biological pretreatment can increase the biogas production; this method was described by Zhong et al. (2011) which led to a 33% increase of biogas production (Zhong et al. 2011). The substrates were exposed to a microbial agent including yeasts, celluleutic bacteria and lactic acid bacteria, which degraded the substrate before the actual start of the anaerobic digestion. A reduction in lignin, cellulose and hemicelluloses content could be observed after 15 days of pretreatment. The following anaerobic digestion showed an increase of biogas yield and methane content (Zhong et al. 2011). Apart from the pretreatment of the single substrates, a mixture of different substrates (co-digestion) or a backmixing of digester effluent can lead to a better performance of the system (Weiland 2010; Sosnowski et al. 2003). Co-digestions can be carried out with mixtures of manure and energy plants or sewage slug and solid wastes and increase the methane production because of stabilizing the C:N ratio within the digester (Ward et al. 2008; Sosnowski et al. 2003). Another optimization method is addition of inorganic particles to the fermentation medium. Addition of nanoparticles of zero-valent iron could enhance the methane production by 28% (Carpenter et al. 2015). An increase in biogas formation could also be observed with magnetic iron oxide particles (Abdelsalam et al. 2017). Other particles include charcoal, silica and mineral salts were investigated (reviewed by Yadvika et al. 2004). The improvement in biogas yield could be due to aggregation of bacteria and methanogens around the particles, leading to a lower washout and higher culture densities; it is also possible that metal particles release electrons to the surrounding medium, which can be used for methane formation, but the exact mechanism remains unclear (Yadvika et al. 2004; Carpenter et al. 2015).

One promising method for biological biogas upgrading in methane content is the conversion of the residual CO_2 to additional methane using hydrogenothrophic methanogens, which are capable of producing methane solely out of CO_2 and H_2 (Bassani et al. 2015). H_2 can either be injected into the anaerobic digester (Luo et al. 2012), or H_2 and biogas can be mixed in a second reactor containing methanogens (Bassani et al. 2015; Luo and Angelidaki 2012) (Fig. 5). If introducing hydrogen to the anaerobic digester, there may be a shift within the methanogenic community: acetoclastic methanogens decrease, while hydrogenotrophic methanogens (especially ***Methanoculleus***) are enriched; also, hydrolyzing and acidifying bacteria decrease, while synthrophic bacteria producing acetate increase (see also "Substrates and metabolism of methanogens" section; Bassani et al. 2015). Technical concepts for the integration of H_2into existing biogas plants and effective new means of process control are necessary to make this process commercially attractive. Therefore, experiments have to be carried out under industrial conditions, i.e. under fluctuating substrate compositions, in reactors with zones of different substrate concentrations, changing microbial consortium and different pressure zones according to a larger reactor height; these conditions will usually not appear in lab scale, unless they are particularly tested.

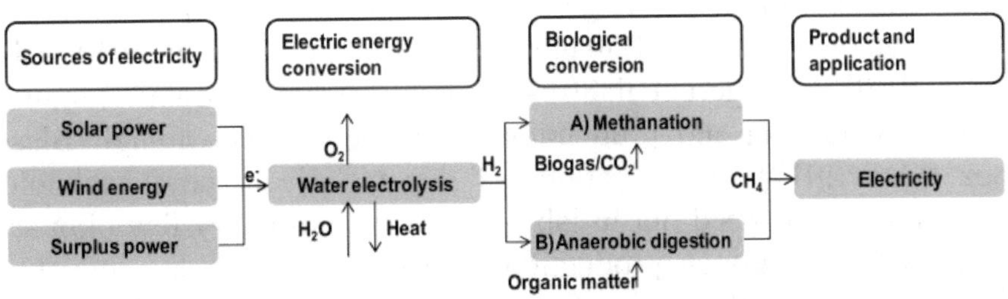

Figure 5: Increasing methane yield by hydrogen addition. H_2 is produced via water electrolyses and (A) fed into the second reactor for the conversion of CO_2 into methane, or (B) feed directly to the anaerobic digester for in situ methane production.

H_2 is usually produced by water electrolysis, a process in which electricity is used to split water and generate oxygen and hydrogen. To couple water electrolysis to anaerobic methanogenesis and provide a constant level of H_2 within the digester, methanogenic bioelectrochemical systems were invented.

Methanogenic Bioelectrochemical Systems

The successful increase of methane production by iron addition leads to the conclusion that methanogens may use inorganic surfaces to boost their metabolism by exchange of electrons with the inorganic material (Carpenter et al. 2015). On the other hand, hydrogen addition could also increase the methane output of a biogas plant. A methanogenic bioelectrochemical system (BES) combines these two improvements for increased methane production (Koch et al. 2015). Here, electrodes are introduced into the reaction medium and an external potential is applied. Methanogens can now either interact directly with the electrode surface to gain electrons (Cheng et al. 2009), and/or hydrogen can be produced at the cathode, which can then be consumed by the methanogens to produce methane (Geppert et al. 2016). The whole process belongs to the field of microbial electrosynthesis (MES), which includes processes that convert a substrate into a desired organic product by using microorganisms and electrical current (Schröder et al. 2015; Lovley 2012; Holtmann et al. 2014). The advantage of a BES system compared to the external production of hydrogen is that short time storage and gassing in of the hardly soluble hydrogen can be avoided (Butler and Lovley 2016).

Notably, the electrode material and size, the membrane material and size and the applied voltage strongly influences the performance of electromethanogenesis, (see Babanova et al. 2017; Krieg et al. 2014; Ribot-Llobet et al. 2013; Siegert et al. 2014 for reviews), but "optimal" conditions for microbial growth and production have not yet been found (Blasco-Gómez et al. 2017). Investigations of this (relatively new) technology have been mostly carried out in lab scale so far, with very few pilot scale approaches (for hydrogen production with methane as side product, see Cusick et al. 2011). Yet, no scale up concept or even well characterized reactor concept exists for electromethanogenesis, whereas various types of bioelectrochemical reactors have been designed (reviewed in Geppert et al. 2016; Krieg et al. 2014; Kadier et al. 2016). Two general modes of integrating electrochemistry into the methanogenic process can be distinguished: first, the electrodes can be integrated into the anaerobic digestion of sewage water or other organic wastes, and secondly, the methanogenic BES can be placed into a second reactor as a stand-alone-process, fed with CO_2, but without additional organic substrates (Fig. 6).

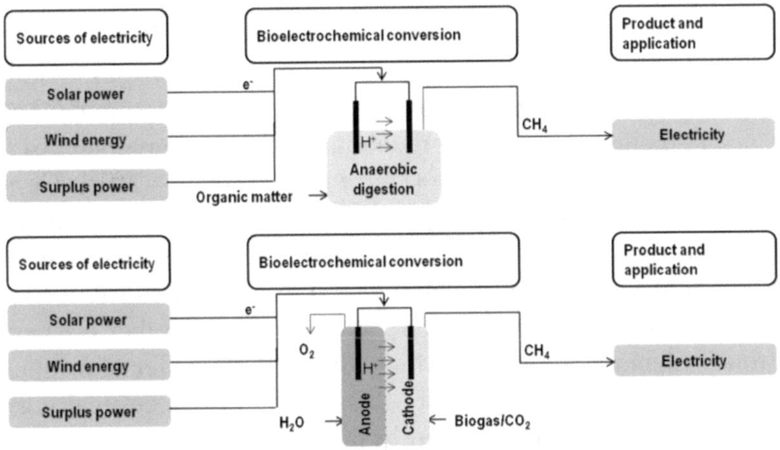

Figure 6: Increasing methane yield by electrode integration. Top: integration of electrodes into the anaerobic digester; bottom: biogas upgrading in an external, separated MES system fed with CO_2 and electricity.

Integration of Electrodes into Waste and Wastewater Treatment

To enhance the production of biogas and increase biogas purity, electrodes can be inserted into the anaerobic digester for in situ biogas upgrading. CO_2, which is produced during the digestion of organic matter, can be converted to methane at the electrodes without an additional reactor (Bo et al. 2014). Therefore, the biogas production can be performed during wastewater treatment (Guo et al. 2017) or sewage sludge treatment (Guo et al. 2013) as well as in a mere biogas producing process (Gajaraj et al. 2017). The methane content within the biogas reached up to 98.1% during the digestion of activated sludge and acetate (Bo et al. 2014). It has been shown that the integration of electrodes alters the microbial consortium within the plant, while it is also possible to use adapted consortia, e.g. for psychrophilic temperature ranges (Koch et al. 2015; Bo et al. 2014; Liu et al. 2016). To achieve a reasonable process, ways of electrode integration into existing treatment plants need to be established.

Bioelectrochemical Systems Fed with CO_2

Methanogenic microbial electrosynthesis can also be carried out in a second reactor, which is equipped with electrodes and fed with CO_2 or a gas containing CO_2. Gas streams rich in CO_2 can be biogas, syngas, or industrial flue gas. The CO_2 contained is often considered a waste component of these gas streams, and since it is also a greenhouse gas, the conversion of CO_2 to more useful chemicals is desirable (Dürre and Eikmanns 2015; Geppert et al. 2016). The conversion of CO_2 by methanogens takes place at the cathode of the system. Since anodic processes like oxygen generation or acid production could inhibit the methanogens, the process can be carried out in a two-chamber system, were anode and cathode chamber are separated by a proton-exchange membrane, which allows the transfer of protons from anode to cathode chamber; this is necessary to allow electrical current in the system and maintain the pH within the cathode chamber (Dykstra and Pavlostathis 2017; Cheng et al. 2009). In this system, it is possible to use a pure methanogenic culture (Beese-Vasbender et al. 2015) or an enriched methanogenic consortium (Dykstra and Pavlostathis 2017) at the cathode, while the anode chamber can be abiotic (water electrolysis) or biotic (degradation of organic matter) (Dykstra and Pavlostathis 2017).

As mentioned, the bioelectrochemical methanogensis is currently still a lab-scale application. To gain a economical technical process, concepts for process characterization and control, reactor balancing, and scale up of reactors have to be developed. To create further progress in this field and also in bioelectrochemical applications, genetic tools might be necessary to create methanogens with higher electron uptake rates, e.g. via the integration of (more) cytochromes into the membrane or the heterologeous secretion of electron shuttles.

CONCLUSIONS

Methanogens are interesting organisms, both from a biological, as well as for a technological, point of view. Research of the last years made it clear that this unique group of microbes is far from being fully understood. During the last years, several reviews on biological aspects of methanogens (Borrel et al. 2016; Goyal et al. 2016), on natural methanogenesis (e.g. Park and Liang 2016; Bao et al. 2016) or on single technical applications, eventually in combination with the very specific biology within the process (Biogas: Wang et al. 2017; Braguglia et al. 2017; Koo et al. 2017; Biogas upgrading and optimization: Neumann et al. 2016; Choong et al. 2016; Romero-Güiza et al. 2016; Bioelectromethanation: Blasco-Gómez et al. 2017; Geppert et al. 2016) have been published. All these review articles are rather specialized to one single aspect of methanogens. This review combines all these aspects, including a review of recently developed tools, to give an overview over the whole field of methanogenic research. Therefore, it makes it possible to understand challenges in industrial applications by giving the biological basics and helps to imagine applications for results from basic research in industry. Industry mainly focused on the production of biogas with methanogens, but other applications, especially when considering electroactivity of methanogens, seem feasible. Newly developed genetic tools for methanogens are useful to design a wider product spectrum, which raises the technical relevance of methanogens. However, most processes possible with methanogens are still not economically feasible, since their strict requirement for anaerobic conditions raises the investment costs and their slow growth leads to long process times. It would be desirable to have further comparable knowledge of the efficiency of different methanogenic strains in terms of space time yield and conversion rates under industrially relevant conditions, for example by performing pure culture studies with fluctuating substrate composition, fluctuating pH and under different substrate concentrations. A major problem here remains the comparability of published data about methanogenic performance in biogas plants as well as in electrochemical systems, since studies have been carried out under various conditions. For some applications, especially microbial electrosynthesis, more research of the methanogenic community and comparisons between pure and mixed cultures have to be done to increase methane yields. Still, process optimization, like the use of CO_2-rich waste gas streams as substrates and intelligent process integration will favor methanogenic processes beyond waste treatment in the future. Scale-up of reactors, e.g., for electromethanogenesis or biogas-upgrading, are a major task for process engineers, while genetic engineering may pave the way to produce higher value products from waste CO_2employing methanogens.

AUTHORS' CONTRIBUTIONS

All authors contributed equally in writing this review article. All authors read and approved the final manuscript.

ACKNOWLEDGEMENTS

The authors thank the DECHEMA Research Institute and the Technical University Dresden for their help in accomplishing this work.

REFERENCES

1. Abdelsalam E, Samer M, Attia YA, Abdel-Hadi MA, Hassan HE, Badr Y. Influence of zero valent iron nanoparticles and magnetic iron oxide nanoparticles on biogas and methane production from anaerobic digestion of manure. Energy. 2017;120:842–853.
2. Amon T, Amon B, Kryvoruchko V, Bodiroza V, Pötsch E, Zollitsch W. Optimising methane yield from anaerobic digestion of manure: effects of dairy systems and of glycerine supplementation. Int Congr Ser. 2006;1293:217–220.
3. Andriani D, Wresta A, Atmaja TD, Saepudin A. A review on optimization production and upgrading biogas through CO_2 removal using various techniques. Appl Biochem Biotechnol. 2014;172:1909–1928.
4. Argyle JL, Tumbula DL, Leigh JA. Neomycin resistance as a selectable marker in ***Methanococcus maripaludis***. Appl Environ Microbiol. 1996;62:4233–4237.
5. Asakawa S, Akagawa-Matsushita M, Morii H, Koga Y, Hayano K. Characterization of ***Methanosarcina mazeii*** TMA isolated from a paddy field soil. Curr Microbiol. 1995;31:34–38.
6. Babanova S, Carpenter K, Phadke S, Suzuki S, Ishii S, Phan T, Grossi-Soyster E, Flynn M, Hogan J, Bretschger O. The effect of membrane type on the performance of microbial electrosynthesis cells for methane production. J Electrochem Soc. 2017;164:H3015–H3023.
7. Balch WE, Fox GE, Magrum LJ, Woese CR, Wolfe RS. Methanogens: reevaluation of a unique biological group. Microbiol Rev. 1979;43:260–296.
8. Bao Y, Huang H, He D, Ju Y, Qi Y. Microbial enhancing coal-bed methane generation potential, constraints and mechanism - a mini-review. J Nat Gas Sci Eng. 2016;35:68–78.
9. Bapteste E, Brochier C, Boucher Y. Higher-level classification of the Archaea: evolution of methanogenesis and methanogens. Archaea. 2005;1:353–363.
10. Bär K, Mörs F, Götz M, Graf F. Vergleich der biologischen und katalytischen Methanisierung für den Einsatz bei PtG-Konzepten. gwf-Gas. 2015;7:1–8.
11. Bassani I, Kougias PG, Treu L, Angelidaki I. Biogas upgrading via hydrogenotrophic methanogenesis in two-stage continuous stirred tank reactors at mesophilic and thermophilic conditions. Environ Sci

Technol. 2015;49:12585–12593.

12. Batlle-Vilanova P, Puig S, Gonzalez-Olmos R, Vilajeliu-Pons A, Balaguer MD, Colprim J. Deciphering the electron transfer mechanisms for biogas upgrading to biomethane within a mixed culture biocathode. RSC Adv. 2015;5:52243–52251.
13. Batstone DJ, Keller J, Angelidaki I, Kalyuzhnyi SV, Pavlostathis SG, Rozzi A, Sanders WT, Siegrist H, Vavilin VA. The IWA anaerobic digestion model no 1 (ADM1) Water Sci Technol. 2002;45:65–73.
14. Bauer F, Persson T, Hulteberg C, Tamm D. Biogas upgrading - technology overview, comparison and perspectives for the future. Biofuels, Bioprod Biorefin. 2013;7:499–511.
15. Becher B, Müller V, Gottschalk G. The methyl-tetrahydromethanopterin: coenzyme M methyltransferase of ***Methanosarcina*** strain Göl is a primary sodium pump. FEMS Microbiol Lett. 1992;91:239–243.
16. Beese-Vasbender PF, Grote J-P, Garrelfs J, Stratmann M, Mayrhofer KJJ. Selective microbial electrosynthesis of methane by a pure culture of a marine lithoautotrophic archaeon. Bioelectrochemistry. 2015;102:50–55.
17. Belay N, Johnson R, Rajagopal BS, Conway de Macario E, Daniels L. Methanogenic bacteria from human dental plaque. Appl Environ Microbiol. 1988;54:600–603.
18. Beneke S, Bestgen H, Klein A. Use of the ***Escherichia coliuidA*** gene as a reporter in ***Methanococcus voltae*** for the analysis of the regulatory function of the intergenic region between the operons encoding selenium-free hydrogenases. Mol Gen Genet. 1995;248:225–228.
19. Blasco-Gómez R, Batlle-Vilanova P, Villano M, Balaguer M, Colprim J, Puig S. On the edge of research and technological application: a critical review of electromethanogenesis. Int J Mol Sci. 2017;18:874.
20. Bo T, Zhu X, Zhang L, Tao Y, He X, Li D, Yan Z. A new upgraded biogas production process: coupling microbial electrolysis cell and anaerobic digestion in single-chamber, barrel-shape stainless steel reactor. Electrochem Commun. 2014;45:67–70.
21. Boccazzi P, Zhang JK, Metcalf WW. Generation of dominant selectable markers for resistance to pseudomonic acid by cloning and mutagenesis of the ***ileS*** gene from the archaeon ***Methanosarcina barkeri*** fusaro. J Bacteriol. 2000;182:2611–2618.
22. Boe K, Angelidaki I. Serial CSTR digester configuration for improving

biogas production from manure. Water Res. 2009;43:166–172.

23. Bond T, Templeton MR. History and future of domestic biogas plants in the developing world. Energy Sustain Dev. 2011;15:347–354.
24. Borrel G, Adam PS, Gribaldo S. Methanogenesis and the Wood–Ljungdahl pathway: an ancient, versatile, and fragile association. Genome Biol Evol. 2016;8:1706–1711.
25. Bougrier C, Battimelli A, Delgenes JP, Carrere H. Combined ozone pretreatment and anaerobic digestion for the reduction of biological sludge production in wastewater treatment. Ozone Sci Eng. 2007;29:201–206.
26. Braguglia Camilla M., Gallipoli Agata, Gianico Andrea, Pagliaccia Pamela. Anaerobic bioconversion of food waste into energy: A critical review. Bioresource Technology. 2018;248:37–56.
27. Bräuer SL, Cadillo-Quiroz H, Yashiro E, Yavitt JB, Zinder SH. Isolation of a novel acidiphilic methanogen from an acidic peat bog. Nature. 2006;442:192–194.
28. Bräuer SL, Cadillo-Quiroz H, Ward RJ, Yavitt JB, Zinder SH. ***Methanoregula boonei*** gen. nov., sp. nov., an acidiphilic methanogen isolated from an acidic peat bog. Int J Syst Evol Microbiol. 2011;61:45–52.
29. Bryant MP, Boone DR. Emended description of strain MST (DSM 800T), the type strain of ***Methanosarcina barkeri***. Int J Syst Bacteriol. 1987;37:169–170.
30. Buckel W, Thauer RK. Energy conservation via electron bifurcating ferredoxin reduction and proton/Na^+ translocating ferredoxin oxidation. Biochim Biophys Acta Bioenerg. 2013;1827:94–113.
31. Burke SA, Krzycki JA. Reconstitution of monomethylamine: coenzyme M methyl transfer with a corrinoid protein and two methyltransferases purified from ***Methanosarcina barkeri***. J Biol Chem. 1997;272:16570–16577.
32. Butler CS, Lovley DR. How to sustainably feed a microbe: strategies for biological production of carbon-based commodities with renewable electricity. Front Microbiol. 2016;7:1879.
33. Cakir FY, Stenstrom MK. Greenhouse gas production: a comparison between aerobic and anaerobic wastewater treatment technology. Water Res. 2005;39:4197–4203.
34. Carpenter AW, Laughton SN, Wiesner MR. Enhanced biogas production

from nanoscale zero valent iron-amended anaerobic bioreactors. Environ Eng Sci. 2015;32:647–655.

35. Cheng S, Xing D, Call DF, Logan BE. Direct biological conversion of electrical current into methane by electromethanogenesis. Environ Sci Technol. 2009;43:3953–3958.

36. Choi O, Sang B-I. Extracellular electron transfer from cathode to microbes: application for biofuel production. Biotechnol Biofuels. 2016;9:11.

37. Choong YY, Norli I, Abdullah AZ, Yhaya MF. Impacts of trace element supplementation on the performance of anaerobic digestion process: a critical review. Bioresour Technol. 2016;209:369–379.

38. Cohen-Kupiec R, Blank C, Leigh JA. Transcriptional regulation in Archaea: in vivo demonstration of a repressor binding site in a methanogen. Proc Natl Acad Sci USA. 1997;94:1316–1320.

39. Cusick RD, Bryan B, Parker DS, Merrill MD, Mehanna M, Kiely PD, Liu G, Logan BE. Performance of a pilot-scale continuous flow microbial electrolysis cell fed winery wastewater. Appl Microbiol Biotechnol. 2011;89:2053–2063.

40. Daniels L, Fuchs G, Thauer RK, Zeikus JG. Carbon monoxide oxidation by methanogenic bacteria. J Bacteriol. 1977;132:118–126.

41. Demirel B, Scherer P. Production of methane from sugar beet silage without manure addition by a single-stage anaerobic digestion process. Biomass Bioenergy. 2008;32:203–209.

42. Demolli S, Geist MM, Weigand JE, Matschiavelli N, Suess B, Rother M. Development of β-lactamase as a tool for monitoring conditional gene expression by a tetracycline-riboswitch in ***Methanosarcina acetivorans***. Archaea. 2014;2014:1–10.

43. Deutzmann JS, Spormann AM. Enhanced microbial electrosynthesis by using defined co-cultures. ISME J. 2017;11:704–714.

44. Deutzmann JS, Sahin M, Spormann AM (2015) Extracellular enzymes facilitate electron uptake in biocorrosion and bioelectrosynthesis. mBio 6(2):e00496–15

45. Doudna JA, Charpentier E. The new frontier of genome engineering with CRISPR-Cas9. Science. 2014;346:1258096.

46. Dridi B, Henry M, El Khéchine A, Raoult D, Drancourt M. High prevalence of ***Methanobrevibacter smithii*** and ***Methanosphaera stadtmanae*** detected in the human gut using an improved DNA

detection protocol. PLoS ONE. 2009;4:e7063.

47. Dridi B, Fardeau M-L, Ollivier B, Raoult D, Drancourt M. ***Methanomassiliicoccus luminyensis*** gen. nov., sp. nov., a methanogenic archaeon isolated from human faeces. Int J Syst Evol Microbiol. 2012;62:1902–1907.

48. Dürre P, Eikmanns BJ. C1-carbon sources for chemical and fuel production by microbial gas fermentation. Curr Opin Biotechnol. 2015;35:63–72.

49. Dykstra CM, Pavlostathis SG. Evaluation of gas and carbon transport in a methanogenic bioelectrochemical system (BES) Biotechnol Bioeng. 2017;114:961–969.

50. Evans PN, Parks DH, Chadwick GL, Robbins SJ, Orphan VJ, Golding SD, Tyson GW. Methane metabolism in the archaeal phylum ***Bathyarchaeota*** revealed by genome-centric metagenomics. Science. 2015;350:434–438.

51. Fantozzi F, Buratti C. Biogas production from different substrates in an experimental Continuously Stirred Tank Reactor anaerobic digester. Bioresour Technol. 2009;100:5783–5789.

52. Feng Y, Zhang Y, Chen S, Quan X. Enhanced production of methane from waste activated sludge by the combination of high-solid anaerobic digestion and microbial electrolysis cell with iron–graphite electrode. Chem Eng J. 2015;259:787–794.

53. Ferrari A, Brusa T, Rutili A, Canzi E, Biavati B. Isolation and characterization of ***Methanobrevibacter oralis*** sp. nov. Curr Microbiol. 1994;29:7–12.

54. Frimmer U, Widdel F. Oxidation of ethanol by methanogenic bacteria. Arch Microbiol. 1989;152:479–483.

55. Fu Q, Kuramochi Y, Fukushima N, Maeda H, Sato K, Kobayashi H. Bioelectrochemical analyses of the development of a thermophilic biocathode catalyzing electromethanogenesis. Environ Sci Technol. 2015;49:1225–1232.

56. Gajaraj S, Huang Y, Zheng P, Hu Z. Methane production improvement and associated methanogenic assemblages in bioelectrochemically assisted anaerobic digestion. Biochem Eng J. 2017;117:105–112.

57. Garcia J-L, Patel BK, Ollivier B. Taxonomic, phylogenetic, and ecological diversity of methanogenic Archaea. Anaerobe. 2000;6:205–226.

58. Gardner WL, Whitman WB. Expression vectors for ***Methanococcus maripaludis***: overexpression of acetohydroxyacid synthase and beta-galactosidase. Genetics. 1999;152:1439–1447.

59. Ge X, Yang L, Sheets JP, Yu Z, Li Y. Biological conversion of methane to liquid fuels: status and opportunities. Biotechnol Adv. 2014;32:1460–1475.

60. Geppert F, Liu D, van Eerten-Jansen M, Weidner E, Buisman C, ter Heijne A. Bioelectrochemical power-to-gas: state of the art and future perspectives. Trends Biotechnol. 2016;34:879–894.

61. Gernhardt P, Possot O, Foglino M, Sibold L, Klein A. Construction of an integration vector for use in the archaebacterium ***Methanococcus voltae*** and expression of a eubacterial resistance gene. Mol Gen Genet. 1990;221:273–279.

62. Gieg LM, Duncan KE, Suflita JM. Bioenergy production via microbial conversion of residual oil to natural gas. Appl Environ Microbiol. 2008;74:3022–3029.

63. Gorby YA, Yanina S, McLean JS, Rosso KM, Moyles D, Dohnalkova A, Beveridge TJ, Chang IS, Kim BH, Kim KS, Culley DE, Reed SB, Romine MF, Saffarini DA, Hill EA, Shi L, Elias DA, Kennedy DW, Pinchuk G, Watanabe K, Ishii S, Logan B, Nealson KH, Fredrickson JK. Electrically conductive bacterial nanowires produced by ***Shewanella oneidensis*** strain MR-1 and other microorganisms. Proc Natl Acad Sci. 2006;103:11358–11363.

64. Gottschalk G, Thauer RK. The Na(+)-translocating methyltransferase complex from methanogenic archaea. Biochim Biophys Acta. 2001;1505:28–36.

65. Goyal N, Zhou Z, Karimi IA. Metabolic processes of ***Methanococcus maripaludis*** and potential applications. Microb Cell Fact. 2016;15:107.

66. Grüber G, Manimekalai MSS, Mayer F, Müller V. ATP synthases from archaea: the beauty of a molecular motor. Biochim Biophys Acta Bioenerg. 2014;1837:940–952.

67. Guo X, Liu J, Xiao B. Bioelectrochemical enhancement of hydrogen and methane production from the anaerobic digestion of sewage sludge in single-chamber membrane-free microbial electrolysis cells. Int J Hydrogen Energy. 2013;38:1342–1347.

68. Guo Z, Thangavel S, Wang L, He Z, Cai W, Wang A, Liu W. Efficient methane production from beer wastewater in a membraneless microbial

electrolysis cell with a stacked cathode: the effect of the cathode/anode ratio on bioenergy recovery. Energy Fuels. 2017;31:615–620.

69. Guss AM, Rother M, Zhang JK, Kulkarni G, Metcalf WW. New methods for tightly regulated gene expression and highly efficient chromosomal integration of cloned genes for ***Methanosarcina***species. Archaea. 2008;2:193–203.

70. Hara M, Onaka Y, Kobayashi H, Fu Q, Kawaguchi H, Vilcaez J, Sato K. Mechanism of electromethanogenic reduction of CO_2 by a thermophilic methanogen. Energy Procedia. 2013;37:7021–7028.

71. Harris JE, Pinn PA. Bacitracin-resistant mutants of a mesophilic ***Methanobacterium*** species. Arch Microbiol. 1985;143:151–153.

72. Hohn MJ, Palioura S, Su D, Yuan J, Söll D. Genetic analysis of selenocysteine biosynthesis in the archaeon ***Methanococcus maripaludis***. Mol Microbiol. 2011;81:249–258.

73. Holm-Nielsen JB, Al Seadi T, Oleskowicz-Popiel P. The future of anaerobic digestion and biogas utilization. Bioresour Technol. 2009;100:5478–5484.

74. Holtmann D, Hannappel A, Schrader J. Encyclopedia of applied electrochemistry. New York: Springer; 2014. Microbial electrosynthesis; pp. 1268–1275.

75. Hook SE, Wright A-DG, McBride BW. Methanogens: methane producers of the rumen and mitigation strategies. Archaea. 2010;2010:1–11.

76. Hummel H, Böck A. Mutations in ***Methanobacterium formicicum*** conferring resistance to anti-80S ribosome-targeted antibiotics. MGG Mol Gen Genet. 1985;198:529–533.

77. Iino T, Tamaki H, Tamazawa S, Ueno Y, Ohkuma M, Suzuki K-I, Igarashi Y, Haruta S. Candidatus ***Methanogranum caenicola***: a novel methanogen from the anaerobic digested sludge, and proposal of ***Methanomassiliicoccaceae*** fam. nov. and ***Methanomassiliicoccales*** ord. nov., for a methanogenic lineage of the class ***Thermoplasmata***. Microbes Environ. 2013;28:244–250.

78. Jiang Y, Su M, Li D. Removal of sulfide and production of methane from carbon dioxide in microbial fuel cells–microbial electrolysis cell (MFCs–MEC) coupled system. Appl Biochem Biotechnol. 2014;172:2720–2731.

79. Jones WJ, Leigh JA, Mayer F, Woese CR, Wolfe RS. ***Methanococcus***

jannaschii sp. nov., an extremely thermophilic methanogen from a submarine hydrothermal vent. Arch Microbiol. 1983;136:254–261.

80. Kadier A, Simayi Y, Abdeshahian P, Azman NF, Chandrasekhar K, Kalil MS. A comprehensive review of microbial electrolysis cells (MEC) reactor designs and configurations for sustainable hydrogen gas production. Alexandria Eng J. 2016;55:427–443.
81. Kamagata Y, Kawasaki H, Oyaizu H, Nakamura K, Mikami E, Endo G, Koga Y, Yamasato K. Characterization of three thermophilic strains of ***Methanothrix*** ("Methanosaeta") ***thermophila*** sp. nov. and rejection of ***Methanothrix*** ("Methanosaeta") ***thermoacetophila***. Int J Syst Bacteriol. 1992;42:463–468.
82. Kaparaju P, Ellegaard L, Angelidaki I. Optimisation of biogas production from manure through serial digestion: lab-scale and pilot-scale studies. Bioresour Technol. 2009;100:701–709.
83. Karakashev D, Batstone DJ, Angelidaki I. Influence of environmental conditions on methanogenic compositions in anaerobic biogas reactors. Appl Environ Microbiol. 2005;71:331–338.
84. Kern T, Linge M, Rother M. ***Methanobacterium aggregans*** sp. nov., a hydrogenotrophic methanogenic archaeon isolated from an anaerobic digester. Int J Syst Evol Microbiol. 2015;65:1975–1980.
85. Kern T, Fischer Martin A, Deppenmeier U, Schmitz RA, Rother M. ***Methanosarcina flavescens*** sp. nov., a methanogenic archaeon isolated from a full-scale anaerobic digester. Int J Syst Evol Microbiol. 2016;66:1533–1538.
86. Kern T, Theiss J, Röske K, Rother M. Assessment of hydrogen metabolism in commercial anaerobic digesters. Appl Microbiol Biotechnol. 2016;100:4699–4710.
87. Kobayashi H, Saito N, Fu Q, Kawaguchi H, Vilcaez J, Wakayama T, Maeda H, Sato K. Bio-electrochemical property and phylogenetic diversity of microbial communities associated with bioelectrodes of an electromethanogenic reactor. J Biosci Bioeng. 2013;116:114–117.
88. Koch C, Harnisch F. What is the essence of microbial electroactivity? Front Microbiol. 2016;7:1890.
89. Koch C, Kuchenbuch A, Kretzschmar J, Wedwitschka H, Liebetrau J, Müller S, Harnisch F. Coupling electric energy and biogas production in anaerobic digesters - impacts on the microbiome. RSC Adv. 2015;5:31329–31340.

90. Koo T, Shin SG, Lee J, Han G, Kim W, Cho K, Hwang S. Identifying methanogen community structures and their correlations with performance parameters in four full-scale anaerobic sludge digesters. Bioresour Technol. 2017;228:368–373.

91. Krieg T, Sydow A, Schröder U, Schrader J, Holtmann D. Reactor concepts for bioelectrochemical syntheses and energy conversion. Trends Biotechnol. 2014;32:645–655.

92. Kurr M, Huber R, König H, Jannasch HW, Fricke H, Trincone A, Kristjansson JK, Stetter KO. ***Methanopyrus kandleri***, gen. and sp. nov. represents a novel group of hyperthermophilic methanogens, growing at 110 °C. Arch Microbiol. 1991;156:239–247.

93. Lay J-J, Li Y-Y, Noike T. Influences of pH and moisture content on the methane production in high-solids sludge digestion. Water Res. 1997;31:1518–1524.

94. Leadbetter JR, Breznak JA. Physiological ecology of ***Methanobrevibacter cuticularis*** sp. nov. and ***Methanobrevibacter curvatus*** sp. nov., isolated from the hindgut of the termite ***Reticulitermes flavipes***. Appl Environ Microbiol. 1996;62:3620–3631.

95. Lebuhn M, Liu F, Heuwinkel H, Gronauer A. Biogas production from mono-digestion of maize silage-long-term process stability and requirements. Water Sci Technol. 2008;58:1645–1651.

96. Lessner DJ, Lhu L, Wahal CS, Ferry JG (2010) An engineered methanogenic pathway derived from the domains ***Bacteria*** and ***Archaea***. mBio 1(5):e00243–10.

97. Li Y, Park SY, Zhu J. Solid-state anaerobic digestion for methane production from organic waste. Renew Sustain Energy Rev. 2011;15:821–826.

98. Lie TJ, Leigh JA. A novel repressor of nif and glnA expression in the methanogenic archaeon ***Methanococcus maripaludis***. Mol Microbiol. 2003;47:235–246.

99. Liu Y, Whitman WB. Metabolic, phylogenetic, and ecological diversity of the methanogenic archaea. Ann NY Acad Sci. 2008;1125:171–189.

100. Liu G, Zhang R, El-Mashad HM, Dong R. Effect of feed to inoculum ratios on biogas yields of food and green wastes. Bioresour Technol. 2009;100:5103–5108.

101. Liu D, Zhang L, Chen S, Buisman C, ter Heijne A. Bioelectrochemical enhancement of methane production in low temperature anaerobic

digestion at 10 °C. Water Res. 2016;99:281–287.

102. Lohner ST, Deutzmann JS, Logan BE, Leigh J, Spormann AM. Hydrogenase-independent uptake and metabolism of electrons by the archaeon ***Methanococcus maripaludis***. ISME J. 2014;8:1673–1681.

103. Lovley DR. Electromicrobiology. Annu Rev Microbiol. 2012;66:391–409.

104. Lucas R, Kuchenbuch A, Fetzer I, Harms H, Kleinsteuber S. Long-term monitoring reveals stable and remarkably similar microbial communities in parallel full-scale biogas reactors digesting energy crops. FEMS Microbiol Ecol. 2015;91:3.

105. Luo G, Angelidaki I. Integrated biogas upgrading and hydrogen utilization in an anaerobic reactor containing enriched hydrogenotrophic methanogenic culture. Biotechnol Bioeng. 2012;109:2729–2736.

106. Luo G, Johansson S, Boe K, Xie L, Zhou Q, Angelidaki I. Simultaneous hydrogen utilization and in situ biogas upgrading in an anaerobic reactor. Biotechnol Bioeng. 2012;109:1088–1094.

107. Lupa B, Hendrickson EL, Leigh JA, Whitman WB. Formate-dependent H_2 production by the mesophilic methanogen ***Methanococcus maripaludis***. Appl Environ Microbiol. 2008;74:6584–6590.

108. Lyu Z, Jain R, Smith P, Fetchko T, Yan Y, Whitman WB. Engineering the autotroph ***Methanococcus maripaludis*** for geraniol production. ACS Synth Biol. 2016;5:577–581.

109. Mah RA. Isolation and characterization of ***Methanococcus mazei***. Curr Microbiol. 1980;3:321–325.

110. Martin I, Pidou M, Soares A, Judd S, Jefferson B. Modelling the energy demands of aerobic and anaerobic membrane bioreactors for wastewater treatment. Environ Technol. 2011;32:921–932.

111. Mathrani IM, Boone DR, Mah RA, Fox GE, Lau PP. ***Methanohalophilus zhilinae*** sp. nov., an alkaliphilic, halophilic, methylotrophic methanogen. Int J Syst Bacteriol. 1988;38:139–142.

112. Mayer F, Müller V. Adaptations of anaerobic archaea to life under extreme energy limitation. FEMS Microbiol Rev. 2014;38:449–472.

113. Mayumi D, Mochimaru H, Tamaki H, Yamamoto K, Yoshioka H, Suzuki Y, Kamagata Y, Sakata S. Methane production from coal by a single methanogen. Science. 2016;354:222–225.

114. McAnulty MJ, Poosarla VG, Li J, Soo VWC, Zhu F, Wood TK. Metabolic engineering of ***Methanosarcina acetivorans*** for lactate

production from methane. Biotechnol Bioeng. 2017;114:852–861.

115. McInerney MJ, Struchtemeyer CG, Sieber J, Mouttaki H, Stams AJ, Schink B, Rohlin L, Gunsalus RP. Physiology, ecology, phylogeny, and genomics of microorganisms capable of syntrophic metabolism. Ann NY Acad Sci. 2008;1125:58–72.

116. Metcalf WW, Zhang JK, Shi X, Wolfe RS. Molecular, genetic, and biochemical characterization of the ***serC*** gene of ***Methanosarcina barkeri*** Fusaro. J Bacteriol. 1996;178:5797–5802.

117. Metcalf WW, Zhang JK, Apolinario E, Sowers KR, Wolfe RS. A genetic system for Archaea of the genus ***Methanosarcina***: liposome-mediated transformation and construction of shuttle vectors. Proc Natl Acad Sci USA. 1997;94:2626–2631.

118. Miller TL, Lin C. Description of ***Methanobrevibacter gottschalkii*** sp. nov., ***Methanobrevibacter thaueri*** sp. nov., ***Methanobrevibacter woesei*** sp. nov. and ***Methanobrevibacter wolinii*** sp. nov. Int J Syst Evol Microbiol. 2002;52:819–822.

119. Miller TL, Wolin MJ, Conway de Macario E, Macario AJ. Isolation of ***Methanobrevibacter smithii***from human feces. Appl Environ Microbiol. 1982;43:227–232.

120. Mochimaru H, Tamaki H, Hanada S, Imachi H, Nakamura K, Sakata S, Kamagata Y. ***Methanolobus profundi*** sp. nov., a methylotrophic methanogen isolated from deep subsurface sediments in a natural gas field. Int J Syst Evol Microbiol. 2009;59:714–718.

121. Mondorf S, Deppenmeier U, Welte C. A novel inducible protein production system and neomycin resistance as selection marker for ***Methanosarcina mazei***. Archaea. 2012;2012:1–8.

122. Moore BC, Leigh JA. Markerless mutagenesis in ***Methanococcus maripaludis*** demonstrates roles for alanine dehydrogenase, alanine racemase, and alanine permease. J Bacteriol. 2005;187:972–979.

123. Mūller B., Allmansberger R., Klein A. Termination of a transcription unit comprising highly expressed genes in the archaebacteriumMethanococcus voltae. Nucleic Acids Research. 1985;13(18):6439–6445.

124. Mussgnug JH, Klassen V, Schlüter A, Kruse O. Microalgae as substrates for fermentative biogas production in a combined biorefinery concept. J Biotechnol. 2010;150:51–56.

125. Nayak DD, Metcalf WW. Cas9-mediated genome editing in the

methanogenic archaeon ***Methanosarcina acetivorans***. Proc Natl Acad Sci. 2017;114:2976–2981.

126. Neumann P, Pesante S, Venegas M, Vidal G. Developments in pre-treatment methods to improve anaerobic digestion of sewage sludge. Rev Environ Sci Biotechnol. 2016;15:173–211.
127. Niu Q, Kobayashi T, Takemura Y, Kubota K, Li Y-Y. Evaluation of functional microbial community's difference in full-scale and lab-scale anaerobic digesters feeding with different organic solid waste: effects of substrate and operation factors. Bioresour Technol. 2015;193:110–118.
128. Nizami A-S, Murphy JD. What type of digester configurations should be employed to produce biomethane from grass silage? Renew Sustain Energy Rev. 2010;14:1558–1568.
129. O'Brien JM, Wolkin RH, Moench TT, Morgan JB, Zeikus JG. Association of hydrogen metabolism with unitrophic or mixotrophic growth of ***Methanosarcina barkeri*** on carbon monoxide. J Bacteriol. 1984;158:373–375.
130. Oelgeschläger E, Rother M. In vivo role of three fused corrinoid/methyl transfer proteins in ***Methanosarcina acetivorans***. Mol Microbiol. 2009;72:1260–1272.
131. Ollivier B. ***Methanoplanus petrolearius*** sp. nov., a novel methanogenic bacterium from an oil-producing well. FEMS Microbiol Lett. 1997;147:51–56.
132. Oren A. Microbial life at high salt concentrations: phylogenetic and metabolic diversity. Saline Syst. 2008;4:2.
133. Park SY, Liang Y. Biogenic methane production from coal: a review on recent research and development on microbially enhanced coalbed methane (MECBM) Fuel. 2016;166:258–267.
134. Park DH, Laivenieks M, Guettler MV, Jain MK, Zeikus JG. Microbial utilization of electrically reduced neutral red as the sole electron donor for growth and metabolite production. Appl Environ Microbiol. 1999;65:2912–2917.
135. Patil SA, Hägerhäll C, Gorton L. Electron transfer mechanisms between microorganisms and electrodes in bioelectrochemical systems. Bioanal Rev. 2012;4:159–192.
136. Pflüger K, Baumann S, Gottschalk G, Lin W, Santos H, Müller V. Lysine-2,3-aminomutase and beta-lysine acetyltransferase genes

of methanogenic archaea are salt induced and are essential for the biosynthesis of N(epsilon)-acetyl-beta-lysine and growth at high salinity. Appl Environ Microbiol. 2003;69:6047–6055.

137. Pflüger K, Wieland H, Müller V. Osmoadaptation in methanogenic archaea: recent insights from a genomic perspective. In: Gunde-Cimerman N, Oren A, Plemenitas A, editors. Adaptation of life at high salt concentrations in ***Archaea***, ***Bacteria***, and ***Eukarya***. Dordrecht: Springer; 2005. pp. 241–251.

138. Plöchl M, Heiermann M (2006) Biogas farming in central and northern Europe: a strategy for developing countries? Agric Eng Int CIGR Ejournal VIII:1–15

139. Porat I, Whitman WB. Tryptophan auxotrophs were obtained by random transposon insertions in the ***Methanococcus maripaludis*** tryptophan operon. FEMS Microbiol Lett. 2009;297:250–254.

140. Pritchett MA, Zhang JK, Metcalf WW. Development of a markerless genetic exchange method for ***Methanosarcina acetivorans*** C2A and its use in construction of new genetic tools for methanogenic archaea. Appl Environ Microbiol. 2004;70:1425–1433.

141. Rajeshwari K, Balakrishnan M, Kansal A, Lata K, Kishore VV. State-of-the-art of anaerobic digestion technology for industrial wastewater treatment. Renew Sustain Energy Rev. 2000;4:135–156.

142. Ravichandran M, Munisamy P, Varadharaju C, Natarajan S. Methanogenic archaea: a multipotent biological candidate focusing toward realizing future global energy. Int J Curr Microbiol Appl Sci. 2015;4:785–793.

143. Ren T, Patel M, Blok K. Steam cracking and methane to olefins: energy use, CO_2 emissions and production costs. Energy. 2008;33:817–833.

144. Ribot-Llobet E, Nam J-Y, Tokash JC, Guisasola A, Logan BE. Assessment of four different cathode materials at different initial pHs using unbuffered catholytes in microbial electrolysis cells. Int J Hydrog Energy. 2013;38:2951–2956.

145. Rincón B, Borja R, González JM, Portillo MC, Sáiz-Jiménez C. Influence of organic loading rate and hydraulic retention time on the performance, stability and microbial communities of one-stage anaerobic digestion of two-phase olive mill solid residue. Biochem Eng J. 2008;40:253–261.

146. Rittmann BE, Lee H, Zhang H, Alder J, Banaszak JE, Lopez R. Full-

scale application of focused-pulsed pre-treatment for improving biosolids digestion and conversion to methane. Water Sci Technol. 2008;58:1895.

147. Robichaux M, Howell M, Boopathy R. Methanogenic activity in human periodontal pocket. Curr Microbiol. 2003;46:53–58.

148. Roeßler M, Pflüger K, Flach H, Lienard T, Gottschalk G, Müller V. Identification of a salt-induced primary transporter for glycine betaine in the methanogen ***Methanosarcina mazei*** Gö1. Appl Environ Microbiol. 2002;68:2133–2139.

149. Romero-Güiza MS, Vila J, Mata-Alvarez J, Chimenos JM, Astals S. The role of additives on anaerobic digestion: a review. Renew Sustain Energy Rev. 2016;58:1486–1499.

150. Rotaru A-E, Shrestha PM, Liu F, Markovaite B, Chen S, Nevin KP, Lovley DR. Direct interspecies electron transfer between ***Geobacter metallireducens*** and ***Methanosarcina barkeri***. Appl Environ Microbiol. 2014;80:4599–4605.

151. Rother M, Metcalf WW. Anaerobic growth of ***Methanosarcina acetivorans*** C2A on carbon monoxide: an unusual way of life for a methanogenic archaeon. Proc Natl Acad Sci. 2004;101:16929–16934.

152. Rother M, Boccazzi P, Bose A, Pritchett MA, Metcalf WW. Methanol-dependent gene expression demonstrates that methyl-coenzyme M reductase is essential in ***Methanosarcina acetivorans*** C2A and allows isolation of mutants with defects in regulation of the methanol utilization pathway. J Bacteriol. 2005;187:5552–5559.

153. Ryckebosch E, Drouillon M, Vervaeren H. Techniques for transformation of biogas to biomethane. Biomass Bioenergy. 2011;35:1633–1645.

154. Sahlström L. A review of survival of pathogenic bacteria in organic waste used in biogas plants. Bioresour Technol. 2003;87:161–166.

155. Sakai S, Imachi H, Hanada S, Ohashi A, Harada H, Kamagata Y. ***Methanocella paludicola*** gen. nov., sp. nov., a methane-producing archaeon, the first isolate of the lineage "Rice Cluster I", and proposal of the new archaeal order ***Methanocellales*** ord. nov. Int J Syst Evol Microbiol. 2008;58:929–936.

156. Sander R. Compilation of Henry's law constants (version 4.0) for water as solvent. Atmos Chem Phys. 2015;15:4399–4981.

157. Sarmiento Felipe B., Leigh John A., Whitman William B. Methods in Methane Metabolism, Part A.2011. Genetic Systems for

Hydrogenotrophic Methanogens; pp. 43–73.

158. Sasaki D, Sasaki K, Watanabe A, Morita M, Igarashi Y, Ohmura N. Efficient production of methane from artificial garbage waste by a cylindrical bioelectrochemical reactor containing carbon fiber textiles. AMB Express. 2013;3:17.

159. Sattler C, Wolf S, Fersch J, Goetz S, Rother M. Random mutagenesis identifies factors involved in formate-dependent growth of the methanogenic archaeon ***Methanococcus maripaludis***. Mol Genet Genomics. 2013;288:413–424.

160. Saum R, Mingote A, Santos H, Müller V. A novel limb in the osmoregulatory network of ***Methanosarcina mazei*** Gö1: N(epsilon)-acetyl-beta-lysine can be substituted by glutamate and alanine. Environ Microbiol. 2009;11:1056–1065.

161. Schink B, Ward JC, Zeikus JG. Microbiology of wetwood: importance of pectin degradation and clostridium species in living trees. Appl Environ Microbiol. 1981;42:526–532.

162. Schiraldi C, Giuliano M, De Rosa M. Perspectives on biotechnological applications of archaea. Archaea. 2002;1:75–86.

163. Schlegel K, Müller V. Evolution of Na^+ and H^+ bioenergetics in methanogenic archaea. Biochem Soc Trans. 2013;41:421–426.

164. Schlegel K, Leone V, Faraldo-Gomez JD, Müller V. Promiscuous archaeal ATP synthase concurrently coupled to Na^+ and H^+ translocation. Proc Natl Acad Sci. 2012;109:947–952.

165. Schlegel K, Welte C, Deppenmeier U, Müller V. Electron transport during aceticlastic methanogenesis by ***Methanosarcina acetivorans*** involves a sodium-translocating Rnf complex. FEBS J. 2012;279:4444–4452.

166. Schnürer A, Nordberg A. Ammonia, a selective agent for methane production by syntrophic acetate oxidation at mesophilic temperature. Water Sci Technol. 2008;57:735–740.

167. Schröder U, Harnisch F, Angenent LT. Microbial electrochemistry and technology: terminology and classification. Energy Environ Sci. 2015;8:513–519.

168. Siegert M, Yates MD, Call DF, Zhu X, Spormann A, Logan BE. Comparison of nonprecious metal cathode materials for methane production by electromethanogenesis. ACS Sustain Chem Eng. 2014;2:910–917.

169. Solli L, Håvelsrud OE, Horn SJ, Rike AG. A metagenomic study of the microbial communities in four parallel biogas reactors. Biotechnol Biofuels. 2014;7:146.
170. Soo VWC, McAnulty MJ, Tripathi A, Zhu F, Zhang L, Hatzakis E, Smith PB, Agrawal S, Nazem-Bokaee H, Gopalakrishnan S, Salis HM, Ferry JG, Maranas CD, Patterson AD, Wood TK. Reversing methanogenesis to capture methane for liquid biofuel precursors. Microb Cell Fact. 2016;15:11.
171. Sorokin DY, Abbas B, Merkel AY, Rijpstra WIC, Damste JSS, Sukhacheva MV, van Loosdrecht MCM. ***Methanosalsum natronophilum*** sp. nov., and ***Methanocalculus alkaliphilus*** sp. nov., haloalkaliphilic methanogens from hypersaline soda lakes. Int J Syst Evol Microbiol. 2015;65:3739–3745.
172. Sorokin DY, Makarova KS, Abbas B, Ferrer M, Golyshin PN, Galinski EA, Ciordia S, Mena MC, Merkel AY, Wolf YI, van Loosdrecht MCM, Koonin EV. Discovery of extremely halophilic, methyl-reducing euryarchaea provides insights into the evolutionary origin of methanogenesis. Nat Microbiol. 2017;2:17081.
173. Sosnowski P, Wieczorek A, Ledakowicz S. Anaerobic co-digestion of sewage sludge and organic fraction of municipal solid wastes. Adv Environ Res. 2003;7:609–616.
174. Stadtman TC, Barker HA. Studies on the methane fermentation. X. A new formate-decomposing bacterium, ***Methanococcus vannielii***. J Bacteriol. 1951;62:269–280.
175. Stolze Y, Zakrzewski M, Maus I, Eikmeyer F, Jaenicke S, Rottmann N, Siebner C, Pühler A, Schlüter A. Comparative metagenomics of biogas-producing microbial communities from production-scale biogas plants operating under wet or dry fermentation conditions. Biotechnol Biofuels. 2015;8:14.
176. Strong PJ, Xie S, Clarke WP. Methane as a resource: can the methanotrophs add value? Environ Sci Technol. 2015;49:4001–4018.
177. Sun J, Klein A. A lysR-type regulator is involved in the negative regulation of genes encoding selenium-free hydrogenases in the archaeon ***Methanococcus voltae***. Mol Microbiol. 2004;52:563–571.
178. Sundberg C, Al-Soud WA, Larsson M, Alm E, Yekta SS, Svensson BH, Sorensen SJ, Karlsson A. 454 pyrosequencing analyses of bacterial and archaeal richness in 21 full-scale biogas digesters. FEMS Microbiol

Ecol. 2013;85:612–626.

179. Sydow A, Krieg T, Mayer F, Schrader J, Holtmann D. Electroactive bacteria - molecular mechanisms and genetic tools. Appl Microbiol Biotechnol. 2014;98:8481–8495.

180. Tabatabaei M, Rahim RA, Abdullah N, Wright AG, Shirai Y, Sakai K, Sulaiman A, Hassan MA. Importance of the methanogenic archaea populations in anaerobic wastewater treatments. Process Biochem. 2010;45:1214–1225.

181. Thauer RK, Kaster A-K, Seedorf H, Buckel W, Hedderich R. Methanogenic archaea: ecologically relevant differences in energy conservation. Nat Rev Microbiol. 2008;6:579–591.

182. Timmers PHA, Welte CU, Koehorst JJ, Plugge CM, Jetten MSM, Stams AJM. Reverse methanogenesis and respiration in methanotrophic archaea. Archaea. 2017;2017:1–22.

183. Tumbula DL, Bowen TL, Whitman WB. Characterization of pURB500 from the archaeon ***Methanococcus maripaludis*** and construction of a shuttle vector. J Bacteriol. 1997;179:2976–2986.

184. Valentine DL, Blanton DC, Reeburgh WS. Hydrogen production by methanogens under low-hydrogen conditions. Arch Microbiol. 2000;174:415–421.

185. Van Lier JB, Mahmoud N, Zeeman G. Anaerobic wastewater treatment. In: Henze M, van Loosdrecht MCM, Ekama GA, Brdjanovic D, editors. Biological wastewater treatment: principles modelling and design. London: IWA Publishing; 2008. pp. 401–442.

186. Vanegas C, Bartlett J. Anaerobic digestion of laminaria digitata: the effect of temperature on biogas production and composition. Waste Biomass Valorization. 2013;4:509–515.

187. Vanwonterghem I, Evans PN, Parks DH, Jensen PD, Woodcroft BJ, Hugenholtz P, Tyson GW. Methylotrophic methanogenesis discovered in the archaeal phylum Verstraetearchaeota. Nat Microbiol. 2016;1:16170.

188. Vavilin VA, Fernandez B, Palatsi J, Flotats X. Hydrolysis kinetics in anaerobic degradation of particulate organic material: an overview. Waste Manag. 2008;28:939–951.

189. Villano M, Aulenta F, Ciucci C, Ferri T, Giuliano A, Majone M. Bioelectrochemical reduction of CO_2 to CH_4 via direct and indirect extracellular electron transfer by a hydrogenophilic methanogenic

culture. Bioresour Technol. 2010;101:3085–3090.

190. Voegeli Y, Lohri C, Kassenga G, Baier U, Zurbrügg C (2009) Technical and biological performance of the Arti compact biogas plant for kitchen waste - case study from Tanzania. In: Proceedings Sardinia, 2009, twelfth international waste management and landfill symposium. CISA Publisher, Italy, S. Margherita die Pula, Cagliari

191. Voordouw G. Production-related petroleum microbiology: progress and prospects. Curr Opin Biotechnol. 2011;22:401–405.

192. Wagner T, Ermler U, Shima S. The methanogenic CO_2 reducing-and-fixing enzyme is bifunctional and contains 46 [4Fe–4S] clusters. Science. 2016;354:114–117.

193. Wang P, Wang H, Qiu Y, Ren L, Jiang B. Microbial characteristics in anaerobic digestion process of food waste for methane production - a review. Bioresour Technol. 2017;17:31056–31058.

194. Ward AJ, Hobbs PJ, Holliman PJ, Jones DL. Optimisation of the anaerobic digestion of agricultural resources. Bioresour Technol. 2008;99:7928–7940.

195. Weiland P. Biogas production: current state and perspectives. Appl Microbiol Biotechnol. 2010;85:849–860.

196. Welander PV, Metcalf WW. Loss of the ***mtr*** operon in ***Methanosarcina*** blocks growth on methanol, but not methanogenesis, and reveals an unknown methanogenic pathway. PNAS. 2005;102:10664–10669.

197. Welander PV, Metcalf WW. Mutagenesis of the C1 oxidation pathway in ***Methanosarcina barkeri***: new insights into the Mtr/Mer bypass pathway. J Bacteriol. 2008;190:1928–1936.

198. Welte C, Deppenmeier U. Bioenergetics and anaerobic respiratory chains of aceticlastic methanogens. Biochim Biophys Acta Bioenerg. 2014;1837:1130–1147.

199. Westerholm M, Moestedt J, Schnürer A. Biogas production through syntrophic acetate oxidation and deliberate operating strategies for improved digester performance. Appl Energ. 2016;179:124–135.

200. Widdel F. Growth of methanogenic bacteria in pure culture with 2-propanol and other alcohols as hydrogen donors. Appl Environ Microbiol. 1986;51:1056–1062.

201. Wonglertarak W, Wichitsathian B. Alkaline pretreatment of waste activated sludge in anaerobic digestion. J Clean Energy Technol. 2014;2:118–121.

202. Worrell VE, Nagle DP, McCarthy D, Eisenbraun A. Genetic transformation system in the archaebacterium ***Methanobacterium thermoautotrophicum*** Marburg. J Bacteriol. 1988;170:653–656.

203. Yadvika Santosh, Sreekrishnan TR, Kohli S, Rana V. Enhancement of biogas production from solid substrates using different techniques - a review. Bioresour Technol. 2004;95:1–10.

204. Yu HW, Samani Z, Hanson A, Smith G. Energy recovery from grass using two-phase anaerobic digestion. Waste Manag. 2002;22:1–5.

205. Zeikus JG, Bowen VG. Fine structure of ***Methanospirillum hungatii***. J Bacteriol. 1975;121:373–380.

206. Zeikus JG, Henning DL. ***Methanobacterium arbophilicum*** sp. nov. An obligate anaerobe isolated from wetwood of living trees. Antonie Van Leeuwenhoek. 1975;41:543–552.

207. Zhang JK, Pritchett MA, Lampe DJ, Robertson HM, Metcalf WW. In vivo transposon mutagenesis of the methanogenic archaeon ***Methanosarcina acetivorans*** C2A using a modified version of the insect mariner-family transposable element ***Himar1***. Proc Natl Acad Sci USA. 2000;97:9665–9670.

208. Zhang G, Jiang N, Liu X, Dong X. Methanogenesis from methanol at low temperatures by a novel psychrophilic Methanogen, "***Methanolobus psychrophilus***" sp. nov., prevalent in Zoige Wetland of the Tibetan plateau. Appl Environ Microbiol. 2008;74:6114–6120.

209. Zhen G, Kobayashi T, Lu X, Xu K. Understanding methane bioelectrosynthesis from carbon dioxide in a two-chamber microbial electrolysis cells (MECs) containing a carbon biocathode. Bioresour Technol. 2015;186:141–148.

210. Zhen G, Lu X, Kobayashi T, Kumar G, Xu K. Promoted electromethanosynthesis in a two-chamber microbial electrolysis cells (MECs) containing a hybrid biocathode covered with graphite felt (GF) Chem Eng J. 2016;284:1146–1155.

211. Zheng Y, Zhao J, Xu F, Li Y. Pretreatment of lignocellulosic biomass for enhanced biogas production. PECS. 2014;42:35–53.

212. Zhong W, Zhang Z, Luo Y, Sun S, Qiao W, Xiao M. Effect of biological pretreatments in enhancing corn straw biogas production. Bioresour Technol. 2011;102:11177–11182

9

Anaerobic Digestion without Biogas?

Robbert Kleerebezem[1], Bart Joosse[1], Rene Rozendal[2], Mark C. M. Van Loosdrecht[1]

[1]Department of Biotechnology, Delft University of Technology, Julianalaan 67, 2628BC Delft, The Netherlands
[2]Paques BV, T. de Boerstraat 24, 8561EL Balk, The Netherlands

ABSTRACT

Anaerobic digestion for the production of methane containing biogas is the classic example of a resource recovery process that combines stabilization of particulate organic matter or wastewater treatment with the production of a valuable end-product. Attractive features of the process include the production of a single end-product from a heterogeneous feedstock, and in-situ product separation of the gaseous end-product. Despite these intrinsic attractive properties of the process, the economic added value of the biogas produced is limited, enabling the development of alternative processes that yield higher-value end-products. Typically the production of higher value end-products from low value feedstock and industrial wastewater proceeds via intermediate production of organic acids (and carbon dioxide and molecular hydrogen). Optimization of organic acid production from particulate feedstocks and wastewater for development of the organic acid

Citation: Kleerebezem, R., Joosse, B., Rozendal, R. et al. Anaerobic Digestion without Biogas? , Rev Environ Sci Biotechnol (2015) 14: 787. https://doi.org/10.1007/s11157-015-9374-6

based resource recovery route receives significant research attention. The organic acid stream generated as such, has no economic value, but if organic acids can either be concentrated via membrane separation or (bio)converted to an end-product that can easily be separated from the liquid, an attractive biomass processing scheme can be developed. Attractive end-products of organic acid processing include polyhydroxyalkanoates, medium chain length fatty acids, or other organic molecules using bio-electrochemical systems. Overall we suggest that these novel bioprocessing routes for conversion of low value feedstock to higher added value products will contribute to a sustainable future and will change the economic status of organic waste.

Keywords: Biogas, Polyhydroxyalkanoates, PHA, Medium chainlength fatty acids, MCFA, VFA-platform, Anaerobic digestion

INTRODUCTION

Environmental engineering processes traditionally aim for removal of polluting compounds from water, soil, or gas. Herewith the main product of this kind of processes are the production of water, soil, or gas that can be returned to the environment without negative health implications (hygienization) or a negative effect on the natural environment (environmental protection). In recent years the recovery of the polluting compounds has been added as a secondary treatment objective to environmental engineering processes. Resource recovery from waste is widely accepted as a more and more important research theme, stimulated by the increasing awareness of the exhaustion of non-renewable natural resources (Agler et al. 2011; Kleerebezem and van Loosdrecht 2007). The attractive concept of combing waste(water) treatment and the production of valuable compounds from a low a value heterogeneous feedstock does not need to be clarified to those working on the anaerobic digestion process. Anaerobic digestion for the production of methane containing biogas can be regarded as ***the*** classic example of a resource recovery process that combines wastewater treatment or solids stabilization with effective conversion of biodegradable organic carbon to a valuable product: methane containing biogas. Methane containing biogas can directly be used for electricity and/or heat production, or upgraded to natural gas quality. Other resources that can be recovered from waste(water) are nutrients like nitrogen and phosphorus, and specific trace metals, but in this paper we will focus on organic carbon recovery.

Anaerobic digestion for the production of methane containing biogas is a worldwide accepted technology for treatment of numerous streams rich in organic carbon. Feedstocks that are currently being treated using the anaerobic digestion process include streams with a high solid content [manure, sewage sludge, organic fraction of municipal solid waste (OFMSW), energy crops, etc], as well as wastewaters with primarily water soluble organic carbon (agroindustrial wastewater, sewage, chemical industry wastewater, etc). A novel feedstock that receives more and more research attention is algae biomass obtained from phototrophic nutrient recovery systems integrated in sewage treatment (Dominguez Cabanelas et al. 2013; Montingelli et al. 2015; Ward et al. 2014). In the past decades bioreactor concepts have been developed for both high and low solids type of feedstock: High-solid bioreactors range from slurry reactors (<5 % solids) to true solid reactors (<40 % solids) with a variable extent of plug flow in the system (Kleerebezem 2015). Anaerobic wastewater treatment has been revolutionized by the effective uncoupling of the solid and liquid retention time in Upflow Anaerobic Sludge Bed (UASB) reactors and derivatives thereof. In the past decades the application window of the process has been extended to a wide range of operational conditions: the process is implemented at psychrophilic, mesophilic, and thermophilic temperatures, and even extreme conditions like high salt concentrations can currently effectively be tolerated in anaerobic digesters provided that adequate operational measures are taken. Overall it is evident that anaerobic digestion can be considered a mature technology that is widely applied for organic waste valorisation.

The production of methane containing biogas makes the anaerobic digestion process a player in the renewable biomass to bioenergy field, competing with biomass based bioethanol and biodiesel production. Methane containing biogas can be applied directly for electricity and heat production in a combined heat and power (CHP) plant or upgraded to natural gas quality by removal of water vapour and carbon dioxide (and hydrogen sulphide) and introduced in the natural gas grid (Andriani et al. 2014; Niesner et al. 2013). Biogas can be applied on different scales with small scale domestic biogas application for cooking, or large scale industrial application with biogas use for electricity production. Upgraded biofuel can furthermore be used as transport fuel as widely implemented in Sweden (Borjesson and Ahlgren 2012; Tilche and Galatola 2008).

Even though it is evident that the production of methane containing biogas provides a clear added value of the anaerobic digestion process over other environmental technologies, it remains unclear to which extent biogas production is a main driver of the process compared to other arguments. In recent years, it has furthermore been recognized that an alternative approach to anaerobic processing of biomass is to aim for production of organic acids and/or alcohols instead of methane containing biogas. Direct recovery of these water soluble products of biomass fermentation or post-processing to obtain other molecules (e.g. polyhydroxyalkanoates, or medium chain length fatty acids) may result in the production of more valuable end-products of the resource recovery process. Alternatively, biomass processing can aim for other gaseous products like molecular hydrogen or direct generation of electricity in microbial fuel cells.

In this paper we will discuss first to which extent methane containing biogas production is an important driver for application of the anaerobic digestion process in relation to other arguments for application of anaerobic digestion instead of alternative technologies. Based on these considerations we will elaborate where the most prosperous niche remains for production of other compounds from waste.

METHANE CONTAINING BIOGAS AS PROCESS DRIVER

Anaerobic digestion can both be considered as a biomass based bioenergy production process or as an environmental engineering process that is capable of wastewater treatment or waste biomass stabilization, enabling agricultural reuse of organic waste after treatment. There are two main arguments that clarify why methane containing biogas as end product of the process makes the anaerobic digestion process a highly attractive bioenergy production process from a low value feedstock.

In Situ Production Separation

The fact that methane is a poorly water soluble compound directly contributes to the attractiveness of the process. No or limited downstream processing is required to enable the utilization of methane containing biogas for electricity and heat production in a combined heat and power plant. Technologies are furthermore available for upgrading of the biogas to natural gas quality for introduction in the natural gas grid. Even though liquid end-products offer intrinsic advantages as well (see below), they require energy intensive downstream processing for water removal through distillation in case of bioethanol production, or other biorefinery based product recovery technologies as required for example for lipid recovery from algae.

A second advantage related to the production of a gaseous end product is the lower energy requirements for bioreactor operation compared to active aeration in aerobic processing of waste. In anaerobic wastewater treatment reactors such as the UASB reactor or anaerobic slurry reactors, no mechanical mixing is required since adequate mixing is established through biogas production. Furthermore, oxygen supply through aeration adds significantly to the electricity consumption for aerobic treatment process: composting of solids or aerobic wastewater treatment.

Thermodynamic Driver of the Process

In a thermodynamically closed systems—in absence of an external electron acceptor or energy source e.g. light—microorganisms catalyse redox reactions towards an eventual state of thermodynamic equilibrium. For organic carbon conversions this implies that the state of thermodynamic equilibrium is achieved when the organic carbon compound with the lowest Gibbs energy change per electron is produced. The Gibbs energy change per electron for oxidation of any organic compound to carbon dioxide in standard conditions but corrected for a pH of 7 (ΔG01eΔGe01) can be calculated from tabulated Gibbs energy of formation values (Kleerebezem and Van Loosdrecht 2010). For methane this results in the following equation:

$$-1 \cdot CH_4(g) - 2 \cdot H_2O + 1 \cdot CO_2(g) + 8 \cdot H^{+1} + 8 \cdot e^{-1}, \quad \Delta G^{01} = -23 \frac{kJ}{emol}$$

From all organic compounds that participate in biological systems methane has the lowest ΔG01eΔGe01-value as can be seen from Fig. 1 that compares the ΔG01eΔGe01-values for a wide range of organic compounds. These data suggest that in all microbial fermentations—where organic carbon is both electron donor and acceptor of the redox reaction—Gibbs energy can be harvested if methane and carbon dioxide are the end-products of the conversion. This highly specific property provides the anaerobic digestion process a principal advantage over other bioenergy producing processes because irrespective of the origin and heterogeneity of the substrate a homogeneous end-product can be generated. Other fermentation end-products can only dominate the process if the final steps of the anaerobic digestion process can be effectively inhibited. This is for example the case in corn or sugar cane based bioethanol production where the production of methane containing biogas is prevented by very high ethanol concentrations that effectively inhibit methanogenic archaea.

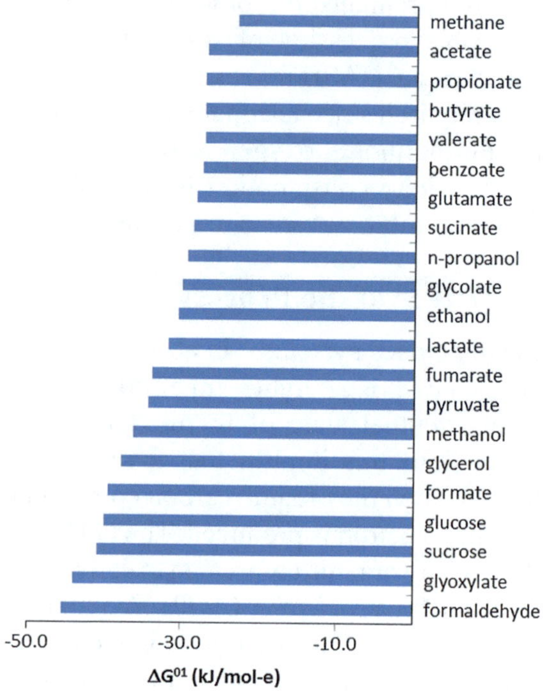

Figure 1: Gibbs energy change per electron (kJ/mol e) upon oxidation to carbon dioxide for a wide range of organic compounds that are relevant in biological systems.

From a bioenergetics perspective one may wonder if by conversion of organic matter into methane containing biogas not a lot of energy is dissipated and therewith not available for electricity production upon combustion. This is however to a limited extent the case, because upon combustion with molecular oxygen the majority of the energy is obtained through oxygen reduction to water and not organic carbon oxidation:

$$-1 \cdot O_2(g) - 4 \cdot H^{+1} - 4 \cdot e^{-1} + 2 \cdot H_2O,$$
$$\Delta G^{01} = -78.7 \frac{kJ}{emol}$$

These two arguments are the main reasons why anaerobic digestion is highly attractive as a bioenergy production process in terms of energy (kJ) production per unit of biomass processed. Typically, net Gibbs energy yields per unit of biomass processed are two to three times higher than other bioenergy process like bioethanol or biodiesel production. This is due to (1) the low energy consumption during processing of biomass, (2) the fact that ***all*** biodegradable organic matter in biomass is converted to methane containing biogas whereas bioethanol and biodiesel only use a fraction of the organic substrates, and (3) biomass yields are very low resulting in a high methane yield during anaerobic digestion. Despite these intrinsic advantages of anaerobic digestion for methane containing biogas production it is a much smaller production process compared to bioethanol and biodiesel. This is due to one big disadvantage that limits its application: economy. Natural gas is a very cheap fossil fuel due to the seemingly endless supply from non-renewable resources. The price of natural gas has dropped in recent years due to exploration of novel shale-gas reservoirs, further complicating the implementation of anaerobic digestion as bioenergy process. Liquid biofuels like bioethanol and biodiesel can readily be mixed with current transport fuel based infrastructure and therefore have a much higher price per amount of energy generated which is (almost) linearly related to the price per electron (Table 1). In summary we suggest that anaerobic digestion as bioenergy process can only be implemented in economically viable way when adequate subsidies are supplied to overcome the competition with natural gas.

Table 1: Approximate price (Euro) per kg and per kmol electrons obtained upon combustion to carbon dioxide for different organic compounds (June 2013)

		EUR/kg	**EUR/kmol-e**
Coal	C	0.05	0.2
Methane (US June 2013)	CH_4	0.20	0.4
Methane (Europe June 2013)	CH_4	0.40	0.8
Oil (June 2013)	CH_2	0.64	1.5
Hydrogen	H_2	2.0	2.0

Sugar (June 2013)	$C_6H_{12}O_6$	0.28	2.1
Ethanol (2013)	C_2H_6O	0.52	2.0
Hexanoic acid	$C_6H_{12}O_2$	1.00	3.6
PHB	CH_3O	2.00	12.4

Another argument against methane containing biogas as end-product of the process is the potential emission of methane into the atmosphere if biogas is not handled adequately. The global warming potential of methane is approximately twenty three times carbon dioxide, suggesting that if only five percent of the methane generated during anaerobic digestion is emitted into the atmosphere, the overall positive impact of anaerobic digestion in terms of greenhouse gas emissions is diminished. The negative impact of methane emissions is particularly important when considering low-strength wastewater at low temperatures (e.g. sewage) where 50 % or more of the methane is dissolved in the bioreactor effluent.

Despite this intrinsic limitation of biogas as end product of organic matter processing, there are other arguments that make anaerobic digestion for waste(water) treatment a highly attractive process. To compare the anaerobic and aerobic treatment, different arguments can be used for wastewater and particulate waste treatment:

Anaerobic Wastewater Treatment

A global comparison of anaerobic and aerobic wastewater treatment processes is presented in Fig. 2. Main advantages of anaerobic treatment are the lower energy demand due to the absence of aeration requirements and the lower biomass production and associated nutrient requirements. Lower nutrient uptake can of course also be a disadvantage if for example nitrogen removal via nitrification-denitrification needs to be established. Biomass is generally considered as an unwanted side-product of wastewater treatment processes and associated with high costs for treatment or disposal. For sewage treatment using the activated sludge process in The Netherlands for example, approximately 50 % of all treatment costs are related to sludge disposal and processing. A ten times lower biomass production per unit of organic carbon converted therefore is an evident advantage of anaerobic wastewater treatment over aerobic wastewater treatment.

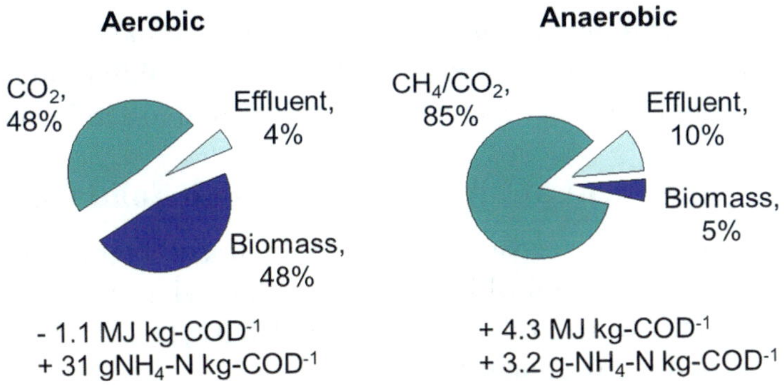

Figure 2: Generalized comparison between aerobic and anaerobic wastewater treatment in terms of the fate of organic carbon [expressed as chemical oxygen demand (COD)] and energy production/consumption and nutrient requirements (expressed as N-requirements) [adopted from van Lier et al. (2008)].

Volumetric treatment capacities for organic carbon removal in aerobic processes are either restricted by the mass transfer capacity for oxygen or the biomass concentrations that can be established in the activated sludge process. Taking into account that the maximum oxygen transfer rate in a bubble column typically amounts 5 kgO_2/m^3/day (Garcia-Ochoa and Gomez 2009), the volumetric COD-removal capacity is limited to approximately 7 $kgCOD/m^3$/day in an activated sludge system. Volumetric treatment capacities in aerobic biofilm systems can even be significantly lower depending on the biofilm surface area available in the system. In activated sludge based sewage treatment systems, the COD-loading rate normally does not exceed 1–2 $kgCOD/m^3$/day.

In anaerobic wastewater treatment systems with adequate biomass retention—such as the granular sludge based UASB-reactor—the volumetric treatment capacities are independent of gas to liquid mass transfer of substrate and therefore much higher treatment capacities can be established. In high-rate anaerobic processes typical volumetric treatment capacities are approximately 30 $kgCOD/m^3$/day or more, provided that good granular biomass can be obtained. The consequence of this is that bioreactor volumes are significantly lower for high-rate anaerobic wastewater treatment processes. Herewith it is evident that one of the main challenges for implementation of anaerobic wastewater treatment is to achieve adequate biomass retention. It should be noted that because the maximum specific growth rate of crucial steps in the anaerobic digestion process like aceticlastic methanogenesis are much lower than aerobic growth rates, suggesting that the time needed for start-up of the process or for recovery after a process failure in anaerobic wastewater treatment plants, is much longer compared to aerobic processes.

Anaerobic Treatment of Slurries with a High Solid Content

Also for treatment of feedstocks with a high solid content such as manure, sewage sludge, OFMSW, or agro-industrial residues, anaerobic digestion competes with aerobic composting. Both type of processes have disadvantages and disadvantages:

- Energy consumption for active aeration of the composting process makes the energy demand of the aerobic process significantly higher,
- Off-gas treatment of aerobic composting is required to minimize ammonia emissions,
- Feedstock subjected to anaerobic digestion need to be increased in temperature to 70 °C for some time for hygienization purposes if reuse of the digestate in agriculture is required. The temperature increase required in composting is achieved through oxygen respiration lowering the energy demand of aerobic composting process,
- Volume reduction through drying of the material in the aerobic composting process is an asset of this process. Depending on the water content of the feedstock, anaerobically digested material often requires energy intensive dewatering before transport.

Overall it is evident that aerobic composting versus anaerobic digestion is a trade-off between different advantages and disadvantages. The added value of production of methane containing biogas in the anaerobic digestion process is one of the arguments that favour anaerobic digestion, but in the Netherlands anaerobic digestion only was implemented on a relatively large scale when green gas subsidies were guaranteed.

In summary we conclude that the production of methane containing biogas is normally not the main reason for choosing for anaerobic digestion for waste organic carbon processing. Only in case "green gas" subsidies are implemented cost-effective industrial scale anaerobic processing of biomass can be implemented. This is also reflected by the observation that in a number of anaerobic wastewater treatment plants biogas is just being flared. This conclusion opens up the question if there are no other methods for organic carbon processing that may yield higher added value products. In the next section we will describe recent initiatives in that direction that aim for feedstock that is traditionally used for the anaerobic digestion process.

VFA AS CENTRAL INTERMEDIATE

Most of the initiatives to produce higher value end-products from heterogeneous low value feedstock are based on initial anaerobic fermentation of organic matter to short chain organic acids and alcohols (Agler et al. 2011; Kleerebezem and van Loosdrecht 2007). These short chain fatty acids include volatile fatty acids (mainly acetate, propionate, butyrate), but also lactate, and are referred to as ***carboxylates***. Anaerobic organic acid production relies on the fermentation of carboydrates and protein derived amino acids and depends furthermore on the effective inhibition of conversion of organic acids to methane containing biogas (Fig. 3). In absence of hydrogenotrophic methanogens long chain fatty acids cannot be oxidized to short chain organic acids for thermodynamic reasons and will therefore not be converted in a mixed culture fermentation process. During the production of organic acids, gaseous molecular hydrogen and carbon dioxide are normally produced as well. Fermentative production of molecular hydrogen from biomass has been investigated intensively as part of a two-stage anaerobic digestion process that aims for hydrogen (and organic acids) production in the first stage, and the production of methane containing biogas in the second stage.

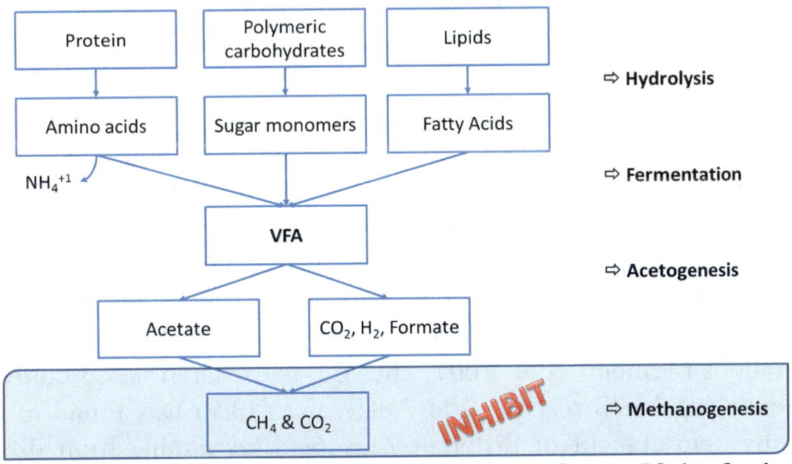

Figure 3: Anaerobic organic carbon degradation scheme. If the final methanogenic steps are fully inhibited, the end product of the process is VFA (and carbon dioxide and hydrogen).

The selective production of organic acids from various feedstocks is an emerging field of research. Within this field the following research challenges can be identified:

Controlling the Product Spectrum in Anaerobic Fermentations

When speaking about organic acid production in a mixed culture fermentation process, it remains an open question ***which*** organic acids are produced. Carbohydrate fermentation products include short and medium chain length fatty acids, lactate, alcohols, hydrogen, and carbon dioxide. Whereas the product spectrum of most amino acid fermentations are relatively well predictable based on general Strickland reaction pathways, the factors that determine the product spectrum of mixed culture fermentations remain to be identified. Consequently, which fermentation products are produced during carbohydrate fermentation in which conditions and by which microorganisms is a research topic that receives significant attention in recent years. It is evident that depending on the subsequent VFA recovery or processing steps it is of vital importance to have a proper control on the fermentation pattern obtained. An additional objective for optimization of the organic acid production process is the minimization of molecular hydrogen for maximization of the organic acid yield per substrate. Control and regulation of anaerobic carbohydrate fermentation furthermore receives a lot of research attention from the field of food microbiology as well as health related human and animal microbiome related research areas.

Numerous environmental factors determine the product spectrum of mixed culture fermentations and the microorganisms dominating the system. Factors that do play a role include pH, temperature, dilution rate in chemostat systems, type of carbohydrate substrate and concentration, product concentration, feeding pattern (continuous versus batch wise feeding), etc. (Temudo et al. 2007, 2008, 2009; Zoetemeyer et al. 1982a, b, d). At moderately acid to around neutral pH-values (4.5–7.0), glucose fermentation at mesophilic temperatures typically is dominated by a mixed butyrate/acetate fermentation pattern typically associated with ***Clostridium*** fermentations (Temudo et al. 2007, 2008). Potato starch based cultivation in a sequencing batch reactor at low pH-values (3.9) was found to result in selective enrichment of different ***Lactobacillus*** strains from different inocula and predominant lactate production (Liang et al. 2015). At higher pH-values (7.0–8.5) and thermophilic temperatures ethanol is often found to be a dominant product during continuous carbohydrate fermentation (Temudo et al. 2007; Zhang et al. 2014).

A pioneering modelling effort that intended to explain the product spectrum of mixed culture fermentation was conducted by Mosey (1983). This author assumed that the actual oxidation state of the NADH/NAD electron carrier was determined by the actual partial pressure of molecular hydrogen in the system. The NADH/NAD ratio subsequently determined which volatile fatty acid was produced, with more reduced products formed at increasing hydrogen partial pressures. Whereas Mosey focussed on fermentation patterns in biogas producing anaerobic digesters, others used a comparable starting point for modelling mixed culture carbohydrate fermentations in absence of methanogenesis (Kleerebezem et al. 2008; Rodriguez et al. 2006). Thermodynamic analysis of product formation pathways combined with product transport considerations enabled these authors to identify the optimal product spectrum in terms of bioenergy conservation potential. Based on experimental evidence (de Kok et al. 2013) electron carriers other than NADH/NAD were included in the second version of the model (Kleerebezem et al. 2008), resulting in dominant production of butyrate and acetate mixtures as fermentation products. This fermentation pattern typically is associated with Clostridium fermentations that indeed are often found to dominate at acid and around neutral pH-values and mesophilic temperatures. Since then the model has been extended to include electron bifurcation as electron transfer mechanism, resulting in a different product spectrum (Zhang et al. 2013).

Both the experimental and modelling efforts clearly demonstrate that not all mechanisms that determine the product spectrum of carbohydrate fermentations have been identified. More research is needed to elucidate what the selective mechanisms are for selection of product formation pathways and corresponding microorganisms in mixed culture fermentation processes.

Bioreactor Development for High-rate Organic Acid Production from Wastewater

Partial pre-acidification of organic substrates is a widely applied pre-treatment step for high-rate anaerobic wastewater treatment. This process is typically conducted in large mixed tanks because the primary aim of this pre-treatment step is equilibration of the wastewater composition and concentration (Cohen et al. 1985, 1979). Partial pre-acidification enhances the formation of methanogenic granular biomass and therewith has a positive impact on subsequent methane production from the organic acids produced.

Pre-acidification is particularly important for wastewaters with relatively high concentrations of carbohydrates and other readily fermentable organic matter.

For optimization of organic acid production in a VFA-platform based biorefinery concept, the VFA production step needs to fulfil other objectives to facilitate either direct VFA recovery, or post-processing to other products (see next section). Objectives for a wastewater treatment process aiming for VFA-production therewith are: (1) the extent of substrate acidification needs to be maximized, (2) the VFA composition should be controlled, (3) production of methane (or hydrogen) containing biogas should be minimized, (4) biomass concentrations in the bioreactor effluent should be minimized, and (5) compact bioreactors will need to be developed in order minimize the footprint and investment costs.

In our laboratory we are working on the development of the Granular Sludge process for volatile Fatty Acid production (GSFA) that can achieve these objectives. Granular sludge technology is successfully applied for both aerobic and anaerobic wastewater treatment, and also a first feasibility of the GSFA process has been demonstrated (Zoetemeyer et al. 1982c).

Bioprocess Development and Pre-treatment for Organic Acid Production from Biomass

There are a number of challenges to overcome for optimization of VFA production from biomass. Factors determining the efficiency of VFA production are explained based on Fig. 4. A two-step process is assumed consisting of (1) a lumped hydrolysis reaction catalysed by extracellular hydrolytic enzymes, and subsequent fermentation of the carbohydrate monomers to VFA, and (2) the production of methane containing biogas from VFA. For determining the end-product composition as a function of the retention time in the system, the process has been implemented in a continuous stirred tank reactor (CSTR). Critical factors that can be identified in the process are (1) the fraction of biodegradable organic matter, (2) the value for the hydrolysis rate constant (K_h), (3) and the critical retention time for VFA production.

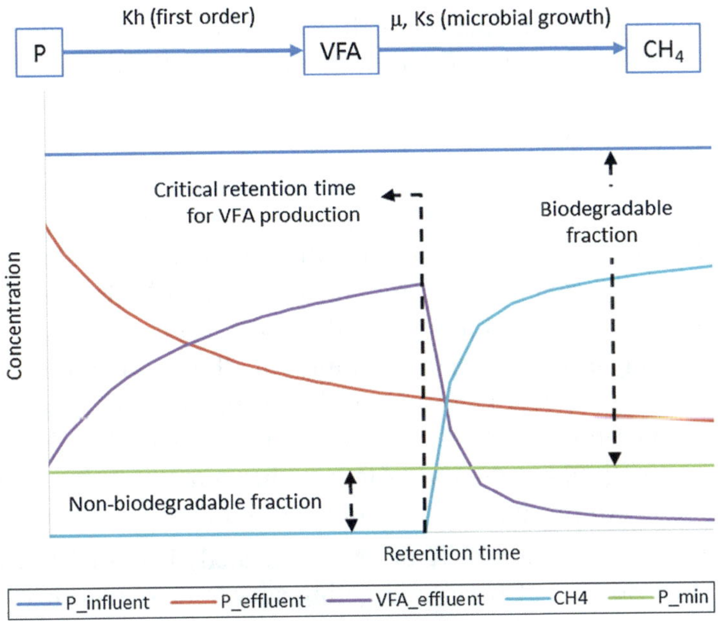

Figure 4: Simplified representation of anaerobic digestion as a two-step process consisting of first order (particulate) substrate (***P***) hydrolysis and fermentation to volatile fatty acids (***VFA***) and subsequent microbial conversion of VFA to methane containing biogas (CH_4) characterised by a maximum growth (μ) and substrate affinity constant (K_S). The lines show the products obtained (not to scale) as a function of the retention time in a CSTR-type bioreactor. The feedstock is characterized by a degradable and non-biodegradable fraction.

Bioprocess optimization for VFA production from biomass typically aim for manipulating these critical factors:

- Optimization of the operational conditions for VFA production can both aim for increasing the hydrolysis rate (increase Kh) through optimization of the operational pH or temperature, or by increasing the critical retention time required for biogas production by inhibiting methanogenesis. Effective inhibition of methane containing biogas production was shown to be feasible through operation at unfavourable pH-values or low solid retention times. It has for example been demonstrated that operational pH-values below 6 and elevated VFA-concentrations effectively inhibit methanogenic activity in the system enabling operation at significantly higher solid retention times. For feedstocks that are characterized by a very low hydrolysis rate constant and a low biodegradable fraction such as manure and secondary sewage sludge, it is evident that at around neutral pH-values it will be hard to separate VFA production and methanogenesis in space or time (Kleerebezem 2015).

- Feedstock pre-treatment methods include thermal, mechanical, chemical (acid or alkaline) or biological (enzymatic) means. (Alzate et al. 2012; Montingelli et al. 2015). In relation to Fig. 4 these methods typically aim for increasing the biodegradable fraction of the biomass and/or increasing the value for the hydrolysis rate constant by improving the enzymatic accessibility of the substrate (Sanders et al. 2000). There is a large amount of literature available on pre-treatment of feedstock for biogas production, but only very few studies aimed for determining the impact of pre-treatment on the VFA production potential.
- Operation as a CSTR or as a (repeated) batch (SBR) or plug-flow process also has a large impact on the extent that VFA and biogas production can be separated effectively in time or space. In a sequence of a two-step process in which the first step that is first order in substrate and the second a microbial growth process, a (sequencing) batch or plug-flow process is strongly preferred since the average hydrolysis rate is higher and the critical retention time for VFA production (Fig. 4) is significantly prolonged (Kleerebezem 2015). This effect is even more pronounced when taking into account product inhibition of the hydrolysis-fermentation process by the organic acids produced (Veeken et al. 2000).
- Bioprocess development for VFA production from a feedstock with a high solid content can only be conducted in CSTR-type slurry reactors through addition of water, and subsequent solid–liquid separation for production of a VFA-rich liquid stream. Mixing in these slurry reactors and the solid liquid separation steps are energy intensive and consequently other bioreactor concepts have been developed for VFA leaching from a feedstock with a high solid content (>20 %) such as the heap leaching process (Cysneiros et al. 2008, 2012; Stella Cadavid-Rodriguez and Horan 2014). A limiting factor for the heap leaching process for VFA production is the mass transfer of organic acids from the solid to the liquid phase. Preferential flow (channelling) in the leach bed may result in strong concentration gradients and

local inhibition of hydrolysis by organic acids and consequently a decrease in the observed hydrolysis rate achieved. The development of bioreactor concepts for more efficient VFA-leaching from a high solid feedstock is a potentially interesting but to date largely untouched field of research and development. One could think for example of a washing machine inspired VFA production reactor (Fig. 5), but a prerequisite of the bioreactor to be developed is that it can be operated in anaerobic conditions which may be difficult to achieve.

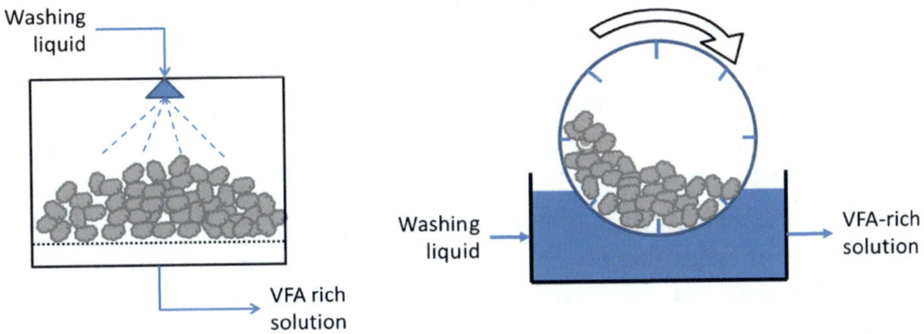

Figure 5: Process schemes for VFA production from particulate substrate: the ***left*** figure shows the heap leaching process and the ***right*** figure is a partially submerged rotating drum type of reactor.

VALORISATION OF VFA

The primary fermentation within the VFA platform produces a mixture of short-chain carboxylates from organic-rich waste streams. These valuable carboxylates can be concentrated from the fermentation broth, for example by the use of nanofiltration (NF) membranes (Zacharof and Lovitt 2014), liquid/liquid extraction (Li et al. 2002), and anion exchange (Cao et al. 2002).

Besides direct concentration of the carboxylates from the fermentation broth, secondary (bio)conversion of the carboxylates facilitates product recovery. The production of biopolymers, such as polyhydroxyalkanoates (PHAs) that are produced by bacteria as intracellular storage compounds, simplifies product separation by solid–liquid separation. Secondary fermentation reactions, occurring either in the same reactor or in separate processes, enable production of more hydrophobic compounds that can be separated from the fermentation broth. Secondary fermentations include chain elongation to medium chain fatty acids (MCFAs) and electrosynthesis using bioelectrochemical systems (BES).

Membranes

Membrane separation of VFAs from the fermentation broth is a widely used process (Lee et al. 2008; Zacharof and Lovitt 2014). Membrane separation can be applied in-situ and has the big advantage of being easily scalable. Fermentation broth usually contains high concentrations of solids like biomass and salts that can have an adverse effect on the performance of the membrane and causes membrane fouling (Le-Clech et al. 2006). To overcome this problem extensive pretreatment is needed, including dilution, sieving and microfiltration (MF). Current research on nanofiltration (NF) of the pretreated liquid achieving retention ratios of 75% at an optimal pH of 7 (Zacharof and Lovitt 2014). The neutral pH during nanofiltration is needed since only acids in their dissociated form are being rejected by nanofiltration membranes. For anaerobic digestion processes with high acid concentrations performed at low pH this can be an economically unattractive condition (Fig. 6).

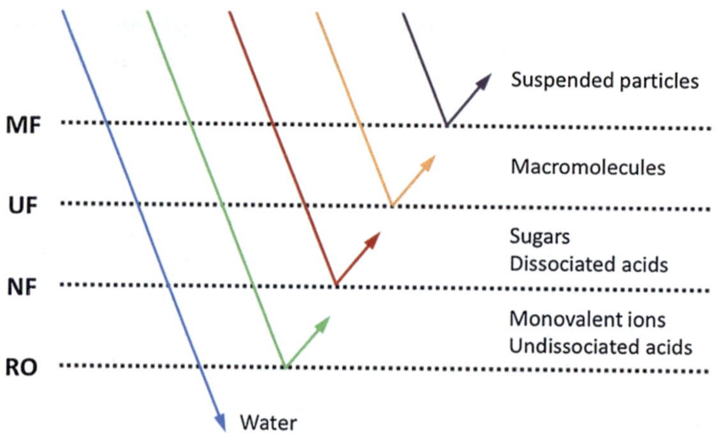

Figure 6: Principle of membrane separation using microfiltration (MF), ultrafiltration (UF), nanofiltration (NF) and reverse osmosis (RO).

Anion Exchange and Direct Recovery

Acid removal using anion exchange is already extensively used for waste vegetable oil (Maddikeri et al. 2012). The possibility to use this process for lactate recovery was investigated by Cao et al. (2002) at a low pH of 2.0

with water as eluent, achieving a maximum recovery yield of 92 %. Salt ions did not affect the performance of the adsorption process, which is important since fermentation broth contains significant amounts of salts. Whether this process can be used for VFA recovery depends on the desired purity of the VFA since for the described adsorption method it is impossible to achieve a high product purity.

Recent studies focused on succinic acid recovery via anion exchange followed by ester formation (López-Garzón et al. 2014). In the catalysis dimethyl carbonate (DMC) is used as a solvent and reagent resulting in a high dimethyl succinate yield of 96%. Formation of methyl esters from VFAs is ongoing research with the focus on integration of the recovery and catalysis step to improve ester production and feasibility of the process itself (Fig. 7).

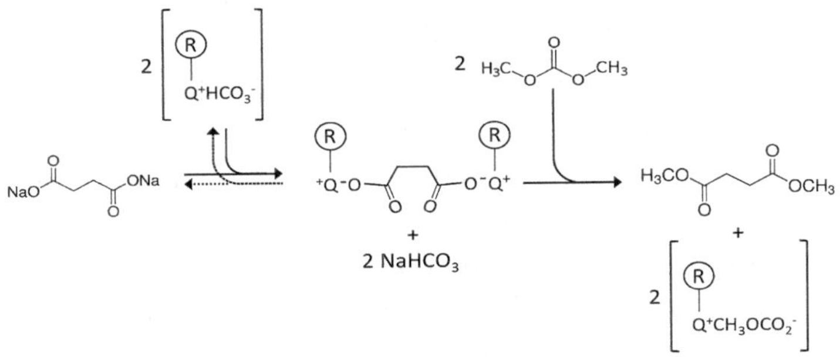

Figure 7: Succinic acid recovery via anion exchange and subsequent methylation using dimethyl carbonate (DMC) as being described by López-Garzón et al. (2014). The anion exchange resin (R) is in its bicarbonate form and contains a quaternary ammonium functional group (Q^+) that catalyzes the reaction.

Traditional bioelectrochemical systems (BES) were used to gain electric power from the oxidation of organic substrates (mostly acetate and other short-chain carboxylates) in a microbial fuel cell (MFC) (Rabaey et al. 2007). However, in more recent studies it was suggested that converting the organic substrates into chemicals, for example by microbial electrosynthesis (MES), had significant environmental and economic benefits (Foley et al. 2010; Rabaey and Rozendal 2010).

PHA Production

Polyhydroxyalkanoates (PHAs) are storage polymers that are accumulated by many types of bacteria as an internal energy and carbon reserve. PHAs have properties comparable to petrochemical plastics and can also serve as a biofuel or building block for the synthesis of all kinds of chemicals (Chen 2009). Preferred substrates for PHA synthesis are butyrate, lactate and, to a lesser extent, acetate (Marang et al. 2013) which are major products from the primary fermentation of the VFA-platform (Agler et al. 2011). Since PHA storing bacteria extract the VFAs from the liquid phase and accumulate them intracellular, product recovery is greatly improved compared to direct VFA recovery.

Lab-scale experiments with synthetic substrates (i.e. acetate or lactate) showed a maximum PHA content of 0.90 gPHA $gVSS^{-1}$ (Jiang et al. 2011; Johnson et al. 2009). Studies with wastewater achieved lower PHA contents of 0.75 gPHA $gVSS^{-1}$ using fermented molasses (Albuquerque et al. 2010) and 0.77 gPHA $gVSS^{-1}$ using fermented paper mill wastewater (Jiang et al. 2012). Recent pilot-scale experiments showed a comparable PHA content of 0.70 ± 0.05 gPHA $gVSS^{-1}$ (Tamis et al. 2014).

Solids in the fermented wastewater and possible non-storing side-populations can influence the production rates of the system and should therefore be minimized. Other challenges remain in the field of cost-effective downstream processing without compromising the polymer quality, and identification of the best product utilization route. In this respect it is worthwhile to note that PHA may also serve as feedstock for biofuel (3-hydroxybutyrate methyl ester, HAME) production, or the production of specific chemical such acrylate and propene (Gao et al. 2011; Spekreijse et al. 2012; Zhang et al. 2009) (Fig. 8)

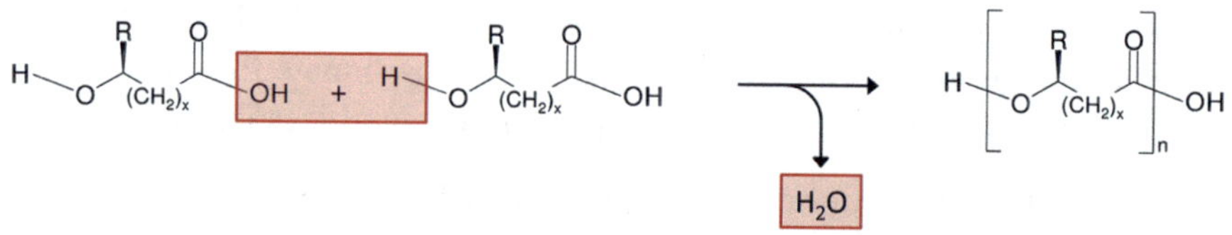

Figure 8: Simplified mechanism of polyhydroxyalkanoate (PHA) formation from two monomers. The R group is a hydrogen atom, alkyl, or alkenyl.

MCFA Production

Chain elongation is an anaerobic fermentation process in which ethanol and short-chain VFAs (mostly acetate) are converted into medium-chain VFAs (Steinbusch et al. 2011). Due to the longer carbon chain of the molecules they become more apolar and thus easier to extract from the fermentation broth.

Chain elongation based on synthetic medium (i.e. acetate and ethanol) achieved promising rates of 57.4 g l^{-1} day^{-1} with an MCFA selectivity above 80% at neutral pH and an HRT of 4 h (Grootscholten et al. 2013). The ***n***-caproate and ***n***-caprylate concentrations (9.3 and 0.3 g/l, respectively) are still below the solubility concentrations (***n***-caproate: 10.19 g/l and ***n***-caprylate: 0.79 g/l) but an increase in HRT caused an increase in concentrations to 12.0 and 0.9 g/l for ***n***-caproate and ***n***-caprylate, respectively.

More realistic substrates for the chain elongation process include the use of effluent streams from syngas (Vasudevan et al. 2014) fermentations and yeast-fermentation beer (Agler et al. 2012). These fermentations were carried out at a lower pH of 5.5 to suppress methanogenesis. At pH 5.5 a larger fraction of the acids are in their undissociated form and thus become more toxic to the cell. In-line liquid/liquid extraction was used to remove the undissociated acids from the broth, which resulted in a selectivity of 80% (Agler et al. 2012).

The low solubility and thus easy extraction of MCFAs gives MCFA fermentation a big advantage over ethanol distillation, which has a high energy demand. Given the fact that high concentrations of MCFAs are toxic to the cell, in-line MCFA extraction is needed to ensure high-rate MCFA production (Fig. 9).

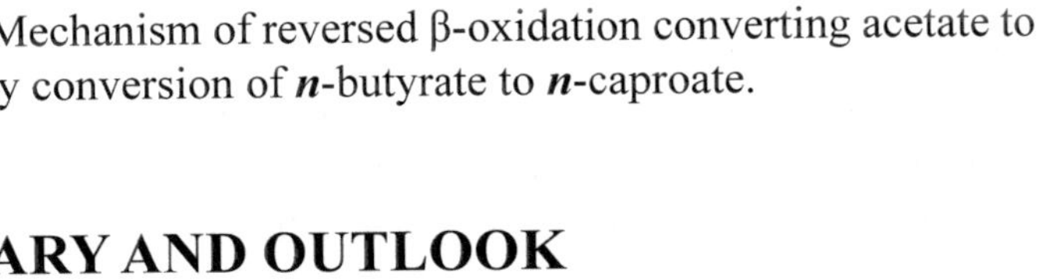

Figure 9: Mechanism of reversed β-oxidation converting acetate to ***n***-butyrate$^-$ followed by conversion of ***n***-butyrate to ***n***-caproate.

SUMMARY AND OUTLOOK

A future bio-based society will rely on effective valorisation of agroindustrial resources. Even though anaerobic digestion for the production of methane containing biogas is a process that may contribute significantly to a sustainable biobased society, it may to some extent be replaced by (bio) processes aiming for production of biopolymers, medium chain length fatty acids, and other higher added value products from heterogeneous feedstocks formerly known as waste. An example that demonstrates the economic potential of one of these alternative biomass processing routes is shown below. The suitability of a feedstock for either biogas production or the production of higher added value products via the VFA platform will depend primarily on the fraction of readily fermentable organic matter. Feedstock with a high fraction of readily fermentable organic carbon—such as agro-industrial wastewaters, papermill wastewater, or the organic fraction of municipal solid waste) provide opportunities for production of higher added value products. A feedstock with a low fraction of readily fermentable organic matter—such as manure, secondary sewage sludge, or spent biomass from industrial fermentations—is more suitable for anaerobic digestion to methane containing biogas. Pre-treatment methods that are capable of significantly increasing the fraction of fermentable organic carbon can furthermore change the status of the feedstock to become more suitable for production of higher added value products.

Another aspect of production processes that use waste as feedstock, is that waste is a side-product of (agro)industrial production and therefore is not generated in a scalable amount. This means that whereas the scale of industrial production typically can be optimized economically, this is not possible for waste based production processes. In many cases this implies that the amounts of products that can be generated from a waste will be much smaller than those normally established during industrial production. This is for example the case for fertilizer production through nutrient recovery processes from wastewater treatment facilities that produce many orders of magnitude smaller amounts of product compared to Haber–Bosch or mining based artificial fertilizer production. This often implies that for effective product utilization a specific (local) niche market needs to be identified that can guarantee effective marketing of the product. The most favourable condition in this respect is achieved if the product of the waste valorisation process can be used in the infrastructure where the waste is generated. In this way a very short value chain can be established with a clear added value.

Example 1

PHA versus Methane containing biogas production from paper mill wastewater

We will consider a cardboard production facility generating 5000 m^3/day (***Q***) wastewater in a closed water cycle. The wastewater contains volatile fatty acids at a concentration of 8 gCOD/L ($\boldsymbol{VFA}_i$). The organic acids need to be removed from the wastewater to avoid VFA incorporation in the cardboard, yielding a bad smell when the cardboard gets wet. Both anaerobic digestion and PHA-production are considered for VFA removal from the recirculation water. PHA-production is based on the enrichment-accumulation process according to Johnson (Johnson et al. 2009). Taken a number of generalized parameters we can calculate the product revenues from both wastewater treatment processes assuming for simplicity reasons full VFA removal:

Process	**Anaerobic digestion**	**PHA production**
Parameters	$Y_{CH4/VFA} = 0.9$ ***gCOD/gCOD*** $COD_{CH4} = 4$ ***gCOD/g***	$Y_{PHA/VFA} = 0.7$ ***gCOD/gCOD*** $Y_{X/VFA} = 0.6$ ***gCOD/gCOD*** ***Final COD–PHA content*** = ***80 %*** ($f_{PHA} = 0.8$) $COD_{PHA} = 1.67$ ***gCOD/g*** υ = ***fraction substrate for PHA production*** $PHA^m = Y_{PHA/VFA} \cdot VFA_i,$ $X^m = Y_{PHA/VFA} \cdot Y_{X/PHA} \cdot VFA_i,$
Price end-product	$€_{CH4} = 0.4$ ***€/kg***	$€_{PHA} = 2.0$ ***€/kg***
Revenue calculation	$Rev = \frac{Q \cdot VFA_i \cdot Y_{CH4/VFA} \cdot €_{CH4}}{COD_{CH4}}$	$\vartheta = \frac{f_{PHA} \cdot PHA^m}{PHA^m - f_{PHA} \cdot (PHA^m - X^m)}$ $Rev = \frac{Q \cdot VFA_i \cdot \vartheta \cdot Y_{PHA/VFA} \cdot €_{PHA}}{COD_{CH4}}$
Revenue	3.6 k€/day	20.2 k€/day

It is evident that the product revenues that can be obtained from PHA production are significantly higher than those obtained from anaerobic digestion. This does not directly mean that PHA-production is economically a much more attractive process option because the calculations do not take into account (1) the costs associated with bioprocess operation, (2) downstream processing required for product recovery, nor (3) potential subsidies for green gas production. Still, it is evident that there is significant economic room for establishment of PHA production in comparison with biogas production and research efforts in this field therefore aim for cost-effective process implementation and product recovery and utilization.

ACKNOWLEDGMENTS

The financial support from the Dutch Applied Science foundation (STW) and Paques BV through the VFA-platform program (Project No. 13002) is gratefully acknowledged.

REFERENCES

1. Agler MT, Wrenn BA, Zinder SH, Angenent LT (2011) Waste to bioproduct conversion with undefined mixed cultures: the carboxylate platform. Trends Biotechnol 29:70–78
2. Agler MT, Spirito CM, Usack JG, Werner JJ, Angenent LT (2012) Chain elongation with reactor microbiomes: upgrading dilute ethanol to medium-chain carboxylates. Energy Environ Sci 5:8189–8192
3. Albuquerque MGE, Torres CAV, Reis MAM (2010) Polyhydroxyalkanoate (PHA) production by a mixed microbial culture using sugar molasses: effect of the influent substrate concentration on culture selection. Water Res 44:3419–3433
4. Alzate ME, Muñoz R, Rogalla F, Fdz-Polanco F, Pérez-Elvira SI (2012) Biochemical methane potential of microalgae: Influence of substrate to inoculum ratio, biomass concentration and pretreatment. Bioresour Technol 123:488–494
5. Andriani D, Wresta A, Atmaja TD, Saepudin A (2014) A review on optimization production and upgrading biogas through CO_2 removal using various techniques. Appl Biochem Biotechnol 172:1909–1928
6. Borjesson M, Ahlgren EO (2012) Cost-effective biogas utilisation—a modelling assessment of gas infrastructural options in a regional energy system. Energy 48:212–226
7. Cao X, Yun HS, Koo YM (2002) Recovery of l-(+)-lactic acid by anion exchange resin Amberlite IRA-400. Biochem Eng J 11:189–196
8. Chen G-Q (2009) A microbial polyhydroxyalkanoates (PHA) based bio- and materials industry. Chem Soc Rev 38:2434–2446
9. Cohen A, Zoetemeyer RJ, Vandeursen A, Vanandel JG (1979) Anaerobic digestion of glucose with separated acid production and methane formation. Water Res 13:571–580
10. Cohen A, Breure AM, Schmedding DJM, Zoetemeyer RJ, Vanandel JG (1985) Significance of partial pre-acidification of glucose for methanogenesis in an anaerobic-digestion process. Appl Microbiol Biotechnol 21:404–408
11. Cysneiros D, Banks CJ, Heaven S (2008) Anaerobic digestion of maize in coupled leach-bed and anaerobic filter reactors. Water Sci Technol 58:1505–1511
12. Cysneiros D, Banks CJ, Heaven S, Karatzas K-AG (2012) The effect of pH control and 'hydraulic flush' on hydrolysis and volatile fatty acids

(VFA) production and profile in anaerobic leach bed reactors digesting a high solids content substrate. Bioresour Technol 123:263–271
13. de Kok S, Meijer J, van Loosdrecht MCM, Kleerebezem R (2013) Impact of dissolved hydrogen partial pressure on mixed culture fermentations. Appl Microbiol Biotechnol 97:2617–2625
14. Dominguez Cabanelas IT, Ruiz J, Arbib Z, Chinalia FA, Garrido-Perez C, Rogalla F, Nascimento IA, Perales JA (2013) Comparing the use of different domestic wastewaters for coupling microalgal production and nutrient removal. Bioresour Technol 131:429–436
15. Foley JM, Rozendal RA, Hertle CK, Lant PA, Rabaey K (2010) Life cycle assessment of high-rate anaerobic treatment, microbial fuel cells, and microbial electrolysis cells. Environ Sci Technol 44:3629–3637
16. Gao X, Chen JC, Wu Q, Chen GQ (2011) Polyhydroxyalkanoates as a source of chemicals, polymers, and biofuels. Curr Opin Biotechnol 22:768–774
17. Garcia-Ochoa F, Gomez E (2009) Bioreactor scale-up and oxygen transfer rate in microbial processes: an overview. Biotechnol Adv 27:153–176
18. Grootscholten TIM, Steinbusch KJJ, Hamelers HVM, Buisman CJN (2013) Improving medium chain fatty acid productivity using chain elongation by reducing the hydraulic retention time in an upflow anaerobic filter. Bioresour Technol 136:735–738
19. Jiang Y, Marang L, Kleerebezem R, Muyzer G, van Loosdrecht MCM (2011) Polyhydroxybutyrate production from lactate using a mixed microbial culture. Biotechnol Bioeng 108:2022–2035
20. Jiang Y, Marang L, Tamis J, van Loosdrecht MCM, Dijkman H, Kleerebezem R (2012) Waste to resource: converting paper mill wastewater to bioplastic. Water Res 46:5517–5530
21. Johnson K, Jiang Y, Kleerebezem R, Muyzer G, van Loosdrecht MCM (2009) Enrichment of a mixed bacterial culture with a high polyhydroxyalkanoate storage capacity. Biomacromolecules 10:670–676
22. Kleerebezem R (2015) Biochemical conversion: anaerobic digestion. In: De Jong W, Van Ommen J (eds) Biomass as a sustainable energy source for the future. Wiley, New Jersey, pp 441–467
23. Kleerebezem R, van Loosdrecht MCM (2007) Mixed culture biotechnology for bioenergy production. Curr Opin Biotechnol

18:207–212

24. Kleerebezem R, Van Loosdrecht MCM (2010) A generalized method for thermodynamic state analysis of environmental systems. Crit Rev Environ Sci Technol 40:1–54
25. Kleerebezem R, Rodriguez J, Temudo MF, van Loosdrecht MCM (2008) Modeling mixed culture fermentations; the role of different electron carriers. Water Sci Technol 57:493–497
26. Le-Clech P, Chen V, Fane TAG (2006) Fouling in membrane bioreactors used in wastewater treatment. J Membr Sci 284:17–53
27. Lee JW, Kim JO, Jung JT (2008) Concentration of organic acid from fermented food wastes using a MF/RO process. Mater Sci Forum 569:281–284
28. Li Z, Qin W, Dai Y (2002) Liquid – liquid equilibria of acetic, propionic, butyric, and valeric acids with trioctylamine as extractant. J Chem Eng Data 47:843–848
29. Liang S, Gliniewicz K, Mendes-Soares H, Settles ML, Forney LJ, Coats ER, McDonald AG (2015) Comparative analysis of microbial community of novel lactic acid fermentation inoculated with different undefined mixed cultures. Bioresour Technol 179:268–274
30. López-Garzón CS, van der Wielen LAM, Straathof AJJ (2014) Green upgrading of succinate using dimethyl carbonate for a better integration with fermentative production. Chem Eng J 235:52–60
31. Maddikeri GL, Pandit AB, Gogate PR (2012) Adsorptive removal of saturated and unsaturated fatty acids using ion-exchange resins. Ind Eng Chem Res 51:6869–6876
32. Marang L, Jiang Y, van Loosdrecht MCM, Kleerebezem R (2013) Butyrate as preferred substrate for polyhydroxybutyrate production. Bioresour Technol 142:232–239
33. Montingelli ME, Tedesco S, Olabi AG (2015) Biogas production from algal biomass: a review. Renew Sustain Energy Rev 43:961–972
34. Mosey FE (1983) Mathematical-modeling of the anaerobic-digestion process-regulatory mechanisms for the formation of short-chain volatile acids from glucose. Water Sci Technol 15:209–232
35. Niesner J, Jecha D, Stehlik P (2013) Biogas upgrading technologies: state of art review in European Region. In: 16th international conference

on process integration, modelling and optimisation for energy saving and pollution reduction (Pres'13) 35, pp 517–522

36. Rabaey K, Rozendal RA (2010) Microbial electrosynthesis—revisiting the electrical route for microbial production. Nat Rev Microbiol 8:706–716
37. Rabaey K, Rodriguez J, Blackall LL, Keller J, Gross P, Batstone D, Verstraete W, Nealson KH (2007) Microbial ecology meets electrochemistry: electricity-driven and driving communities. ISME J 1:9–18
38. Rodriguez J, Kleerebezem R, Lema JM, van Loosdrecht MCM (2006) Modeling product formation in anaerobic mixed culture fermentations. Biotechnol Bioeng 93:592–606
39. Sanders WTM, Geerink M, Zeeman G, Lettinga G (2000) Anaerobic hydrolysis kinetics of particulate substrates. Water Sci Technol 41:17–24
40. Spekreijse J, Le Notre J, van Haveren J, Scott EL, Sanders JPM (2012) Simultaneous production of biobased styrene and acrylates using ethenolysis. Green Chem 14:2747–2751
41. Steinbusch KJJ, Hamelers HVM, Plugge CM, Buisman CJN (2011) Biological formation of caproate and caprylate from acetate: fuel and chemical production from low grade biomass. Energy Environ Sci 4:216–224
42. Stella Cadavid-Rodriguez L, Horan NJ (2014) Production of volatile fatty acids from wastewater screenings using a leach-bed reactor. Water Res 60:242–249
43. Tamis J, Lužkov K, Jiang Y, Loosdrecht MCMV, Kleerebezem R (2014) Enrichment of Plasticicumulans acidivorans at pilot-scale for PHA production on industrial wastewater. J Biotechnol Part A 192:161–169
44. Temudo MF, Kleerebezem R, van Loosdrecht M (2007) Influence of the pH on (open) mixed culture fermentation of glucose: a chemostat study. Biotechnol Bioeng 98:69–79
45. Temudo MF, Muyzer G, Kleerebezem R, van Loosdrecht MCM (2008) Diversity of microbial communities in open mixed culture fermentations: impact of the pH and carbon source. Appl Microbiol Biotechnol 80:1121–1130

46. Temudo MF, Mato T, Kleerebezem R, van Loosdrecht MCM (2009) Xylose anaerobic conversion by open-mixed cultures. Appl Microbiol Biotechnol 82:231–239
47. Tilche A, Galatola M (2008) The potential of bio-methane as bio-fuel/bio-energy for reducing greenhouse gas emissions: a qualitative assessment for Europe in a life cycle perspective. Water Sci Technol 57:1683–1692
48. van Lier JB, Mahmoud N, Zeeman G (2008) Anaerobic wastewater treatment. In: Henze M, Van Loosdrecht MCM, Ekama GA, Brdjanovic D (eds) Biological watewater treatment principles, modelling and design. IWA Publishing, London, pp 401–442
49. Vasudevan D, Richter H, Angenent LT (2014) Upgrading dilute ethanol from syngas fermentation to ***n***-caproate with reactor microbiomes. Bioresour Technol 151:378–382
50. Veeken A, Kalyuzhnyi S, Scharff H, Hamelers B (2000) Effect of pH and VFA on hydrolysis of organic solid waste. J Environ Eng ASCE 126:1076–1081
51. Ward AJ, Lewis DM, Green B (2014) Anaerobic digestion of algae biomass: a review. Algal Res Biomass Biofuels Bioprod 5:204–214
52. Zacharof MP, Lovitt RW (2014) Recovery of volatile fatty acids (VFA) from complex waste effluents using membranes. Water Sci Technol 69:495–503
53. Zhang XJ, Luo RC, Wang Z, Deng Y, Chen GQ (2009) Application of (R)-3-hydroxyalkanoate methyl esters derived from microbial polyhydroxyalkanoates as novel biofuels. Biomacromolecules 10:707–711
54. Zhang F, Zhang Y, Chen M, van Loosdrecht MCM, Zeng RJ (2013) A modified metabolic model for mixed culture fermentation with energy conserving electron bifurcation reaction and metabolite transport energy. Biotechnol Bioeng 110:1884–1894
55. Zhang F, Chen Y, Dai K, Zeng RJ (2014) The chemostat study of metabolic distribution in extreme-thermophilic (70 A degrees C) mixed culture fermentation. Appl Microbiol Biotechnol 98:10267–10273
56. Zoetemeyer RJ, Arnoldy P, Cohen A, Boelhouwer C (1982a) Influence of temperature on the anaerobic acidification of glucose in a mixed culture forming part of a 2-stage digestion process. Water Res 16:313–321

57. Zoetemeyer RJ, Matthijsen A, Cohen A, Boelhouwer C (1982b) Product inhibition in the acid forming stage of the anaerobic-digestion process. Water Res 16:633–639

58. Zoetemeyer RJ, Matthijsen A, Vandenheuvel JC, Cohen A, Boelhouwer C (1982c) Anaerobic acidification of glucose in an upflow reactor. Biomass 2:187–199

59. Zoetemeyer RJ, Vandenheuvel JC, Cohen A (1982d) Ph influence on acidogenic dissimilation of glucose in an anaerobic digester. Water Res 16:303–311

10

Biological Pretreatment Strategies for Second-Generation Lignocellulosic Resources to Enhance Biogas Production

Andreas Otto Wagner, Nina Lackner, Mira Mutschlechner, Eva Maria Prem, Rudolf Markt , and Paul Illmer

Department of Microbiology, Universität Innsbruck, Technikerstraße 25d, A-6020 Innsbruck, Austria

ABSTRACT

With regard to social and environmental sustainability, second-generation biofuel and biogas production from lignocellulosic material provides considerable potential, since lignocellulose represents an inexhaustible, ubiquitous natural resource, and is therefore one important step towards independence from fossil fuel combustion. However, the highly heterogeneous structure and recalcitrant nature of lignocellulose restricts its commercial utilization in biogas plants. Improvements therefore rely on effective pretreatment methods to overcome structural impediments, thus facilitating the accessibility and digestibility of (ligno)cellulosic substrates during anaerobic digestion. While chemical and physical pretreatment

Citation: Wagner AO, Lackner N, Mutschlechner M, Prem EM, Markt R, Illmer P. Biological Pretreatment Strategies for Second-Generation Lignocellulosic Resources to Enhance Biogas Production. Energies. 2018; 11(7):1797.

strategies exhibit inherent drawbacks including the formation of inhibitory products, biological pretreatment is increasingly being advocated as an environmentally friendly process with low energy input, low disposal costs, and milder operating conditions. Nevertheless, the promising potential of biological pretreatment techniques is not yet fully exploited. Hence, we intended to provide a detailed insight into currently applied pretreatment techniques, with a special focus on biological ones for downstream processing of lignocellulosic biomass in anaerobic digestion.

Keywords: Pretreatment, biological pretreatment, anaerobic digestion, biogas, methane

INTRODUCTION

Although it is known that CO_2 production from fossil fuel combustion is a major contributor to global warming, these energy carriers are still the most important resources for global energy generation [1]. Great efforts have been devoted to increasing energy production from nonfossil fuels and to replacing climate-change-relevant energy sources by renewable ones. Hydropower, wind, and solar energy are probably the most promising alternative energy resources but can exhibit limitations concerning flexible energy production, storage and/or backup, transportation, and land requirements [2].

Biogas production from anaerobic digestion (AD) processes is considered as an attractive source for green energy [3,4] and, therefore, endeavors have been made to increase the share of biogas in global energy production. During anaerobic digestion, organic feedstocks are converted into biogas containing methane (CH_4) as a valuable end-product. The energy input for biogas production is calculated to be lower than in current ethanol production, leading to a higher energy output-to-input ratio [5]. However, the expanded production of biogas was often achieved by the utilization of energy crops directly competing with food crop farming (first-generation biofuels). Therefore, the exploration of lignocellulosic materials (second-generation biofuels) for bio-methane production was substantially accelerated during the past years, thus offering ecological as well as economic advantages [6]. However, lignin resists (complete) degradation under anaerobic conditions, posing a challenge regarding the overall degradability of lignocellulose in AD. In this context, enhancing the substrate conversion to overcome the degradation resistance of lignocellulosic resources is of utmost importance to achieving environmentally friendly and economically feasible processes

[7,8]. Hence, effective pretreatment methods are needed, particularly because lignocellulosic biomass has been evaluated as an attractive renewable energy source due to its inexhaustible, ubiquitous character [2,9].

The main objectives of this work are, therefore, (i) to present a short update on the currently available pretreatment strategies for enhanced disintegration of lignocellulosic resources and their application, and (ii) to review biological pretreatments currently applied for enhanced biogas production.

LIGNOCELLULOSES

Lignocellulose is the most abundant renewable biomass [10], with a worldwide annual production of an estimated 1000 Gt [11], including wheat straw, sugarcane bagasse, corn stalks, rye straw, rice straw, and barley straw as well as various types of organic waste (fractions). For data on the composition of different feedstocks, please refer to, e.g., Dahadha et al. [12], or Paudel et al. [13].

Lignocellulose contains up to 45% cellulose as the main component, 30% hemicellulose, and 25% lignin, although the composition varies considerably among different plants [14,15]. With about from 50% to 80% of organic material deriving from photosynthetic processes, lignocellulose represents one of the main components of global biomass [16,17]. Therefore, lignocellulose plays a major role as a constituent of biological resources and represents the most abundantly available raw material for the generation of renewable primary products and energy [18].

In the following section the chemical structure and characteristics of the most important fibrous components of lignocellulosic resources are described briefly.

Cellulose

Cellulose is the major component of plant matter and, therefore, a valuable source of biomass storing an enormous quantity of energy conserved by photosynthesis. It is a fibrous, hard, and water-insoluble substance that can be found in the wooden part of plant tissue. As a linear polymer, it comprises from 3000 to 14,000 glucose monomers, which are linked via β-1,4 glycosidic bonds. Approximately 60–70 of those cellulose polymers are interconnected by hydrogen bonds, forming so-called elementary fibrils, which themselves build up microfibrils. Multiple of those collocated chains can form a network of stable supramolecular fibers, with high tensile strength and a partially crystalline structure [19]. In plants, cellulose molecules are synthesized individually, which then undergo immediate self-assembly [20] probably regulated by hemicelluloses [21]. An important feature of cellulose is its crystalline structure, with the degree of crystallinity being highly variable depending on the type of plant tissue [19]. While the crystalline structure of cellulose fibers hinders degradation, various types of irregularities—pits, pores, and capillaries—increase the surface area of cellulose molecules [22]. This results in at least partially hydrated areas when being immersed in water, thus permitting access for enzymatic attack, including cellulases.

Hemicellulose

Hemicellulose is the term for branched heteropolysaccharides—mostly matrix polysaccharides—including monomers like glucose, mannose, galactose, xylose, and arabinose. Although similar enzymes are involved in cellulose and hemicellulose decomposition, complete hemicellulose degradation requires more enzymes due to its greater chemical and structural heterogeneity [23]. Hemicellulose is degraded to monomeric sugars and acetic acid [18], with the latter being of special interest for anaerobic digestion, representing the dominant methane precursor [24].

Lignin

Lignin is an aromatic polymer synthesized of phenylpropanoid precursors. Lignin is predominantly found in combination with cellulose (and hemicellulose), the so-called lignocellulose. Therein, lignin is encrusting both cellulose and hemicellulose, forming a physical seal, and is an impenetrable barrier in the plant cell wall. This polymer is synthesized by the generation of free radicals, which are released in the peroxide-mediated dehydrogenation of three phenylpropionic alcohols [18]. Lignin breakdown is necessary to facilitate the access to cellulose and hemicellulose but can, however, only occur via co-metabolism [18].

BIODEGRADATION OF LIGNOCELLULOSE

Bioconversion of lignocellulosic residues is predominantly carried out by fungi. Because of the insolubility of cellulose, hemicellulose, and lignin, it occurs exocellularly, in association with the outer cell envelope, or extracellularly [18]. Two types of fungal enzymes are known to break down lignocellulose: (i) the hydrolytic system that produces hydrolases responsible for polysaccharide degradation and (ii) a unique oxidative and extracellular ligninolytic system degrading lignin by opening phenyl rings [18,25].

The ability to digest cellulose is widely distributed among many genera in the domain of Bacteria and in fungal groups within the domain of Eukarya [26], whereas cellulolytic organisms in the domain of Archaea have not (yet) been identified. Specialized groups of fungi are further able to attack lignin-encrusted cellulose. Generally, a 10- to 100-fold higher productivity of fungal compared with bacterial enzymes was assessed for cellulases [25].

Concerning the eubacteria, the ability to decompose cellulose is widespread in bacteria within the predominantly aerobic order Actinomycetales (Actinobacteria) and the anaerobic order Clostridiales (Firmicutes) [19,27]. Mechanisms of bacterial decomposition differ significantly from those of their fungal counterparts. Within cellulolytic clostridia, the breakdown of cellulose is organized in the so-called cellulosome [28,29], which is attached to the cell surface, contains all necessary enzymes, and forms a bridge between the cell and the insoluble cellulose components [16]. In anaerobic digestion systems, cellulose-degrading bacteria play an important role regarding the interaction between several groups of organisms, resulting in a complete conversion into carbon dioxide, methane, and water [30]. However, due to the small amount of energy that can be preserved in anaerobic processes and the lower productivity of bacterial cellulases compared with fungal ones [25], the degradation of cellulose is significantly slower under anoxic than under oxic conditions.

A specialized group within the Neocallimastigomycota called "anaerobic fungi", commonly found in ruminants, is able to degrade cellulose and hemicellulose under strictly anaerobic conditions [31,32]. The use of anaerobic fungi for an improved anaerobic digestion was taken into account, e.g., by Dollhofer et al. [33] or Leis et al. [34]. Also, Nakashimada et al. [35] investigated methane production from cellulose as a substrate with defined mixed cultures using the cellulolytic ***Neocallimastix frontalis*** and methanogens.

In contrast to anaerobic fungi, the direct application of aerobically growing fungi in anaerobic systems is completely hampered by their oxygen demand. Among fungi, there are a number of representatives, e.g., of the genera Fusarium and Chaetomonium that also target lignin-encrusted cellulose. In particular, so-called white rot fungi can effectively degrade lignin using an oxidative process with phenol oxidases as key enzymes [36], including ***Phanerochaete chrysosporium*** and ***Trametes versicolor***, representing the most extensively studied members [37]. As the degradation of lignin is hardly possible under anoxic conditions, aerobic pretreatment prior to anaerobic digestion is of special interest [38,39,40].

CONCEPTS OF PRETREATMENT

Pretreatment strategies commonly comprise physical, chemical, and biological methods [40], and are applied in various fields of bioenergy and biofuel generation including biogas, bioethanol, biohydrogen, and hythane (H_2 + CH_4) production. Since lignocellulose materials represent the largest fraction of waste generated by modern society, increasing scientific interest is orientated towards combined cellulose waste management and energy resources [41]. Factors for ecological and economical feasible application of pretreatment strategies include low capital and energy investments, applicability over a wide variety of substrates, and high product yields to enhance revenues along with low waste treatment costs [7].

Among all bioconversion technologies for energy production, anaerobic digestion seems to be the most cost-effective that has been implemented worldwide for commercial production of electricity, heat, and compressed natural gas [40]. Anaerobic digestion has been adopted for bioenergy production from different organic feedstocks, such as forestry and agricultural residues, animal manures, organic fractions of municipal solid wastes, food wastes, and energy crops [42], answering the increasing demand for renewable energy sources. For recalcitrant substrates such as lignocellulosic resources, conventional anaerobic digestion cannot maximize the substrate conversion into biogas [43]. Thus, the application of biological pretreatments has gained significant importance in the past few years because of (i) the complex composition of lignocellulosic resources persistent in anaerobic environments; (ii) the desire to reduce hydraulic retention times; and (iii) the wish to increase the net carbon conversion rates. The latter is characterized by an enhanced total biogas and methane yield, representing the ultimate goal for any pretreatment strategy.

Physical and Chemical Pretreatment

Physical and chemical pretreatments are the most widespread strategies to improve the substrate quality designated for anaerobic digestion. They are often designed to improve the general digestibility and do not specifically target a certain compound of the substrate matrix. Physical strategies comprise comminution, heat and/or pressure treatment, steam explosion, liquid hot water, extrusion, and irradiation as well as ultrasonic and microwave technologies. Chemical ones include the use of acids or bases, catalyzed steam explosion, ozonization, oxidation, organosolve methods, and ionic liquid extraction [40]. A combination of physical and/or chemical methods is often applied. The main disadvantages of physical/chemical pretreatments are high energy and/or chemical demands with possible quality reductions of the digestion residues, thus hampering the subsequent use as biological fertilizer, accompanied by increasing costs for their disposal [44,45]. For a detailed description of physical and chemical pretreatment strategies, please refer to the respective review papers.

Biological Pretreatment

An effective biological pretreatment requires no preceding mechanical size reduction, preserves the pentose (hemicellulose) fractions, avoids the formation of degradation products that inhibit growth of fermentative microorganisms, minimizes energy demands, and limits costs. Therefore, a major objective of biological pretreatment is to break and remove the lignin seal and to disrupt the crystalline structure of cellulose to make it (more) susceptible to an enzymatic or microbial attack, while minimizing the loss of carbohydrates for anaerobic digestion [40,46,47]. The delignification and the decomposition of hemicellulose enhance the availability of cellulose and resultant monomers, which can boost the overall anaerobic digestion process. The choice of application is mainly dependent on the chemical composition of the substrate; however, in practice, structural and economic factors like available facilities or excess energy can often play an equally important role. Biological pretreatment techniques for enhanced biogas production have mainly focused on the use of fungal and bacterial strains or microbial consortia under both aerobic and anaerobic conditions, as well as on enzymes, with the latter being less important [40]. Therefore, this review is focusing on pretreatment strategies using active microorganisms.

The advantages of biological pretreatments compared with nonbiological procedures are the potential production of useful by-products, reduced formation of inhibitory substances due to milder operation conditions, the minimization of applied chemicals and energy input, and lower costs for waste deposit [44,45]. Beside the hydrolysis of (ligno)cellulose during pretreatment, microorganisms can further be used to upgrade the quality of certain substrates by removing undesired, potentially inhibitory substances.

However, the efficiency of biological pretreatment is limited by the rate of microbial growth and the utilization of readily available sugars by the engaged organisms [48].

Micro-Aerobic Pretreatment

Micro-aeration during anaerobic processes is known to increase microbial activity during the initial hydrolysis phase [49]. Pretreatments using different doses of oxygen during anaerobic digestion can also be ascribed to biological pretreatments since the oxygen input alters the microbial community; however, these methods are mainly applied in waste water treatment plants [49,50]. The goal of micro-aeration is to stimulate microbial growth and activity during hydrolysis, the rate-limiting step in anaerobic digestion [51]. Up to now, this method has been successfully applied for brown water and/or food waste [52,53,54] as well as for energy crops [43] and agricultural residues [55,56].

Ensiling, Composting

Another microbiological pretreatment strategy originated from the necessity to store and stabilize lignocellulosic resources to guarantee a whole year's substrate supply for anaerobic digestion facilities. Cui et al. [57] investigated a wet storage technique via ensiling with simultaneous chemical and fungal pretreatment which could increase glucose and xylose yields 2.9- and 3.9-fold, respectively. Sugar beet pulp silage in different maturity stages was evaluated concerning its methane potential by Heidarzadeh et al. [58], indicating a positive trend but also the risk of energy loss if the ensiling was not conducted properly. Papinagsorn et al. [59] successfully tested ensiled napier grass in combination with chemical pretreatment for co-digestion with cow dung yielding up to 8.34 $kJ \cdot g^{-1}$ VS. Vervaeren et al. [60] investigated maize silage additives for enhanced biogas production and could verify up to a 14.7% increase in biogas production for certain additives. Wagner et al. [61] used composting as a treatment strategy to enhance methane production from digestate and showed a positive impact of composting with increased biogas and methane yields in a subsequent anaerobic digestion process.

Physical Separation of Digestion Phases or Microbial Consortia

Efforts have been made earlier to separate the different phases of anaerobic digestion to increase total biogas and methane yields. These methods are often referred to special digestion systems but not to pretreatment technologies. Since almost all methods are at least some kind of upstream treatment prior to anaerobic digestion per se, they can also be seen as pretreatment methods. As this type of technology does not aim to increase degradation rates of one or more specific components of substrates, the mode of action is rather unspecific. The physical separation of the hydrolytic and methanogenic phase can further be used to apply suitable conditions (e.g., temperature, pH) for each step. Quin et al. [62] investigated the effect of a thermophilic and hyperthermophilic anaerobic treatment prior to mesophilic anaerobic digestion. In this context, a preceding hyperthermophilic step increased the hydrolytic activity of the engaged microorganisms and resulted in a higher organic solids reduction rate [63,64]. Thermophilic or hyperthermophilic conditions are further beneficial to pathogen removal [65,66,67]. During aerobic hyperthermophilic pretreatment, ***Geobacillus stearothermophilus***, for instance, turned out to be important for downstream processing [68].

Aerobic Pretreatment with Defined Fungal Cultures

Due to their powerful enzymatic capabilities, fungi offer great potential for biotechnological application. Lignocellulose-degrading fungi can be classified into white rot, brown rot, and soft rot fungi. White and soft rot fungi are known to attack lignin and, to a certain extent, also cellulose, while brown rots mainly target cellulose [69]. Nevertheless, white rot fungi are the preferred pretreatment organisms as they mandate highly efficient delignification enzyme equipment [70,71], with basidiomycetes being supremely effective [18,72]. The degree of delignification using white rots varies among applied fungi and depends on various factors such as pH and available N sources [70]. In contrast to anaerobic fungi, higher decomposition rates can be achieved due to aerobic conditions, which lead to higher energy yields for the engaged microorganisms and can therefore result in faster turnover rates. However, the application of fungi to pretreat substrates for anaerobic digestion is rather new [70]. Nevertheless, aerobic pretreatment using fungi was described for various species with diverse outcomes and is summarized in Table 1. A comparison concerning the effectivity and efficiency is rather difficult due to different experimental setups, substrates, inocula, inoculation rates, incubation times, and conditions, etc.; however, the table is intended to give a helpful overview. However, in more than 60 percent of the reviewed publications dealing with fungal pretreatment, white rot fungi were applied with pretreatment periods extending over approximately 3–4 weeks, which is notably longer than for other fungal pretreatments. In most cases, different organic waste fractions were used as substrate; to a minor extent, energy crops were also applied.

Table 1: Comparison of different fungal pretreatment strategies for enhanced biogas production

Pretreatment Organism (Type of Fungus [1])	**Substrate**	**Pretreatment Incubation Conditions [2]**	**Additional Information on Fungal Pretreatment Process [3]**	**Anaerobic Digestion Conditions [4]**	**Impact of Pretreatment on Substrate**	**Impact of Pretreatment on Biogas Production**	**Reference**
Ceriporiopsis subvermispora(wrf)	Japanese cedar wood	8 weeks a 28 °C b 70% c	orig, hyphal biomass grown on agar added, substrate supplemented with 10% wheat bran.	batch, mp, t 60	28% lignin removal in initial substrate, ~75%cleavage of ß-O-4 aryl ether	35% and 5% conversion of holocellulose to methane with and without pretreatment, respectively	[77]
Ceriporiopsis subvermispora(wrf)	Albizia biomass (forestry waste)	48 days a 28 °C b 60% c	e, autoc	batch, mp, ssAD, t 58	24% lignin removal of initial substrate, 4-fold increase in xylose and glucose production after 72 h of enzymatic hydrolysis	3.7-fold increase in methane production	[74]
Ceriposiopsis subvermispora(wrf)	Hazel and acacia branches, barley straw, and sugarcane bagasse	28 days a 28 °C b	e, autoc, grinded substrate	batch, mp	2- to 4-fold increase in enzymatic cellulose degradability for hazel and bagasse, decrease for straw and acacia	Increase of biomethane potential (BMP) for hazel (60%), loss of BMP for acacia (34%), straw and sugarcane bagasse	[73]
Phanerochaete chrysosporium(wrf)	Corn stover silage	30 days a 28 °C b Stable ambient d	f, autoc, washed substrate	batch, mp, t 30	39% lignin removal of initial substrate, improved degradation of substrate cell wall components	19.6–32.6% increase in methane production compared with controls	[86]

Fusarium sp. (wrf)	Paddy straw	10 days a 30 °C b 70% c	g, orig	batch, mp, t 35	17.1% decrease in lignin content, 10.8% decrease in silica content compared with controls	53.8% increase in biogas production	[78]
Trametes versicolor (wrf)	Corn silage	7 days a 27 °C b 70–80% c	g, orig	cont, mp, co-digestion with cow manure	70% increase in lignin degradation compared with control approach	Increased pH stability and biogas productivity, enhanced anaerobic degradation	[87]
Ceriporiopsis subvermispora(wrf)	Yard trimmings	30 days a 28 °C b 60% c	e, autoc	batch, mp, ssAD, t 40	20.9% degradation of initial lignin content	54% increase in methane production compared with controls, increased cellulose degradation	[83]
Polyporus brumalis (wrf)	Wheat straw	12.5 to 20 days a 20–30 °C b wet weight to initial solid ratio of 2.1 to 4.5	e, autoc, addition of metal supplement solution	batch, mp, t 57		Decrease in methane production compared with the control. Within fungal pretreatment, best methane production after 12.5 days incubation at 30 °C at 3.7 ww/ts ratio	[79]
Pleurotus ostreatus (wrf) ***Trichoderma reesei*** (srf [5])	Rice straw	20 days a 28 °C b 75% c	g, autoc	batch, mp, ssAD, t 45	33% lignin removal of initial substrate with wrf and 23.6% with brf Lignin-to-cellulose ratio after treatment: wrf 4.2, brf 2.88	20% increase in methane production with wrf and 21.7% decrease for brf treatment	[81]
CCHT-1 (wrf) ***Trichoderma reesei*** (srf [5])	Sisal leaf decortication residues	4 + 8 days a 28 °C b	g, orig, two fungal stages: wrf followed by brf	batch, mp, t 84	22.5%. decrease in neutral detergent fiber content, 21% increase in cellulose content	30–101% increase in biogas production compared to control	[75]

Sporotrichum*sp.** ***Aspergillus sp. ***Fusarium*** sp. ***Penicillium*** sp.	Orange processing waste	3 days a 30 °C b 65% c	g, orig, mixed culture pretreatment.	cont, mp, t 25	Reduction in inhibitory limonene content in the substrate.	Pretreatment leads to higher possible organic loading rates that improve overall productivity	[88]
Trichoderma viride (srf [5])	Organic waste	4 days a 25 °C b	e, orig	batch, tp, t 18	Increased cellulase activity during pretreatment compared with controls	Up to 400% increase in methane production compared with controls	[85]
Trichoderma viride (srf [5])	Organic waste	10 days a 22 °C b 70% c	f, orig	batch, tp, t 14	Increased cellulase and dehydrogenase activity compared to control	More than 2-fold increase in methane production	[84]

[1]**wrf**: white rot fungi; **brf**: brown rot fungi; **srf**: soft rot fungi; [2]**a**: incubation period, **b**: incubation temperature, **c**: moisture content in %, **d**: humidity in %; [3] inoculation with **e**: submerged fungal culture, **f**: fungal spores, **g**: autoclaved substrate overgrown by fungal mycelia; **autoc**: autoclaved substrate, **orig**: original, unmodified substrate; [4] **batch**: batch system, **cont**: continuous system, **ssAD**: solid-state anaerobic digestion, **mp**: mesophilic conditions, **tp**: thermophilic conditions, **t**: anaerobic incubation period or hydraulic retention time in days; [5]according to Klein and Eveleigh [90].

Liu et al. [73] found a positive impact of pretreatment on methane production potential using forest residues as co-substrate inoculated with ligninolytic ***Ceriporiopsis subvermispora***, but the effect was dependent on the basic substrate used. Ge et al. [74] incubated Albizia (silk tree) biomass, a forestry waste, with the same organism and were also able to increase the cumulative methane yield. From sisal (agave) leaf decortication residues pretreated with two fungal strains including ***Trichoderma reesei***, an increased methane production was observed [75]. In comparison with other strategies, Take et al. [76] found a positive effect of biological pretreatment with ***Cyathus stercoreus*** and ***Trametes hirsute*** on subsequent biogas production using cedar wood chips as substrate, whereas the pretreatment with ***Ceriporiopsis subvermispora*** positively influenced the methane yield in another study using the same substrate but with a nutritional supplement [77]. Phutela et al. [78] pretreated paddy straw with ***Fusarium***sp. and observed decreased lignin and cellulose contents in the substrate and an improved digestibility. However, the application of a facultative pathogenic microorganism seems to be problematic. Wheat straw was incubated with ***Polyporus brumalis*** BRFM 985, a white rot fungi, and in combination with metal amendment with the result that the treatment positively influenced the methane potential [79]. In another study by Vasmara et al. [80], 4- and 10-week incubation of wheat straw with 7 different fungal isolates was investigated regarding increased methane yields in a subsequent anaerobic digestion step. A positive effect was found enhancing the methane yield by an optimized treatment up to 16% for the 4-week and up to 37% for the 10-week pretreatment. ***Pleurotus ostreatus*** and ***Trichoderma reesei*** pretreatment of rice straw resulted in a 120% increase in methane yields in a study by Mustafa et al. [81]. Moreover, Mustafa et al. [82] found increased delignification and methane production from rice straw by pretreatment with the fungus ***Pleurotus ostreatus***. Yard trimmings were subjected to a pretreatment using the white rot ***Ceriporiopsis subvermispora*** [83]; the enhanced methane production was attributed to an increased delignification by the fungus. Mutschlechner et al. [84] inoculated a similar substrate containing high portions of grass and tree cut with ***Trichoderma viride*** and could secure increased methane production. This organism was also used to pretreat raw bio-waste with a positive effect on the methane production potential of the substrate [85]. ***Phanerochaete chrysosporium*** was used during solid-stage fermentation of corn stover to successfully enhance methane production in a subsequent anaerobic digestion step [86]. Tisma et al. [87] observed a positive effect of ***Trametes versicolor*** pretreatment on the biogas productivity of corn silage. Pretreating orange processing waste with strains of ***Sporotrichum***, ***Aspergillus***, ***Fusarium***, and ***Penicillium***, Srilatha et al. [88] observed a positive effect on biogas production and biogas potential. Mackul'ak et al. [89] inoculated sweet chestnut leaves and hay with the fungus ***Auricularia auricula-judae*** and observed an increase in methane productivity. Parthiba Karthikeyan et al. [48] found that biological pretreatments are not yet available for food wastes and demand urgent need for further research.

BY-PRODUCT FORMATION

Pretreatment of recalcitrant material enhances the availability of substrates but can also result in the formation of various inhibitory or even toxic substances. While biological pretreatments are less rigid, physico-chemical ones can be problematic. For example, although high glucose yields are achieved by acid treatment of lignocellulosic biomass, this procedure also leads to hydroxymethylfurfural (HMF) formation, one of the most unwanted pretreatment by-products [10,91]. Moreover, toxic and highly corrosive heavy metal ions like copper, nickel, chromium, and iron are released due to acid application [10]. In contrast, biological pretreatments apply milder conditions, tend to be less corrosive, and release fewer inhibitory substances [92]. Inhibitory substances introduced with the initial substrate can even be degraded during biological pretreatment, leading to an increase in substrate quality. In this context, lignocellulose pretreatment for biofuel production using the fungus ***Coniochaeta ligniaria***NRRL30616 led to a degradation of various undesired by-products including phenolic compounds, furfural, and HMF along with an assimilation of these inhibitory substances in the cells and/or a release of less toxic intermediates into the liquid phase [93]. In another study, pretreatment of oil palm mill effluent using thermophilic bacteria resulted in the removal of unwanted phenols coevally improving the anaerobic digestion performance [94,95].

Inhibitory substances can have an adverse effect on the engaged microorganisms involved in biological pretreatment and downstream anaerobic degradation [96] or on pretreatment facilities by corrosion [91,97]. Synergistic toxic effects are known for lignocellulosic hydrolysates, meaning that the toxicity of two or more toxic substances combined (on yeasts) can be higher than their sum [98].

However, a more profound knowledge of inhibitory substances is urgently needed for anaerobic degradation processes including ethanol/ biofuel production, waste water treatment, and, especially, biogas production.

CLOSING REMARKS—CONCLUSIONS

Various pretreatment strategies—physical, chemical, and biological—have been developed to overcome the inherent resistance of lignocellulose to anaerobic degradation. Biological pretreatment strategies, however, outcompete other pretreatments due to the application of milder conditions, and lower by-product formation and corrosiveness. The variety of applied techniques comprises micro-aerobic treatments, ensiling or composting, the separation of digestion stages, and pretreatments using various fungi. Fungal pretreatments have achieved particular success using various white, brown, and soft rot fungi, or a combination of these. Pretreatment processes applying white rot fungi from the genera Ceripoioposis, Phanerochaete, Fusarium, Trametes, Polyporus, and Pleurotus target cellulose as well as lignin, allowing the use of recalcitrant, second-generation substrates for biogas production. Therefore, biological pretreatment strategies offer great potential to improve the digestibility of different biogas substrates; however, detailed investigations of the mode of action, the application of different substrates, full-scale implementation, and possible by-product formation are still needed.

AUTHOR CONTRIBUTIONS

N.L., M.M., E.M.P. and R.M. contributed equally.

REFERENCES

1. Sawatdeenarunat, C.; Surendra, K.C.; Takara, D.; Oechsner, H.; Khanal, S.K. Anaerobic digestion of lignocellulosic biomass: Challenges and opportunities. ***Bioresour. Technol.*** 2015, ***178***, 178–186.
2. Ellabban, O.; Abu-Rub, H.; Blaabjerg, F. Renewable energy resources: Current status, future prospects and their enabling technology. ***Renew. Sustain. Energy Rev.*** 2014, ***39***, 748–764.
3. Martínez-Gutiérrez, E. Biogas production from different lignocellulosic biomass sources: Advances and perspectives. ***3 Biotech*** 2018, ***8***, 233–286.
4. Amin, F.R.; Khalid, H.; Zhang, H.; Rahman, S.U.; Zhang, R.; Liu, G.; Chen, C. Pretreatment methods of lignocellulosic biomass for anaerobic digestion. ***AMB Express*** 2017, *7*, 72.
5. Börjesson, P.; Mattiasson, B. Biogas as a resource-efficient vehicle fuel. ***Trends Biotechnol.*** 2008, ***26***, 7–13.
6. Zhen, G.; Lu, X.; Kato, H.; Zhao, Y.; Li, Y.-Y. Overview of pretreatment strategies for enhancing sewage sludge disintegration and subsequent anaerobic digestion: Current advances, full-scale application and future perspectives. ***Renew. Sustain. Energy Rev.*** 2017, ***69***, 559–577.
7. Wyman, C.E.; Dale, B.E.; Elander, R.T.; Holtzapple, M.; Ladisch, M.R.; Lee, Y.Y. Coordinated development of leading biomass pretreatment technologies. ***Bioresour. Technol.*** 2005, ***96***, 1959–1966.
8. Capolupo, L.; Faraco, V. Green methods of lignocellulose pretreatment for biorefinery development. ***Appl. Microbiol. Biotechnol.*** 2016, ***100***, 9451–9467.
9. Zieminski, K.; Romanowska, I.; Kowalska, M. Enzymatic pretreatment of lignocellulosic wastes to improve biogas production. ***Waste Manag.*** 2012, ***32***, 1131–1137.
10. Alvira, P.; Tomás-Pejó, E.; Ballesteros, M.; Negro, M.J. Pretreatment technologies for an efficient bioethanol production process based on enzymatic hydrolysis: A review. ***Bioresour. Technol.*** 2010, ***101***, 4851–4861.
11. Hermosilla, E.; Schalchli, H.; Mutis, A.; Diez, M.C. Combined effect of enzyme inducers and nitrate on selective lignin degradation in wheat straw by ***Ganoderma lobatum***. ***Environ. Sci. Pollut. Res. Int.*** 2017, ***24***, 21984–21996.
12. Dahadha, S.; Amin, Z.; Bazyar Lakeh, A.A.; Elbeshbishy, E. Evaluation

of Different Pretreatment Processes of Lignocellulosic Biomass for Enhanced Biomethane Production. ***Energy Fuels*** 2017, ***31***, 10335–10347.

13. Paudel, S.R.; Banjara, S.P.; Choi, O.K.; Park, K.Y.; Kim, Y.M.; Lee, J.W. Pretreatment of agricultural biomass for anaerobic digestion: Current state and challenges. ***Bioresour. Technol.*** 2017, ***245***, 1194–1205.
14. Howard, R.I.; Abotsi, E.; van Jansen Regensburg, E.L.; Howard, S. Lignocellulose biotechnology: Issues of bioconversion and enzyme production. ***Afr. J. Biotechnol.*** 2003, ***2***.
15. Prassad, S.; Singh, A.; Joshi, H.C. Ethanol as an alternative fuel from agricultural, industrial and urban residues. ***Resour. Conserv. Recycl.*** 2007, ***50***, 1–15.
16. Pérez, J.; Muñoz-Dorado, J.; de La Rubia, T.; Martínez, J. Biodegradation and biological treatments of cellulose, hemicellulose and lignin: An overview. ***Int. Microbiol. Off. J. Span. Soc. Microbiol.*** 2002, ***5***, 53–63.
17. Tuomela, M.; Vikman, M.; Hatakka, A.; Itvaara, M. Biodegradation of lignin in a compost environment: A review. ***Bioresour. Technol.*** 1999, ***72***, 169–178.
18. Sánchez, C. Lignocellulosic residues: Biodegradation and bioconversion by fungi. ***Biotechnol. Adv.*** 2009, ***27***, 185–194.
19. Lynd, L.R.; Weimer, P.J.; van Zyl, W.H.; Pretorius, I.S. Microbial cellulose utilization: Fundamentals and biotechnology. ***Microbiol. Mol. Biol. Rev.*** 2002, ***66***, 506–577.
20. Brown, R.M., Jr.; Saxena, I.M. Cellulose biosynthesis: A model for understanding the assembly of biopolymers. ***Plant Physiol. Biochem.*** 2000, ***38***, 57–67.
21. Atalla, R.H.; Hackney, J.M.; Uhlin, I.; Thompson, N.S. Hemicelluloses as structure regulators in the aggregation of native cellulose. ***Int. J. Biol. Macromol.*** 1993, ***15***, 109–112.
22. Fan, L.T.; Lee, Y.-H.; Beardmore, D.H. Mechanism of the enzymatic hydrolysis of cellulose: Effects of major structural features of cellulose on enzymatic hydrolysis. ***Biotechnol. Bioeng.*** 1980, ***22***, 177–182.
23. Malherbe, S.; Cloete, T.E. Lignocellulose biodegradation: Fundamentals and applications. ***Rev. Environ. Sci. Biotechnol.*** 2002, ***1***, 105–114.
24. Gerardi, M.H. ***The Microbiology of Anaerobic Digesters***; Wiley: New York, NY, USA, 2003.

25. Adney, W.S.; Rivard, C.J.; Ming, S.A.; Himmel, M.E. Anaerobic digestion of lignocellulosic biomass and wastes: Cellulases and related enzymes. ***Appl. Biochem. Biotechnol.*** 1991, ***30***, 165–183.
26. Dashtban, M.; Schraft, H.; Qin, W. Fungal Bioconversion of Lignocellulosic Residues; Opportunities & Perspectives. ***Int. J. Biol. Sci.*** 2009, ***5***, 578–595.
27. Shrestha, S.; Fonoll, X.; Khanal, S.K.; Raskin, L. Biological strategies for enhanced hydrolysis of lignocellulosic biomass during anaerobic digestion: Current status and future perspectives. ***Bioresour. Technol.*** 2017, ***245***, 1245–1257.
28. Desvaux, M. The cellulosome of Clostridium cellulolyticum. ***Enzym. Microbiol. Technol.*** 2005, ***37***, 373–385.
29. Vodovnik, M.; Marinsek-Logar, R. Cellulosomes—Promising supramolecular machines of anaerobic cellulolytic microorganisms. ***Acta Chem. Slov.*** 2010, ***57***, 767–774.
30. Leschine, S.B. Cellulose degradiation in anaerobic environment. ***Annu. Rev. Microbiol.*** 1995, ***49***, 399–426.
31. Orpin, C.G. Studies on rumen flagellate ***Neocallimastix frontalis***. ***J. Gen. Microbiol.*** 1975, ***91***, 249–262.
32. Teunissen, M.J.; Dencamp, H.J.M.O.; Orpin, C.G.; Veld, J.H.; Vogels, G.D. Comparsion of growth-characteristics of anaerobic fungi isolated from ruminant and non-ruminant herbivores during cultivation in a defined medium. ***J. Gen. Microbiol.*** 1991, ***137***, 1401–1408.
33. Dollhofer, V.; Podmirseg, S.M.; Callaghan, T.M.; Griffith, G.W.; Fliegerova, K. Anaerobic Fungi and Their Potential for Biogas Production. ***Adv. Biochem. Eng. Biotechnol.*** 2015, ***151***, 41–61.
34. Leis, S.; Dresch, P.; Peintner, U.; Fliegerová, K.; Sandbichler, A.M.; Insam, H.; Podmirseg, S.M. Finding a robust strain for biomethanation: Anaerobic fungi (Neocallimastigomycota) from the Alpine ibex (***Capra ibex***) and their associated methanogens. ***Anaerobe*** 2014, ***29***, 34–43.
35. Nakashimada, Y.; Srinivasan, K.; Murakami, M.; Nishio, N. Direct conversion of cellulose to methane by anaerobic fungus ***Neocallimastix frontalis*** and defined methanogens. ***Biotechnol. Lett.*** 2000, ***22***, 223–227.
36. Rabinovich, M.L.; Bolobova, A.V.; Vasilchenko, N. Fungal decomposition of natural aromatic structures and xenobiotics: A review. ***Appl. Biochem. Microbiol.*** 2004, ***40***, 1–17.

37. Manzanares, P.; Fajardo, S.; Martin, C. Production of ligniolytic activities when treating paper pulp effluents by Trymes versicolor. ***J. Biotechnol.*** 1995, ***43***, 125–132.

38. Brown, M.E.; Chang, M.C.Y. Exploring bacterial lignin degradation. ***Curr. Opin. Chem. Biol.*** 2014, ***19***, 1–7.

39. Ko, J.-J.; Shimizu, Y.; Ikeda, K.; Kim, S.-K.; Park, C.-H.; Matsui, S. Biodegradation of high molecular weight lignin under sulfate reducing conditions: Lignin degradability and degradation by-products. ***Bioresour. Technol.*** 2009, ***100***, 1622–1627.

40. Zheng, Y.; Zhao, J.; Xu, F.; Li, Y. Pretreatment of lignocellulosic biomass for enhanced biogas production. ***Prog. Energy Combust. Sci.*** 2014, ***42***, 35–53.

41. Bayer, E.A.; Lamed, R.; Himmel, M.E. The potential of cellulases and cellulosomes for cellulosic waste management. ***Curr. Opin. Biotechnol.*** 2007, ***18***, 237–245.

42. Sawatdeenarunat, C.; Sung, S.; Khanal, S.K. Enhanced volatile fatty acids production during anaerobic digestion of lignocellulosic biomass via micro-oxygenation. ***Bioresour. Technol.*** 2017, ***237***, 139–145.

43. Kaparaju, P.; Serrano, M.; Thomsen, A.B.; Kongjan, P.; Angelidaki, I. Bioethanol, biohydrogen and biogas production from wheat straw in a biorefinery concept. ***Bioresour. Technol.*** 2009, ***100***, 2562–2568.

44. Zhong, W.; Zhang, Z.; Luo, Y.; Sun, S.; Qiao, W.; Xiao, M. Effect of biological pretreatments in enhancing corn straw biogas production. ***Bioresour. Technol.*** 2011, ***102***, 11177–11182.

45. Zimbardi, F.; Viggiano, D.; Nanna, F.; Demichele, M.; Cuna, D.; Cardinale, G. Steam explosions of straw in batch and continous systems. ***Appl. Biochem. Biotechnol.*** 1999, ***77***, 117–125.

46. Mosier, N.; Wyman, C.; Dale, B.; Elander, R.; Lee, Y.Y.; Holtzapple, M.; Ladisch, M. Features of promising technologies for pretreatment of lignocellulosic biomass. ***Coord. Dev. Lead. Biomass Pretreat. Technol.*** 2005, ***96***, 673–686.

47. Karimi, K.; Taherzadeh, M.J. A critical review of analytical methods in pretreatment of lignocelluloses: Composition, imaging, and crystallinity. ***Bioresour. Technol.*** 2016, ***200***, 1008–1018.

48. Parthiba Karthikeyan, O.; Trably, E.; Mehariya, S.; Bernet, N.; Wong, J.W.C.; Carrere, H. Pretreatment of food waste for methane and hydrogen recovery: A review. ***Bioresour. Technol.*** 2018, ***249***, 1025–1039.

49. Jensen, T.R.; Lastra Milone, T.; Petersen, G.; Andersen, H.R. Accelerated anaerobic hydrolysis rates under a combination of intermittent aeration and anaerobic conditions. ***Water Sci. Technol. J. Int. Assoc. Water Pollut. Res.*** 2017, ***75***, 1944–1951.
50. Montalvo, S.; Vielma, S.; Borja, R.; Huiliñir, C.; Guerrero, L. Increase in biogas production in anaerobic sludge digestion by combining aerobic hydrolysis and addition of metallic wastes. ***Renew. Energy*** 2018, ***123***, 541–548.
51. Henze, M. ***Wastewater Treatment. Biological and Chemical Processes***, 3rd ed.; Springer: Berlin, Germany; London, UK, 2011.
52. Lim, J.W.; Wang, J.-Y. Enhanced hydrolysis and methane yield by applying microaeration pretreatment to the anaerobic co-digestion of brown water and food waste. ***Waste Manag.*** 2013, ***33***, 813–819.
53. Ni, Z.; Liu, J.; Zhang, M. Short-term pre-aeration applied to the dry anaerobic digestion of MSW, with a focus on the spectroscopic characteristics of dissolved organic matter. ***Chem. Eng. J.*** 2017, ***313***, 1222–1232.
54. Sarkar, O.; Venkata Mohan, S. Pre-aeration of food waste to augment acidogenic process at higher organic load: Valorizing biohydrogen, volatile fatty acids and biohythane. ***Bioresour. Technol.*** 2017, ***242***, 68–76.
55. Tsapekos, P.; Kougias, P.G.; Vasileiou, S.A.; Lyberatos, G.; Angelidaki, I. Effect of micro-aeration and inoculum type on the biodegradation of lignocellulosic substrate. ***Bioresour. Technol.*** 2017, ***225***, 246–253.
56. Xu, W.; Fu, S.; Yang, Z.; Lu, J.; Guo, R. Improved methane production from corn straw by microaerobic pretreatment with a pure bacteria system. ***Bioresour. Technol.*** 2018, ***259***, 18–23.
57. Cui, Z.; Shi, J.; Wan, C.; Li, Y. Comparison of alkaline- and fungi-assisted wet-storage of corn stover. ***Bioresour. Technol.*** 2012, ***109***, 98–104.
58. Heidarzadeh Vazifehkhoran, A.; Triolo, J.; Larsen, S.; Stefanek, K.; Sommer, S. Assessment of the Variability of Biogas Production from Sugar Beet Silage as Affected by Movement and Loss of the Produced Alcohols and Organic Acids. ***Energies*** 2016, ***9***, 368.
59. Prapinagsorn, W.; Sittijunda, S.; Reungsang, A. Co-Digestion of Napier Grass and Its Silage with Cow Dung for Methane Production. ***Energies*** 2017, ***10***, 1654.
60. Vervaeren, H.; Hostyn, K.; Ghekiere, G.; Willems, B. Biological

ensilage additives as pretreatment for maize to increase the biogas production. ***Renew. Energy*** 2010, ***35***, 2089–2093.

61. Wagner, A.O.; Janetschek, J.; Illmer, P. Using Digestate Compost as a Substrate for Anaerobic Digestion. ***Chem. Eng. Technol.*** 2018, ***41***, 747–754.
62. Qin, Y.; Higashimori, A.; Wu, L.-J.; Hojo, T.; Kubota, K.; Li, Y.-Y. Phase separation and microbial distribution in the hyperthermophilic-mesophilic-type temperature-phased anaerobic digestion (TPAD) of waste activated sludge (WAS). ***Bioresour. Technol.*** 2017, ***245***, 401–410.
63. Carrère, H.; Dumas, C.; Battimelli, A.; Batstone, D.J.; Delgenès, J.P.; Steyer, J.P.; Ferrer, I. Pretreatment methods to improve sludge anaerobic degradability: A review. ***J. Hazard. Mater.*** 2010, ***183***, 1–15.
64. Carrere, H.; Antonopoulou, G.; Affes, R.; Passos, F.; Battimelli, A.; Lyberatos, G.; Ferrer, I. Review of feedstock pretreatment strategies for improved anaerobic digestion: From lab-scale research to full-scale application. ***Pretreat. Biomass*** 2016, ***199***, 386–397.
65. Lu, J.; Gavala, H.N.; Skiadas, I.V.; Mladenovska, Z.; Ahring, B.K. Improving anaerobic sewage sludge digestion by implementation of a hyper-thermophilic prehydrolysis step. ***J. Environ. Manag.*** 2008, ***88***, 881–889.
66. Wagner, A.O.; Malin, C.; Gstraunthaler, G.; Illmer, P. Survival of selected pathogens in diluted sludge of a thermophilic waste treatment plant and in NaCl-solution under aerobic and anaerobic conditions. ***Waste Manag.*** 2009, ***29***, 425–429.
67. Wagner, A.O.; Gstraunthaler, G.; Illmer, P. Survival of bacterial pathogens during the thermophilic anaerobic digestion of biowaste: Laboratory experiments and in situ validation. ***Anaerobe*** 2008, ***14***, 181–183.
68. Hasegawa, S.; Shiota, N.; Katsura, K.; Akashi, A. Solubilization of organic sludge by thermophilic aerobic bacteria as a pretreatment for anaerobic digestion. ***Water Sci. Technol.*** 2000, ***41***.
69. Rodriguez, C.; Alaswad, A.; Benyounis, K.Y.; Olabi, A.G. Pretreatment techniques used in biogas production from grass. ***Renew. Sustain. Energy Rev.*** 2017, ***68***, 1193–1204.
70. Rouches, E.; Herpoël-Gimbert, I.; Steyer, J.P.; Carrere, H.

Improvement of anaerobic degradation by white-rot fungi pretreatment of lignocellulosic biomass: A review. ***Renew. Sustain. Energy Rev.*** 2016, ***59***, 179–198.

71. Hattaka, A.I. Pretreatment of wheat straw by white-rot fungi for enzymatic saccharification of cellulose. ***Appl. Microbiol. Biotechnol.*** 1983, ***82***, 131–137.
72. Isroi; Millati, R.; Syamsiah, S.; Niklasson, C.; Cahyanto, M.N.; Lundquist, K.; Taherzadeh, M.J. Biological pretreatment of lignocelluloses with white-rot fungi and its applications: A review. ***BioResources*** 2011, ***6***, 5224–5259.
73. Liu, X.; Hiligsmann, S.; Gourdon, R.; Bayard, R. Anaerobic digestion of lignocellulosic biomasses pretreated with ***Ceriporiopsis subvermispora***. ***J. Environ. Manag.*** 2017, ***193***, 154–162.
74. Ge, X.; Matsumoto, T.; Keith, L.; Li, Y. Fungal Pretreatment of Albizia Chips for Enhanced Biogas Production by Solid-State Anaerobic Digestion. ***Energy Fuels*** 2015, ***29***, 200–204.
75. Muthangya, M.; Mshandete, A.M.; Kivaisi, A.K. Two-stage fungal pre-treatment for improved biogas production from sisal leaf decortication residues. ***Int. J. Mol. Sci.*** 2009, ***10***, 4805–4815.
76. Take, H.; Andou, Y.; Nakamura, Y.; Kobayashi, F.; Kurimoto, Y.; Kuwahara, M. Production of methane gas from Japanese cedar chips pretreated by various delignification methods. ***Biochem. Eng. J.*** 2006, ***28***, 30–35.
77. Amirta, R.; Tanabe, T.; Watanabe, T.; Honda, Y.; Kuwahara, M.; Watanabe, T. Methane fermentation of Japanese cedar wood pretreated with a white rot fungus, ***Ceriporiopsis subvermispora***. ***J. Biotechnol.*** 2006, ***123***, 71–77.
78. Phutela, U.G.; Sahni, N. Effect of ***Fusarium*** sp. on Paddy Straw Digestibility and Biogas Production. ***J. Adv. Lab. Res. Biol.*** 2012, ***3***, 9–12.
79. Rouches, E.; Zhou, S.; Sergent, M.; Raouche, S.; Carrere, H. Influence of white-rot fungus ***Polyporus brumalis*** BRFM 985 culture conditions on the pretreatment efficiency for anaerobic digestion of wheat straw. ***Biomass Bioenergy*** 2018, ***110***, 75–79.
80. Vasmara, C.; Cianchetta, S.; Marchetti, R.; Galletti, S. Biogas produciton from wheat straw pre-treated with ligninolytic fungi and

co-digestion with pig slurry. ***Environ. Eng. Manag. J.*** 2015, ***14***, 1751–1760.

81. Mustafa, A.M.; Poulsen, T.G.; Sheng, K. Fungal pretreatment of rice straw with ***Pleurotus ostreatus*** and ***Trichoderma reesei*** to enhance methane production under solid-state anaerobic digestion. ***Appl. Energy***2016, ***180***, 661–671.
82. Mustafa, A.M.; Poulsen, T.G.; Xia, Y.; Sheng, K. Combinations of fungal and milling pretreatments for enhancing rice straw biogas production during solid-state anaerobic digestion. ***Bioresour. Technol.*** 2017, ***224***, 174–182.
83. Zhao, J.; Zheng, Y.; Li, Y. Fungal pretreatment of yard trimmings for enhancement of methane yield from solid-state anaerobic digestion. ***Bioresour. Technol.*** 2014, ***156***, 176–181.
84. Mutschlechner, M.; Illmer, P.; Wagner, A.O. Biological pre-treatment: Enhancing biogas production using the highly cellulolytic fungus ***Trichoderma viride***. ***Waste Manag.*** 2015, ***43***, 98–107.
85. Wagner, A.O.; Schwarzenauer, T.; Illmer, P. Improvement of methane generation capacity by aerobic pre-treatment of organic waste with a cellulolytic ***Trichoderma viride*** culture. ***J. Environ. Manag.*** 2013, ***129***, 357–360.
86. Liu, S.; Li, X.; Wu, S.; He, J.; Pang, C.; Deng, Y.; Dong, R. Fungal pretreatment by ***Phanerochaete chrysosporium*** for enhancement of biogas production from corn stover silage. ***Appl. Biochem. Biotechnol.***2014, ***174***, 1907–1918.
87. Tišma, M.; Planinić, M.; Bucić-Kojić, A.; Panjičko, M.; Zupančič, G.D.; Zelić, B. Corn silage fungal-based solid-state pretreatment for enhanced biogas production in anaerobic co-digestion with cow manure. ***Bioresour. Technol.*** 2018, ***253***, 220–226.
88. Srilatha, H.R.; Nand, K.; Sudhakar, K.; Madhukara, K. Fungal pretreatment of orange processing waste by solid-state fermentation for improved production of methane. ***Process Biochem.*** 1995, ***30***, 327–331.
89. Mackuľak, T.; Prousek, J.; Švorc, Ľ.; Drtil, M. Increase of biogas production from pretreated hay and leaves using wood-rotting fungi. ***Chem. Pap.*** 2012, ***66***, 3498.
90. Klein, D.; Eveleigh, D.E. Ecology of Trichoderma. In ***Trichoderma & Gliocladium, Basic Biology, Taxonomy and Genetics***; Kubicek, C.P.,

Harmna, G.E., Eds.; Taylor and Francis: London, UK, 1998; Volume 1, pp. 57–69. ISBN 0-7484-0572-0.

91. Jönsson, L.J.; Martín, C. Pretreatment of lignocellulose: Formation of inhibitory by-products and strategies for minimizing their effects. ***Bioresour. Technol.*** 2016, ***199***, 103–112.

92. Sindhu, R.; Binod, P.; Pandey, A. Biological pretreatment of lignocellulosic biomass—An overview. ***Bioresour. Technol.*** 2016, ***199***, 76–82.

93. Cao, G.; Ximenes, E.; Nichols, N.N.; Zhang, L.; Ladisch, M. Biological abatement of cellulase inhibitors. ***Bioresour. Technol.*** 2013, ***146***, 604–610.

94. Chantho, P.; Musikavong, C.; Suttinun, O. Removal of phenolic compounds from palm oil mill effluent by thermophilic ***Bacillus thermoleovorans*** strain A2 and their effect on anaerobic digestion. ***Int. Biodeterior. Biodegrad.*** 2016, ***115***, 293–301.

95. Borja, R.; Martin, A.; Alonso, V.; Garcia, I.; Banks, J. Influence of different aerobic pretreatments on the kinetics of anaerobic digestion of olive mill wastewater. ***Water Res.*** 1995, ***29***, 489–495.

96. Panagiotou, G.; Olsson, L. Effect of compounds released during pretreatment of wheat straw on microbial growth and enzymatic hydrolysis rates. ***Biotechnol. Bioeng.*** 2006, ***96***, 250–258.

97. Haghighi Mood, S.; Hossein Golfeshan, A.; Tabatabaei, M.; Salehi Jouzani, G.; Najafi, G.H.; Gholami, M.; Ardjmand, M. Lignocellulosic biomass to bioethanol, a comprehensive review with a focus on pretreatment. ***Renew. Sustain. Energy Rev.*** 2013, ***27***, 77–93.

98. Pienkos, P.T.; Zhang, M. Role of pretreatment and conditioning processes on toxicity of lignocellulosic biomass hydrolysates. ***Cellulose*** 2009, ***16***, 743–762.

11

Combined Biogas and Bioethanol Production: Opportunities and Challenges for Industrial Application

Alessandra Cesaro and Vincenzo Belgiorno

Sanitary Environmental Engineering Division (SEED), Department of Civil Engineering, University of Salerno, via Giovanni Paolo II, 84084 Fisciano (SA), Italy

ABSTRACT

In the last decades the increasing energy requirements along with the need to face the consequences of climate change have driven the search for renewable energy sources, in order to replace as much as possible the use of fossil fuels. In this context biomass has generated great interest as it can be converted into energy via several routes, including fermentation and anaerobic digestion. The former is the most common option to produce ethanol, which has been recognized as one of the leading candidates to substitute a large fraction of the liquid fuels produced from oil. As the economic competitiveness of bioethanol fermentation processes has to be enhanced in order to promote its wider implementation, the most recent trends are directed towards the use of fermentation by-products within anaerobic digestion. The integration

Citation: Cesaro A, Belgiorno V. Combined Biogas and Bioethanol Production: Opportunities and Challenges for Industrial Application. Energies. 2015; 8(8):8121-8144.

of both fermentation and anaerobic digestion, in a biorefinery concept, would allow the production of ethanol along with that of biogas, which can be used to produce heat and electricity, thus improving the overall energy balance. This work aims at reviewing the main studies on the combination of both bioethanol and biogas production processes, in order to highlight the strength and weakness of the integrated treatment for industrial application.

Keywords: anaerobic digestion, biorefinery, digestate, energy, fermentation, stillage

INTRODUCTION

Since the middle of the last century energy demand has significantly increased as a result of both industrial development and population growth. Currently the basic energy requirements are largely met by the use of fossil fuels [1]. It has been recently reported that 80% of the global consumption of primary energy, of which 58% is taken up by the transport sector, is dependent on the use of fossil fuels [2]. As the production of energy from these fuels is associated to both resource depletion and climate change issues, the identification of new and renewable sources of energy has become a matter of priority in many countries [3].

Renewable energy is a resource that is naturally regenerated over a short time scale and derived from the sun, either directly (such as thermal, photochemical, and photoelectric energy) or indirectly (such as wind, hydropower, and photosynthetic energy stored in biomass), as well as from other natural movements and mechanisms of the environment, such as geothermal and tidal energy [4].

Among these renewable sources, biomass represents an essential substitute for fossil fuels and it has attracted great attention worldwide, contributing between 9% and 13% to the total global energy supply [5].

The term biomass refers to the biodegradable fraction of products, wastes and residues from agriculture, forestry and related industries. It also includes dedicated energy crops and trees [6]. In Europe energy from biomass is the fastest growing renewable energy source: its production almost doubled over the last decade and it currently supplies 6% of the total primary energy [7].

Biomass is indeed a versatile resource that can be converted to energy via several routes [8]. Some of the most relevant factors in choosing a specific conversion route are the nature of the feedstock, the availability of a given technology and the demand for a specific energy product [9]. The most common techniques can be broadly divided into thermochemical and biochemical processes.

Thermochemical conversion technologies rely on the thermal breakdown of biomass into fuels and valuable chemicals [10]. The process can occur via direct combustion [11], pyrolysis [12,13], gasification [14], torrefaction [15,16] and liquefaction [17,18].

Biochemical processes are based on different chemical reactions biologically catalysed by microorganisms and/or enzymes that convert fermentable substrates into fuels or other high-value products [19]. The most common types of biochemical processes are fermentation, as defined by Kumar ***et al.*** [20], and anaerobic digestion. The former uses microorganisms and/or enzymes to process a fermentable substrate, converting it into recoverable products: currently the most required fermentation product is ethanol [21]. Anaerobic digestion can be defined as the biological degradation of organic matter into biogas, which is mainly composed of methane and carbon dioxide; the process, occurring in an oxygen-free environment, has been largely applied to treat several kinds of biodegradable materials [22,23].

Bioethanol, which is mainly produced from sugars and starch-rich materials, is one of the leading candidates to replace a large fraction of liquid fuels produced from oil. It can be used in mixtures up to 10% in gasoline without modification of the engines as well as in higher proportion in the so-called flexi-fuel vehicles, able to use up to 85% alcohol in mixtures with gasoline. Moreover, it is possible to use 100% bioethanol in specially designed engines [24,25,26]. Recently bioethanol has also been identified as one of the most promising bio-based raw materials for the chemical industry [27].

The economic feasibility of the bioethanol production process is tightly related to the market of its main by-product—distiller dried grains with solubles (DDGS)—as animal food: if the market of DDGS did not expand like that of ethanol, it would be an obstacle for the future large scale expansion of bioethanol production [28].

On the other hand, anaerobic digestion has been evaluated as one of the most efficient and sustainable processes for energy production [29]. Although it is not readily available as gaseous vehicle fuel, biogas can be easily utilized to produce heat and electricity, with minimal clean up after production [30].

The proper combination of both bioethanol and biogas production processes has been thus regarded as a suitable strategy to improve the competitiveness of fermentation plants, by producing both ethanol and biogas in a biorefinery concept. Such strategy pursues the combination of the material flows of different bio-industries, so that the residue from a bio-industry becomes the input of another one [31,32].

This work aims at reviewing main studies on the combination of bioethanol and biogas production processes. Both techniques are fully described and the feedstock characterization is discussed, in order to highlight its influence on the production of different types of energy. Main experimental results dealing with the use of fermentation by-products as anaerobic digestion substrates are compared, thus pointing out the strength and weakness of the combined process for industrial application.

BIOETHANOL PRODUCTION

Bioethanol has been identified as the most widely used biofuel for transportation worldwide [33]. Its biochemical production involves at least the following stages: hydrolysis, fermentation, distillation and dehydration [34]. The biological degradation of biomass begins during the hydrolysis of complex molecules into simple compounds, which are readily available for the fermentation step. This natural pathway, depending on both operating conditions and raw material, is carried out by microorganisms that ferment sugars into alcohol, lactic acid or other end products. Bioethanol obtained from a fermentation process requires further treatment: to this end, fractional distillation is typically implemented to separate ethanol from water based on their different volatilities. Because the boiling point of water (100°C) is higher than that of ethanol (78.3°C), the ethanol-water mixture is boiled so that ethanol is converted to vapour before water. Then water can be separated via a condensation procedure and ethanol distillate is recaptured at a concentration of 95% [35].

The efficient biochemical conversion of biomass to ethanol can be obtained via different technological options (Figure 1). Thus, the process success requires the combination of the most appropriate options at each step [36]. As a biological process, bioethanol fermentation is highly dependent on several factors, including pH, oxygen and temperature, which can greatly influence the specific rate of growth. The viability of cell populations, the specific rate of fermentation and the sugar uptake rate are all directly related to the desired medium condition [37]. Similarly both process configuration and yeast strain play a fundamental role in the definition of fementation yields [38,39,40,41], along with the feedstock characteristics.

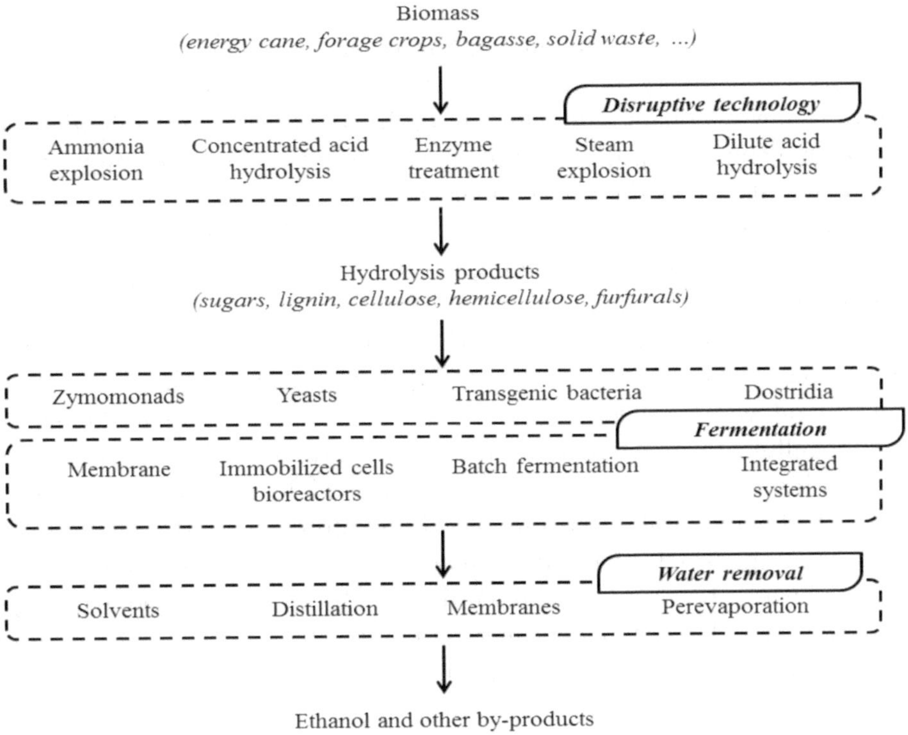

Figure 1: Biomass to ethanol conversion options, adapted by [36].

Bioethanol can be produced from the fermentation of different kinds of biomass. They are broadly classified as first generation materials (including starch and sugar) and second generation materials (lignocellulosic sugar). Currently the former are the most employed [42], but lignocellulosic materials, which are abundant and inexpensive, are gaining great interest [43].

The use of edible crops to produce the so-called first generation bioethanol has rightly been greatly criticized as an unsustainable technique, which poses the issues of both competition with food sources and disruption in the food-to-population ratio [44]. These concerns have driven the development of second-generation bioethanol that is derived from lignocellulosic wastes. More recently third generation bioethanol from alternative biomass such as microalgae and macroalgae has also generated significant interest [45]. The kind of substrate influences the whole bioethanol production process scheme.

For sucrose-based feedstocks such as molasses and sugarcane juice, processes like milling, pretreatment, hydrolysis, and detoxification are not necessary. Conversely most starch-based feedstocks and lignocellulose require some adequate pretreatment to be used for the production of fermentable sugar [46]. Lignocellulosic substrates are indeed composed of three materials: (i) cellulose, a sugar-based polymer; (ii) hemicelluloses, amorphous polysaccharides composed of different kinds of sugars, smaller than the ones of cellulose; (iii) lignin, which is a complex, aromatic polymer, mainly consisting of phenylpropane units [47]. Both hemicellulose and the amorphous portion of the cellulose can be easily hydrolysed, while crystalline cellulose is resistant to bioconversion [48]. This aspect explains the need for pretreatment, which is necessary to provide adequate structural modification of lignocellulosic substrates, in order to increase cellulose enzymatic digestibility yields. Once being accessible, cellulose is hydrolysed into glucose and hemicellulose is converted into both pentoses (C5 sugars) and hexoses (C6 sugars). Differently from anaerobic digestion, the fermentation efficiency varies for these sugar fractions [49,50], so that separate pathways have been studied to optimize techno-economics of ethanol production from lignocellulosic biomass utilization [51].

BIOGAS PRODUCTION

The biochemical degradation of organic substrates to biogas has been a promising option due to its economic and environmental feasibility compared to other biofuels [52]. Anaerobic digestion is a biological process occurring through defined steps. The first stage is the hydrolysis of complex organics (***i.e.***, carbohydrates, lipids and proteins) into soluble compounds; it is recognized as the process rate limiting step, so that it is often promoted by adequate substrate pretreatment [53].

The hydrolysed organic compounds (monomeric sugars, amino-acids and fatty acids) are utilized by acidogenic (acid-forming) bacteria for their growth and accumulate volatile fatty acids (VFAs) during the so-called acidogenesis step. VFAs are then transformed by acetogenic bacteria into simpler products, including acetic acid, hydrogen and carbon dioxide, during the step known as acetogenesis. These products are finally converted into methane by methanogens [54].

Anaerobic digestion yields depend on several factors [55,56], including temperature [57,58,59,60], organic loading rate [61,62,63,64] and substrate to inoculum ratio [65,66,67]. Under similar operating conditions, substrate characteristics play a key role in defining the amount of biogas that can be recovered by anaerobic digestion, as the theoretical gas yield varies with substrate composition in terms of carbohydrates, proteins and fats, as given in Table 1.

Substrates highly rich in lipids and easily-degradable carbohydrates exhibit higher methane potential than lignocellulosic materials [68]. These are more recalcitrant to the biological degradation, due to the presence of lignin (10%–25%), which is the main constituent along with cellulose (35%–50%) and hemicellulose (20%–35%). The amount of lignin, cellulose and hemicellulose varies between different biomass species as well as due to growth conditions and maturation [69]. However, it should be pointed out that the real methane production is generally higher than the one theoretically estimated, because part of the carbon dioxide, which is the main component of biogas along with methane, is solubilized in the digestate.

Table 1: Maximal theoretical methane yields for substrate constituents, adapted from [29]

Constituent	Methane yield [Nm^3/t_{TS}]
Carbohydrates	395–400
Raw protein	497
Raw fat	816–850
Lignin	0

Table 2 reports the composition of selected kinds of biomass in terms of lignocellulosic constituents as well as their methane potential. Despite the relevant presence of lignin, some of these biomasses show methane potentials comparable to those reported for energy crops [70], organic waste [71] and other substrates conventionally used within industrial scale anaerobic digestion facilities. This evidence indicates by far the great energetic potential of these organic materials.

Table 2: Composition and methane yields of selected kinds of biomass

Biomass	Cellulose [%]	Hemicellulose [%]	Lignin [%]	Methane potential [Nm^3_{CH4}/t_{VS}]	Ref.
Bagasse	38.2	27.1	20.2	NA	[69]
Barley straw	37.5–45	24.2–31.7	9.0–26.1	226.4	[69,72,73]
Corn stover	33.7–41.3	22.4	15.2–18.6	80–157.3	[69,74]
Eucalyptus	38.0–45.0	12.0–13.0	25.0–37.0	NA	[69]
Leaves	11.1–12.2	-	22.7–23.1	47–75	[69,74]
Maize	37.5	30.0	10.3	418	[70,75]
Pine wood	26.0–44.15	30.5	24.3–28.3	20	[69,74,76]
Rice straw	32.0–39.6	18.5–24.0	13.0–22.7	302	[69,70,77]
Sorghum	22.2	19.4	21.4	286–319	[69]
Sugarcane	25.0	16.8–32.2	12.0–31.4	278	[69,70,78]
Sunflower stalk	31.0	15.6	29.2	231–297	[69]
Switch grass	31.0–45.0	20.0–31.0	12.0–18.0	125	[69,74]
Wheat straw	32.3–37.9	21.2–25.8	8.3–23.4	130–290	[69,70,74,79]

THE COMBINATION OF BIOETHANOL AND BIOGAS PRODUCTION PROCESSES

Ethanol and biogas are industrially produced from the same kinds of easily degradable biomass resources [80], so that the question of producing either bioethanol or biogas from a given substrate has been already debated. Although the use of sugar-/starch-rich energy crops has been reported to be energetically effective to produce methane rather than ethanol, Moshi ***et al.*** [81] concluded that each of these products may be more appropriate than the other depending on both the area of application and existing infrastructure. However, if anaerobic digestion is a well-established technology, fermentation spread is often limited due to operating costs [82], mainly related to the high energy requirements as well as to the management of by-products. It has been estimated that a typical dry milling ethanol plant can roughly produce 40.2 L of ethanol and 32.1 kg of DDGS per 100 kg of corn used [83].

The proper combination of bioethanol and biogas production processes can be regarded as a suitable option to attain a greater bioenergy potential from biomass, thus improving both the mass and energy balances of single processes and, in particular, that of bioethanol production plants.

Stillage, which is the main fermentation by-product, is indeed characterized by high COD values, variable according to the feedstock used for bioethanol production (Table 3), as well as to the operating conditions of ethanol production process. Such COD values are comparable with the ones of other organic materials typically used as anaerobic digestion substrate [84,85,86], pointing out stillage organic potential to be transformed into methane.

Table 3: Characteristics of stillage from the fermentation of different biomasses

Biomass	pH [-]	TS [%]	COD [g/L]	Soluble COD [g/L]	Ref.
Bagasse	-	6.89	-	38.6	[87]
Barley	4.2	5.97	29.5	-	[88,89]
Beet molasses	4.3–6.7	-	55.5–147	-	[89]
Cassava	4.08	6.05	101.2	32.5	[90]
Corn stover	4.4	6.2-7.7	74.8	73.9	[91,92]
Maize	3.7	3.7–7.5	21.85–59.4	-	[93]
Sugarcane	3.9–4.6	-	31.5–45	-	[94]
Sweet sorghum	4.5	-	79.9	-	[89]
Wheat straw	3.6	12.0	150–154.2	61–76.8	[95,96]

Anaerobic digestion can thus convert over 50% of the by-product COD to biogas, which may be used as fuel within the plant [89]. Moreover, as stillage represents the partially degraded form of the biomass used within the fermentation process, the energy required for its treatment prior to anaerobic digestion can be significantly reduced [97]. Several studies reported the use of bioethanol residues for biogas production. Most of them were carried out to explore the methane potential of stillage originating from corn and grain bioethanol production processes, as reported in Table 4.

Table 4: Anaerobic processing of different bioethanol residues

Stillage	AD operating conditions	Highlights	Reference
Algae	Batch tests (60 mL working volume); T = 35 °C	Methane yields in the range 239–283 L/kg_{VS}, corresponding to energy potential 2.24 times higher than that from the ethanol produced in the main process.	[98]
Algae	Batch tests (100 mL working volume); T = 35 °C; Inoculum as 4.5 and 20 g_{VSS}/L	The inhibition by dilute-acid hydrolysis byproduct was overcome by increasing inoculum concentration. Under optimal tested condition, 84.8% methane conversion rate was achieved.	[99]
Cassava	CSTR batch tests (37 °C and 60 °C) converted into semi-continuous (37 °C)	Under OLR higher than 10 g_{VS}/ L d, the two-phase thermophilic CSTR was stable with hydrogen and methane yields of 56.6 mL_{H2}/g_{VS} and 249 mL_{CH4}/g_{VS}, respectively.	[100]
Cereal	Semi-continuous CSTR (5 L); T = 35 °C. Manure used as co-substrate (15% based on VS)	Anaerobic digestion of whole stillage turned to instability after 120 days of operation. Co-digestion with manure improved process stability and methane yield compared with theoretical values.	[101]
Corn	Anaerobic fluidized bed bioreactor (AFBR) employing zeolite as the carrier media; T = 37 °C	The AFBR showed up to 88% TCOD and 78% TSS removal at OLR of 29 kg_{COD}/m^3 d and HRT of 3.5 days. Methane production rates of up to 40 $L/L_{stillage}$ d was reached at the steady state.	[102]
Corn	BMP assays with alkalinity adjustment; T = 35 °C and 55 °C	Methane potential up to 1 L/g_{VS}, with the best performances under thermophilic conditions.	[103]

?	Semi continuous tests at SRT of 60, 45 and 35 days; T = 35 °C and 55 °C	Thermophilic digester was unable to cope with high organic loading rate at SRT of 60 days, whereas under mesophilic conditions, the digester was stable with 58 $L_{CH4}/kg_{stillage}$.	
Corn	BMP assays with alkalinity adjustment; T = 35 °C; ISR: 3.67–0.46 g/g	At the organic loading rate of 0.46 g/g, degradation was completed in 15–16 days. Methane potential in standard conditions was found in the range 401-458 mL/$g_{VS\ added}$, with organic removals between 76% and 94% in batch mode.	[104]
Corn	Batch test (250 mL); T = 35 °C; ISR: 0.5–3.0 g_{VS}/g_{VS}; Mechanically shaking provided	Thin stillage, treated by centrifugation and lime addition, resulted in rapid production of high levels of biogas (763 mL biogas/$g_{VS\ added}$), under the optimal ISR of 2.0 g_{VS}/g_{VS}.	[105]
Corn	Batch tests; T = 35 °C; Substrate/inoculum ratios: 4–8 g_{TCOD}/g_{VSS}	Two-stage process enabled 18.5% increase in total energy yield.	[106]
Corn	Batch tests (2.8 L working volume); T = 35 °C	Anaerobic digestion of corn stillage was able to improve the overall content utilization and extract a greater yield of lignocellulosic biomass compared to ethanol fermentation alone.	[107]
Grain	Continuous UASB reactor (working volume of 255 mL); HRT = 48 h; T = 55 °C	Maximum methane yield of 155 ml/g_{COD} was obtained at stillage mixtures with water of 25% (v/v) in the feed and at an OLR of 17.1 g_{COD}/L d.	[95]
Grain	CSTR (5 L), FBR (12.9 L) and ASBR (13 L); T = 38 °C; OLR up to 10 g_{VS}/L d	At HRT as low as 6 days, good performances can be achieved by continuous grain anaerobic digestion when trace elements are supplemented, despite the reactor system.	[108]

Grain	Semi continuous CSTR digester; T = 38 °C and 44 °C; OLR up to 6 g_{VS}/L d	Operation at 44 °C was the most successful strategy, resulting in up to 22% higher methane yield compared with the mesophilic reactor.	[109]
Grain	CSTR digester (12 L); T = 35 °C; OLR = 10 g_{VS}/L d; HRT = 7–8.5 days	The depletion of Fe and Ni resulted in a rapid accumulation of volatile fatty acids while Co and W seem to have a long-term effect.	[110]
Grain	Semi continuous CSTR digester (4 L working volume); T = 35 °C; OLR = 2.5 and 4 g_{VS}/L d	The solubility and speciation of supplemented Fe were controlled by precipitation of FeS(s) and formation of the aqueous complexes of Fe-sulfide and Fe-thiol.	[111]
Kitchen garbage	Semi continuous CSTR digester (8 L working volume) with biogas recirculation; T = 53 °C and 60 °C; OLR = 7 g_{VS}/L d	A biogas recirculation ratio of 150 facilitated stable digestion performance and biogas production, while the ammonia removal efficiency increased 1.23-fold when the temperature increased from 53 °C to 60 °C.	[112]
Sugarcane bagasse	BMP assay, 55 °C	Methane potential over 80 days was approximately 10 $mL_{CH4}/mL_{stillage}$, with 85% of ultimate methane yield produced in 50 days.	[87]

Corn and grain can be classified as starch-based materials. Their widespread use in biomass-to-ethanol processes [46,113] results in the availability of great amount of both corn- and grain-based stillage. Eskicioglu ***et al.***[103] evaluated the anaerobic potential of whole stillage from a dry-grind corn-based ethanol plant by continuous-flow digesters under both thermophilic and mesophilic conditions. They found that mesophilic digestion at 60 days of sludge retention time (SRT) provided a methane yield of 58 L/kg stillage, but the best performances were observed under thermophilic conditions. Further studies [104] pointed out that the ultimate methane yield of corn stillage was not affected by inoculum to substrate ratio (ISR), which influenced the content of methane in biogas: the higher was ISR, the lower was the percentage of methane in biogas. Similar results were obtained by Alkan-Ozkaynak and Karthikeyan [105]. They studied the effect of ISR in the range 0.5–3 g_{VS}/g_{VS} on the methane potential of centrifuged and lime-treated corn stillage. The rapid initial biogas production at all the ISR levels tested confirmed that the biomass was well acclimated to the treated-thin stillage. The maximum net biogas production followed then the decreasing order: ISR 1 > ISR 2 > ISR 0.5 > ISR 3. In reactors operating with an ISR of 0.5 g_{VS}/g_{VS}, despite the high initial VS amount added, no significant biogas production was observed after 2.5 days. This outcome was ascribed to the rapid formation of intermediate products, as indicated by the final pH values that were in the acidic range. The best performances both in terms of biogas production (763 $mL/g_{VS\ added}$) and organic matter removal (80.6% COD) were achieved with an ISR of 2 g_{VS}/g_{VS}. The study also highlighted that, under the same operating conditions, pretreated thin stillage resulted in higher biogas production than untreated samples. This evidence was attributed to the removal of some potentially toxic components, such as long-chain fatty acids.

Pretreatments represent by far the most widespread strategy to overcome the limits of the hydrolysis of complex organic molecules to soluble substrates. Their application has thus been broadly studied to enhance anaerobic digestion yields and several research attempts have also dealt with the pre-processing of substrates destined to bio-ethanol production. In this view, the work of Wang ***et al.*** [107] is particularly interesting. The authors treated chopped corn by steam explosion, performed at 2.0 MPa for 5 min. The feedstock was then dried at room temperature up to 92%TS and fed to a 2 L batch reactor for a simultaneous enzymatic saccharification fermentation process. The resulting stillage was used to perform both BMP assays and anaerobic digestion tests. Both pretreatment and hydrolytic enzyme action determined the release of easily digestible compounds. The ready bioavailability of these compounds promoted fast methane production, which reaches its peak value during the first 2 days. The overall product yield was 197 g ethanol and 96 g methane/kg corn stover: thus methane production involved 9.6% improvement in the conversion of corn stover to energy. Consistent results were obtained by using bioethanol fermentation residues from steam exploded oat straw as anaerobic digestion substrate [114]. Biogas production rate from the fermentation residues was found to be higher than the one obtained from the steam exploded oat straw that had not undergone the bioethanol production process. This evidence suggests that the ethanol process offered an additional pretreatment for the biogas process.

Pretreatments can enhance biological process yields by simplifying the molecular structure of organic substrates, as difficult-to degrade fractions obviously provide a lower contribution to either ethanol or biogas production. Tian ***et al.*** [87] tested the products of bagasse stillage screening for their methane potential and compared these results to the methane potential of the whole stillage. The filtrate, which is the stillage fraction passing through 0.5 mm screen, was found to contribute by 70% to the whole stillage methane potential. When fed to a 15 L semi-continuous anaerobic digester, stillage filtrate determined approximately 80% COD removal. Based on these results, it was estimated that the anaerobic digestion of bagasse stillage filtrate can cover 62% of the energy consumed by ethanol distillation. The main drawback of treating only stillage filtrate by anaerobic digestion lays in the need for a proper strategy to manage the filtration residues, which were found to have relatively low methane potential.

This aspect is particularly relevant when the feasibility of an integrated bioethanol/biogas production process is considered. The disposal of residues can indeed result in additional costs which could make the process itself not sustainable. The integrated bioethanol/anaerobic digestion process optimization not only in terms of net energy gain, but also with reference to mass flows is, therefore, fundamental. In this view Zhang ***et al.*** [115] used the stillage from pretreated cassava to feed a thermophilic UASB reactor with 10 L working volume. The resulting digestion liquid was centrifuged and the supernatant was put into a mesophilic UASB reactor (10 L working volume). The digestion liquid flowing out of the mesophilic UASB was finally mixed with the raw materials for next batch of ethanol fermentation. The introduction of the two-stage anaerobic process allowed full stillage recycling to the ethanol fermentation step, with the consequent optimization of the process mass balance. Experimental results showed that the presence of potentially inhibitory substances (***i.e.***, organic compounds, VFAs, ions and colorants) reached a relative steady state after 3–7 batches recycling, without producing any negative effect on ethanol fermentation.

The use of a two-stage anaerobic process is a further option to enhance the recovery of stillage energetic potential. The separation of the acidogenic step from the methanogenic one results in enhanced stability to the different groups of microorganisms as well as in a better process control [116]. Nasr ***et al.*** [106] compared one-stage and two-stage anaerobic digestion processes for the treatment of thin stillage. The separation of acidogenesis and methanogenesis allowed a maximum methane yield of 0.33 L/gCOD added and an overall increase of 18.5% in the energy yield.

In the context of stillage anaerobic digestion optimization, several studies have been performed to assess the influence of reactor configuration [95,100,102]. At full-scale excellent performances and high process stability were achieved by a FBR treating the stillage originating from sweet potato-shochu production [117].

However the comparison of Continuous Stirred Tank Reactor (CSTR), Fixed Bed Reactor (FBR), and Anaerobic Sequencing Batch Reactor (ASBR), operating under mesophilic conditions and with OLR up to 10 g/L d, showed no significant differences [108]. Even the CSTR system, which was different from the other kinds of reactor (Figure 2) as it run without biomass immobilization, was able to provide a stable process at HRT values lower than 10–14 days.

Process stability maintenance is the most important aspect when anaerobic digestion has to be applied on a full scale. This ensures high methane production and the proper economic feasibility of the process.

Bioethanol residues are usually characterized by a prevailing protein content. The anaerobic digestion of these residues can thus suffer from ammonia inhibition [118]. In order to face this drawback, the effect of increasing organic loading rate (up to 6.0 g_{VS}/L d) and simultaneously decreasing hydraulic retention time (down to 24 days) was investigated [109]. The substrate was a mixture of thin stillage and milled grain and it was characterized by an ammonia-nitrogen level ranging between 689 and 984 mg/kg. The anaerobic process was performed in continuously stirred tank laboratory reactors, operating at both 38 °C and 44 °C. The thermophilic process at low HRT proved to be the most successful.

Thermophilic operation would further improve the overall energy balance of the combined bioethanol/anaerobic digestion process because it would not require the cooling of the fed stillage, typically characterized by temperature values around 70 °C.

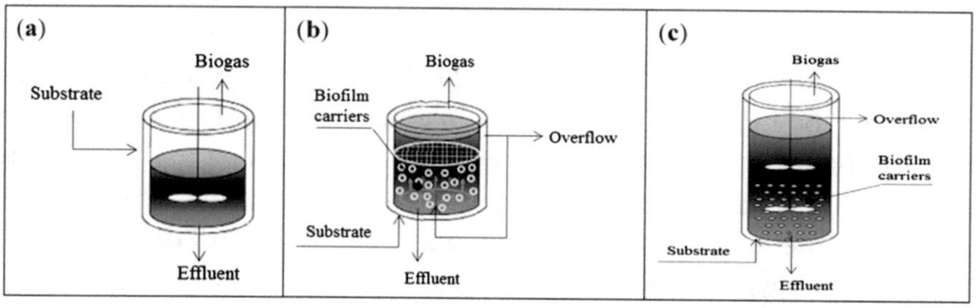

Figure 2: Reactor configuration, adapted from Schmidt ***et al.*** [108]. (a) Continuous Stirred Tank Reactor (CSTR). (b) Fixed Bed Reactor (FBR). (c) Anaerobic Sequencing Batch Reactor (ASBR).

A thermophilic process was investigated along with biogas recirculation as another option to optimize the anaerobic digestion of nitrogen-rich stillage [112], obtained from garbage bio-ethanol production. Anaerobic digestion tests were performed in an 8 L working volume CSTR, with OLR up to 7 g_{VS}/L d, at temperature of 53 °C or 60 °C. Biogas was recirculated into either the headspace of the reactor or the liquid phase of the reactor, varying the recirculation ratio in the range 10–150. Micro-aeration was also provided by continuously supplying ambient air at 3% of the amount of produced biogas in order to reduce the H_2S content in the biogas itself. The best performances were obtained in the system where biogas was recirculated at the highest investigated ratio (150) into the reactor headspace. At 60 °C process stability could be ensured even lowering the recirculation ratio to 10 or 50, thus suggesting that the simplification of system operation could be achieved.

Although ammonia inhibition is a quite common drawback of stillage anaerobic digestion, sulfur presence can also occur as a cause of instability. Sulphur-rich stillage, like the one originating from wheat, can be difficult to digest, as both H_2S production and metal sulfide precipitation in the digester can limit methanogenesis. However, it has been reported that even when sulphur concentrations are higher than metal ones, precipitated metal sulfides can act as a source of trace metals which do not inhibit methanogenesis. Following this statement, Schmidt ***et al.*** [110] developed a dosing strategy of trace metals in wheat stillage anaerobic digestion. To this end, CSTR systems were semi-continuously fed under mesophilic conditions, at an OLR of 10 g_{VS}/L d and HRT varying between 7 and 8.5 days. Trace elements were daily provided to the substrates. Results showed that cobalt (Co) and tungsten (W) had long-term effects on the anaerobic process. Therefore their dosing can be less frequent, with consequent economic benefits. Conversely iron (Fe) and nickel (Ni) depletion resulted in a rapid accumulation of volatile fatty acids and Fe deficiency was found to affect not only methane production but also propionate oxidation. Trace metals need to be properly supplied to limit inhibition phenomena, but their role within anaerobic process also depends on their solubility and bioavailability. Further studies demonstrated that these properties are influenced by complex reactions that involve both inorganic bisulfide (HS^-) and organic thiol moieties [111].

Supplying trace metals can thus play a key role in the anaerobic digestion of stillage. However from a technical point of view, the addition of these substances can prove to be not economically feasible, especially when considering full-scale operation.

A viable alternative can be recognized in stillage anaerobic co-digestion, which is the simultaneous digestion of stillage with two or more substrates [119]. Such process allows the macro- and micro-nutrients equilibrium, moisture balance and promotes the dilution of potentially inhibitory compounds, thus improving methane yields [120]. However the main reason of anaerobic co-digestion success has been recognized in the high OLR that can be applied to the digesters. In this view all kinds of feedstock can be digested within the same industrial process [121].

Whole stillage and manure co-digestion significantly improved biogas productivity and process stability. Methane yield was found to be higher than 20 m^3/m^3 of biomass, which is recognized as the level required for anaerobic digestion to be economically feasible [101]. Similar results were also obtained in other studies [90,122,123,124].

Although it is not as widespread as corn and grain, the use of algae as bioethanol feedstock is gaining increasing attention [125]. The algal ethanol production residue has been reported to be a suitable anaerobic digestion feedstock. It can retain up to 70% of algal biomass as energy basis and it can produce energy up to 2.24 times higher than that of ethanol production in the main process [98]. On the other hand, one of the by-products of algae utilization for ethanol production can limit the subsequent anaerobic digestion process. Inhibitory effects were also observed during the anaerobic digestion of red algal ethanol fermentation residues but, in this case, increasing inoculum concentration was found to be the solution [99].

The use of stillage as anaerobic digestion feedstock is thus a suitable strategy to improve the competitiveness of bioethanol fermentation plants. Several technological solutions can be considered but, among the issue to be solved, there is the management of the residues originating from the combined bioethanol/anaerobic digestion process, so that further research should be performed, in order to address the optimization of both mass and energy balances.

Literature analysis also pointed out that the combination of bioethanol and biogas production processes, at both research and industrial level, mainly rely on the use of corn and grain as feedstocks. Few reports deal with the use of lignocellulosic substrates for the production of both ethanol and methane [126,127,128], but up-scale attempts have already been provided in this field.

Based on previous lab-scale works [129], a biorefinery concept for the production of both second generation ethanol and biogas was developed at pilot-scale. The process included: (i) a wet explosion pretreatment of lignocellulosic substrate; (ii) its enzymatic hydrolysis and fermentation of the C6 sugars, after which lignin was separated, and (iii) a separate C5 sugar fermentation into ethanol. The residual stream from ethanol production became the input flow to an anaerobic process performed in a UASB reactor, under both mesophilic and thermophilic conditions. Results showed that, in the proposed biorefinery concept, the energy from stillage anaerobic digestion accounted for approximately 30% of the overall energy production; further process improvement could be achieved by removing suspended solids from the input stream to the UASB reactor [130]. The frame of this work is not only the simple combination, but rather the integration of both bioethanol and biogas processes, which pursues the optimization of biochemical conversion pathways by lignocellulosic matter fractioning.

Further development of this research resulted in the construction of one of the first integrated bioethanol/biogas plants, which was opened at Kalundborg, in Denmark in late 2009. It was designed to treat 30,000 t/year of biomass. The technology was used to produce a second generation ethanol from either the enzymatic fermentation of C6 sugars or the use of advanced yeast to convert a mixture of both C5 and C6 sugars. Ethanol yields were in the range 200–280 L/ton of dry straw: this product as well as the lignin separated from the incoming substrate were sold as fuels to other companies [131]. The development of this technology will be used, along with an innovative enzymatic product, within the Maabjerg Energy Concept, with the aim of producing second generation bio-ethanol, based on 300,000 tons of straw [132].

The Use of Digestate for Bioethanol Production

A further viable option for the combination of bioethanol fermentation and anaerobic digestion processes consists of the use of the anaerobic effluent as substrate for ethanol production. The anaerobic digestion residue, also known as digestate, is a mixture of partially degraded organic matter, microbial biomass and inorganic compounds. It is characterized by high potential fertiliser value due to its contents of nitrogen, phosphorous, potassium and micronutrients [133]. The agricultural use of the anaerobic digestate has not been clearly regulated yet [134], so that its aerobic processing has become a current practise [135,136] to produce compost, which is commonly used as fertilizer. However the application on soil of waste-produced compost may turn to be not sustainable, due to the risk associated to the presence of potentially toxic compounds [137]. This condition requires the identification of further options to manage anaerobic digestion residues.

Despite its potential as fertilizer, the digestate still contains undigested solids, which can be either used to recover additional methane [138] or to produce ethanol [139]. The use of digestate as bioethanol production substrate appears a suitable strategy to handle the anaerobic residues properly as well as to provide a competitive supply of biomass for biofuel production.

Teater ***et al.*** [140] compared the bioethanol potential of switch-grass, corn stover and anaerobic digestion fiber from a commercial CSTR, treating dairy manure. The digestion fiber was obtained as the solid digestate after liquid/solid separation. The substrates were pretreated using an autoclave at different sodium hydroxide concentrations (0.5–3 wt.%), two retention times (1–3 h), and two temperatures (120 °C and 130 °C). Although comparable ethanol conversion efficiencies were observed, the use of digestate could offer some advantages, including: (i) the possibility of removing the size reduction unit from the bioethanol process; (ii) a low presence of pentose, whose fermentation raises operating issues in bioethanol production process; (iii) the relevant quantity and year round availability of cattle manure that could solve the logistical storage problems associated with annual crops, commonly destined to bioethanol production.

The bioethanol potential of the fiber originating from the anaerobic co-digestion of corn stover and swine manure, at different ratios, was also investigated [141]. At a stover-to-manure ratio of 40:60 the highest performances were observed both in terms of methane (152 g/kg dry raw feed) and ethanol (50 g/kg dry raw feed) yields. The resulting net energy was estimated to be 5.5 MJ/kg of dry raw feed.

Anaerobic digestion effluent can also be used to replace freshwater and nutrients for bioethanol production: when co-fermented with wheat, ethanol concentration can be up to 18% higher in comparison to the production in freshwater. The enhanced bioethanol production was ascribed to the synergistic effect of nutrients, anaerobes, biochemical processing and enzymes in the anaerobic digestion effluent [142].

The outcomes of these studies are particularly interesting as they introduce further options to take greater advantage of the bioenergy potential of biomass feedstock, by integrating both anaerobic digestion and ethanol fermentation process under different operating conditions. However the number of reports is still limited and does not allow the generalization of results, which requires further studies.

CONCLUSIONS

Biomass is a versatile and abundant resource, which can be used to produce energy via different routes, including fermentation and anaerobic digestion. Several kinds of biomass can be used either to produce bioethanol or biogas and, although widely debated in scientific literature, it is not currently possible to state which is the best treatment option for a given substrate. To this end some boundary conditions, including the area of application and the presence of existing infrastructure, have to be previously identified.

As the transport sector accounts for the largest part of the world primary energy consumption, the production of bioethanol from biomass fermentation is gaining great interest. However the process spread on industrial scale is often limited by its low economic competitiveness with energy production processes based on the use of fossil fuels. Such aspect is mainly related to the high energy requirements to treat the biomass before the fermentation step as well as to the management of the stillage originating from the separation of ethanol concentrated solution from water.

This review highlighted that the combination of both bioethanol and anaerobic digestion has been identified as a valuable option to overcome these limits. The anaerobic processing of stillage can indeed solve the issue of managing bioethanol fermentation by-products while producing energy to fulfil the requirements of the facility. Moreover experimental results point out that fermentation acts as biomass pretreatment, so that the energy required to convert stillage into biogas via anaerobic digestion is lower than the one necessary to treat the whole substrate. The synergistic combination of both bioethanol and biogas production processes results in the optimization of the energy balance of single processes as well as in the possibility of taking greater advantage of biomass energetic potential.

Based on the latter statement the use of digestate as bioethanol fermentation substrate has also been studied. In this case the full exploitation of biomass energy was pursued along with the identification of a strategy to handle the anaerobic residues while providing a competitive supply of biomass for biofuel production. However further research is required in this field, with the aim of both enhancing knowledge on the weakness and strengths of the combined anaerobic digestion/fermentation process and identifying a proper option to manage bioethanol by-products from digestate fermentation. Conversely, the inverse process represents a successful treatment option.

The proposed review showed that anaerobic digestion can convert up to 80% of stillage COD into biogas. Organic loading rate higher than 10 g_{VS}/L d as well as hydraulic retention time as high as 6 days can be applied, but either operating strategies or system monitoring has to be considered to limit potentially inhibitory phenomena. In this context common issues are related to both ammonia and sulphur accumulation, which in turn depend on the use of nitrogen- or sulphur-rich stillage, respectively. Although the supply of trace metals can promote the correct development of the anaerobic process, it can turn to be not economically feasible, especially when considering full-scale operation. In order to ensure the stability of the anaerobic process, co-digestion could provide a proper solution. Further research is thus needed to identify the optimal combination of different substrates, taking into account not only the proper balance of chemical-physical properties, but also the simultaneous availability of substrates themselves.

Although both bioethanol fermentation and anaerobic digestion are reliable techniques, already common at full-scale, the integrated process feasibility for industrial application should be more deeply investigated. Reviewed literature highlights uncertain net energy gain as well as process stability issue, mainly dependent on the kind of feedstock. Further research should be thus addressed towards: (i) the characterization of different kinds of stillage according to its production process; (ii) the use of models to predict anaerobic digestion yields from stillage; (iii) the integrated bioethanol/biogas process investigation at larger scale, under continuous feed conditions, aimed at (iv) the development of both mass and energy balance.

Different considerations can be drawn when the use of lignocellulosic feedstocks is considered: the production of second generation ethanol, along with biogas and lignin, sold as fuels to power plants, has already been realised in Denmark. The effective integration of bioethanol and biogas processes was firstly performed in a demo-scale plant and then further developed as industrial project, pointing out the technical and economic feasibility of using waste biomass for a sustainable energy production.

ACKNOWLEDGMENTS

This work was performed at the Department of Civil Engineering of Salerno University under Research Contract n. Rep. 1609 prot. 63565.

AUTHOR CONTRIBUTIONS

Alessandra Cesaro carried out the analysis of scientific literature and prepared the manuscript; Vincenzo Belgiorno supervised the literature analysis and reviewed the final draft.

REFERENCES

1. Maity, S.K. Opportunities, recent trends and challenges of integrated biorefinery: Part I. ***Renew. Sustain. Energy Rev.*** 2015, ***43***, 1427–1445.
2. Nigam, P.S.; Singh, A. Production of liquid biofuels from renewable resources. ***Prog. Energy Combust. Sci.***2011, ***37***, 52–68.
3. Bazmi, A.A.; Zahedi, G. Sustainable energy systems: Role of optimization modeling techniques in power generation and supply—A review. ***Renew. Sustain. Energy Rev.*** 2011, ***15***, 3480–3500.
4. Cucchiella, F.; D'Adamo, I. Issue on supply chain of renewable energy. ***Energy Convers. Manag.*** 2013, ***76***, 774–780.
5. Long, H.; Li, X.; Wang, H.; Jia, J. Biomass resources and their bioenergy potential estimation: A review. ***Renew. Sustain. Energy Rev.*** 2013, ***26***, 344–352.
6. Angelis-Dimakis, A.; Biberacher, M.; Dominguez, J.; Fiorese, G.; Gadocha, S.; Gnansounou, E.; Guariso, G.; Kartalidis, A.; Panichelli, L.; Pinedo, I.; ***et al.*** Methods and tools to evaluate the availability of renewable energy sources. ***Renew. Sustain. Energy Rev.*** 2011, ***15***, 1182–1200.
7. Gabrielle, B.; Bamière, L.; Caldes, N.; De Cara, S.; Decocq, G.; Ferchaud, F.; Loyce, C.; Pelzer, E.; Perez, Y.; Wohlfahrt, J.; ***et al.*** Paving the way for sustainable bioenergy in Europe: Technological options and research avenues for large-scale biomass feedstock supply. ***Renew. Sustain. Energy Rev.*** 2014, ***33***, 11–25.
8. Chadwick, D.T.; McDonnell, K.P.; Brennan, L.P.; Fagan, C.C.; Everard, C.D. Evaluation of infrared techniques for the assessment of biomass and biofuel quality parameters and conversion technology processes: A review. ***Renew. Sustain. Energy Rev.*** 2014, ***30***, 672–681.
9. Fiorese, G.; Catenacci, M.; Bosetti, V.; Verdolini, E. The power of biomass: Experts disclose the potential for success of bioenergy technologies. ***Energy Policy*** 2014, ***65***, 94–114.
10. Tekin, K.; Karagöz, S.; Bektaş, S. A review of hydrothermal biomass processing. ***Renew. Sustain. Energy Rev.*** 2014, ***40***, 673–687.
11. Khan, A.A.; de Jong, W.; Jansens, P.J.; Spliethoff, H. Biomass combustion in fluidized bed boilers: Potential problems and remedies. ***Fuel Process. Technol.*** 2009, ***90***, 21–50.
12. Ward, J.; Rasul, M.G.; Bhuiya, M.M.K. Energy Recovery from Biomass by Fast Pyrolysis. ***Procedia Eng.***2014, ***90***, 669–674.

13. Abnisa, F.; Wan Daud, W.M.A. A review on co-pyrolysis of biomass: An optional technique to obtain a high-grade pyrolysis oil. ***Energy Convers. Manag.*** 2014, ***87***, 71–85.
14. Albertazzi, S.; Basile, F.; Brandin, J.; Einvall, J.; Hulteberg, C.; Fornasari, G.; Rosetti, V.; Sanati, M.; Trifirò, F.; Vaccari, A. The technical feasibility of biomass gasification for hydrogen production. ***Catal. Today*** 2005, ***106***, 297–300.
15. Batidzirai, B.; Mignot, A.P.R.; Schakel, W.B.; Junginger, H.M.; Faaij, A.P.C. Biomass torrefaction technology: Techno-economic status and future prospects. ***Energy*** 2013, ***62***, 196–214.
16. Van der Stelt, M.J.C.; Gerhauser, H.; Kiel, J.H.A.; Ptasinski, K.J. Biomass upgrading by torrefaction for the production of biofuels: A review. ***Biomass Bioenergy*** 2011, ***35***, 3748–3762.
17. Elliott, D.C.; Biller, P.; Ross, A.B.; Schmidt, A.J.; Jones, S.B. Hydrothermal liquefaction of biomass: Developments from batch to continuous process. ***Bioresour. Technol.*** 2015, ***178***, 147–156.
18. Brand, S.; Kim, J. Liquefaction of major lignocellulosic biomass constituents in supercritical ethanol. ***Energy***2015, ***80***, 64–74.
19. Srirangan, K.; Akawi, L.; Moo-Young, M.; Chou, C.P. Towards sustainable production of clean energy carriers from biomass resources. ***Appl. Energy*** 2012, ***100***, 172–186.
20. Kumar, A.; Kumar, N.; Baredar, P.; Shukla, A. A review on biomass energy resources, potential, conversion and policy in India. ***Renew. Sustain. Energy Rev.*** 2015, ***45***, 530–539.
21. Cherubini, F. The biorefinery concept: Using biomass instead of oil for producing energy and chemicals. ***Energy Convers. Manag.*** 2010, ***51***, 1412–1421.
22. Nallathambi Gunaseelan, V. Anaerobic digestion of biomass for methane production: A review. ***Biomass Bioenergy*** 1997, ***13***, 83–114.
23. Ward, A.J.; Lewis, D.M.; Green, F.B. Anaerobic digestion of algae biomass: A review. ***Algal Res.*** 2014, ***5***, 204–214.
24. Morales, M.; Quintero, J.; Conejeros, R.; Aroca, G. Life cycle assessment of lignocellulosic bioethanol: Environmental impacts and energy balance. ***Renew. Sustain. Energy Rev.*** 2015, ***42***, 1349–1361.
25. Manzetti, S.; Andersen, O. A review of emission products from bioethanol and its blends with gasoline. Background for new guidelines for emission control. ***Fuel*** 2015, ***140***, 293–301.

26. Yue, D.; You, F.; Snyder, S.W. Biomass-to-bioenergy and biofuel supply chain optimization: Overview, key issues and challenges. ***Comput. Chem. Eng.*** 2014, ***66***, 36–56.

27. Posada, J.A.; Patel, A.D.; Roes, A.; Blok, K.; Faaij, A.P.C.; Patel, M.K. Potential of bioethanol as a chemical building block for biorefineries: Preliminary sustainability assessment of 12 bioethanol-based products. ***Bioresour. Technol.*** 2013, ***135***, 490–499.

28. Sarkar, N.; Ghosh, S.K.; Bannerjee, S.; Aikat, K. Bioethanol production from agricultural wastes: An overview. ***Renew. Energy*** 2012, ***37***, 19–27.

29. Weiland, P. Biogas production: current state and perspectives. ***Appl. Microbiol. Biotechnol.*** 2010, ***85***, 849–860.

30. Oleskowicz-Popiel, P.; Kádár, Z.; Heiske, S.; Klein-Marcuschamer, D.; Simmons, B.A.; Blanch, H.W.; Schmidt, J.E. Co-production of ethanol, biogas, protein fodder and natural fertilizer in organic farming – Evaluation of a concept for a farm-scale biorefinery. ***Bioresour. Technol.*** 2012, ***104***, 440–446.

31. Parajuli, R.; Dalgaard, T.; Jørgensen, U.; Adamsen, A.P.S.; Knudsen, M.T.; Birkved, M.; Gylling, M.; Schjørring, J.K. Biorefining in the prevailing energy and materials crisis: A review of sustainable pathways for biorefinery value chains and sustainability assessment methodologies. ***Renew. Sustain. Energy Rev.*** 2015, ***43***, 244–263.

32. Martin, M.; Svensson, N.; Fonseca, J.; Eklund, M. Quantifying the environmental performance of integrated bioethanol and biogas production. ***Renew. Energy*** 2014, ***61***, 109–116.

33. Balat, M. Production of bioethanol from lignocellulosic materials via the biochemical pathway: A review. ***Energy Convers. Manag.*** 2011, ***52***, 858–875.

34. Scott, F.; Venturini, F.; Aroca, G.; Conejeros, R. Selection of process alternatives for lignocellulosic bioethanol production using a MILP approach. ***Bioresour. Technol.*** 2013, ***148***, 525–534.

35. Limayem, A.; Ricke, S.C. Lignocellulosic biomass for bioethanol production: Current perspectives, potential issues and future prospects. ***Prog. Energy Combust. Sci.*** 2012, ***38***, 449–467.

36. Saxena, R.C.; Adhikari, D.K.; Goyal, H.B. Biomass-based energy fuel through biochemical routes: A review. ***Renew. Sustain. Energy Rev.*** 2009, ***13***, 167–178.

37. Lin, Y.; Zhang, W.; Li, C.; Sakakibara, K.; Tanaka, S.; Kong, H. Factors affecting ethanol fermentation using Saccharomyces cerevisiae BY4742. ***Biomass Bioenergy*** 2012, ***47***, 395–401.

38. De Souza, C.J.A.; Costa, D.A.; Rodrigues, M.Q.R.B.; dos Santos, A.F.; Lopes, M.R.; Abrantes, A.B.P.; dos Santos Costa, P.; Silveira, W.B.; Passos, F.M.L.; Fietto, L.G. The influence of presaccharification, fermentation temperature and yeast strain on ethanol production from sugarcane bagasse. ***Bioresour. Technol.*** 2012, ***109***, 63–69.

39. Dragone, G.; Silva, D.P.; de Almeida e Silva, J.B. Factors influencing ethanol production rates at high-gravity brewing. ***LWT-Food Sci. Technol.*** 2004, ***37***, 797–802.

40. Wang, P.-M.; Zheng, D.-Q.; Chi, X.-Q.; Li, O.; Qian, C.-D.; Liu, T.-Z.; Zhang, X.-Y.; Du, F.-G.; Sun, P.-Y.; Qu, A.-M.; ***et al.*** Relationship of trehalose accumulation with ethanol fermentation in industrial Saccharomyces cerevisiae yeast strains. ***Bioresour. Technol.*** 2014, ***152***, 371–376.

41. Alkasrawi, M.; Abu Jrai, A.; Al-Muhtaseb, A.H. Simultaneous saccharification and fermentation process for ethanol production from steam-pretreated softwood: Recirculation of condensate streams. ***Chem. Eng. J.***2013, ***225***, 574–579.

42. Fujimoto, S.; Yanagida, T.; Nakaiwa, M.; Tatsumi, H.; Minowa, T. Pinch analysis for bioethanol production process from lignocellulosic biomass. ***Appl. Therm. Eng.*** 2011, ***31***, 3332–3336.

43. Haghighi Mood, S.; Hossein Golfeshan, A.; Tabatabaei, M.; Salehi Jouzani, G.; Najafi, G.H.; Gholami, M.; Ardjmand, M. Lignocellulosic biomass to bioethanol, a comprehensive review with a focus on pretreatment. ***Renew. Sustain. Energy Rev.*** 2013, ***27***, 77–93.

44. Tan, I.S.; Lee, K.T. Enzymatic hydrolysis and fermentation of seaweed solid wastes for bioethanol production: An optimization study. ***Energy*** 2014, ***78***, 53–62.

45. Doan, Q.C.; Moheimani, N.R.; Mastrangelo, A.J.; Lewis, D.M. Microalgal biomass for bioethanol fermentation: Implications for hypersaline systems with an industrial focus. ***Biomass Bioenergy*** 2012, ***46***, 79–88.

46. Vohra, M.; Manwar, J.; Manmode, R.; Padgilwar, S.; Patil, S. Bioethanol production: Feedstock and current technologies. ***J. Environ. Chem. Eng.*** 2014, ***2***, 573–584.

47. Carrier, M.; Loppinet-Serani, A.; Denux, D.; Lasnier, J.-M.; Ham-Pichavant, F.; Cansell, F.; Aymonier, C. Thermogravimetric analysis as a new method to determine the lignocellulosic composition of biomass. ***Biomass Bioenergy*** 2011, ***35***, 298–307.

48. Yang, L.; Xu, F.; Ge, X.; Li, Y. Challenges and strategies for solid-state anaerobic digestion of lignocellulosic biomass. ***Renew. Sustain. Energy Rev.*** 2015, ***44***, 824–834.

49. Bruni, E.; Jensen, A.P.; Pedersen, E.S.; Angelidaki, I. Anaerobic digestion of maize focusing on variety, harvest time and pretreatment. ***Appl. Energy*** 2010, ***87***, 2212–2217.

50. Khoo, H.H. Review of bio-conversion pathways of lignocellulose-to-ethanol: Sustainability assessment based on land footprint projections. ***Renew. Sustain. Energy Rev.*** 2015, ***46***, 100–119.

51. Mariano, A.P.; Dias, M.O.S.; Junqueira, T.L.; Cunha, M.P.; Bonomi, A.; Filho, R.M. Utilization of pentoses from sugarcane biomass: Techno-economics of biogas vs. butanol production. ***Bioresour. Technol.*** 2013, ***142***, 390–399.

52. Divya, D.; Gopinath, L.R.; Merlin Christy, P. A review on current aspects and diverse prospects for enhancing biogas production in sustainable means. ***Renew. Sustain. Energy Rev.*** 2015, ***42***, 690–699.

53. Cesaro, A.; Belgiorno, V. Pretreatment methods to improve anaerobic biodegradability of organic municipal solid waste fractions. ***Chem. Eng. J.*** 2014, ***240***, 24–37.

54. Kothari, R.; Pandey, A.K.; Kumar, S.; Tyagi, V.V.; Tyagi, S.K. Different aspects of dry anaerobic digestion for bio-energy: An overview. ***Renew. Sustain. Energy Rev.*** 2014, ***39***, 174–195.

55. Yan, Z.; Song, Z.; Li, D.; Yuan, Y.; Liu, X.; Zheng, T. The effects of initial substrate concentration, C/N ratio, and temperature on solid-state anaerobic digestion from composting rice straw. ***Bioresour. Technol.*** 2015, ***177***, 266–273.

56. Zhang, C.; Su, H.; Baeyens, J.; Tan, T. Reviewing the anaerobic digestion of food waste for biogas production. ***Renew. Sustain. Energy Rev.*** 2014, ***38***, 383–392.

57. El-Mashad, H.M.; Zeeman, G.; van Loon, W.K.P.; Bot, G.P.A.; Lettinga, G. Effect of temperature and temperature fluctuation on thermophilic anaerobic digestion of cattle manure. ***Bioresour. Technol.*** 2004, ***95***, 191–201.

58. Kim, J.K.; Oh, B.R.; Chun, Y.N.; Kim, S.W. Effects of temperature and hydraulic retention time on anaerobic digestion of food waste. ***J. Biosci. Bioeng.*** 2006, ***102***, 328–332.
59. Deng, L.; Yang, H.; Liu, G.; Zheng, D.; Chen, Z.; Liu, Y.; Pu, X.; Song, L.; Wang, Z.; Lei, Y. Kinetics of temperature effects and its significance to the heating strategy for anaerobic digestion of swine wastewater. ***Appl. Energy*** 2014, ***134***, 349–355.
60. Chae, K.J.; Jang, A.; Yim, S.K.; Kim, I.S. The effects of digestion temperature and temperature shock on the biogas yields from the mesophilic anaerobic digestion of swine manure. ***Bioresour. Technol.*** 2008, ***99***, 1–6.
61. Noutsopoulos, C.; Mamais, D.; Antoniou, K.; Avramides, C.; Oikonomopoulos, P.; Fountoulakis, I. Anaerobic co-digestion of grease sludge and sewage sludge: The effect of organic loading and grease sludge content. ***Bioresour. Technol.*** 2013, ***131***, 452–459.
62. Gou, C.; Yang, Z.; Huang, J.; Wang, H.; Xu, H.; Wang, L. Effects of temperature and organic loading rate on the performance and microbial community of anaerobic co-digestion of waste activated sludge and food waste. ***Chemosphere*** 2014, ***105***, 146–151.
63. Esposito, G.; Frunzo, L.; Panico, A.; Pirozzi, F. Modelling the effect of the OLR and OFMSW particle size on the performances of an anaerobic co-digestion reactor. ***Process Biochem.*** 2011, ***46***, 557–565.
64. Wijekoon, K.C.; Visvanathan, C.; Abeynayaka, A. Effect of organic loading rate on VFA production, organic matter removal and microbial activity of a two-stage thermophilic anaerobic membrane bioreactor. ***Bioresour. Technol.*** 2011, ***102***, 5353–5360.
65. Motte, J.-C.; Escudié, R.; Bernet, N.; Delgenes, J.-P.; Steyer, J.-P.; Dumas, C. Dynamic effect of total solid content, low substrate/inoculum ratio and particle size on solid-state anaerobic digestion. ***Bioresour. Technol.***2013, ***144***, 141–148.
66. Kafle, G.K.; Bhattarai, S.; Kim, S.H.; Chen, L. Effect of feed to microbe ratios on anaerobic digestion of Chinese cabbage waste under mesophilic and thermophilic conditions: Biogas potential and kinetic study. ***J. Environ. Manage.*** 2014, ***133***, 293–301.
67. Kawai, M.; Nagao, N.; Tajima, N.; Niwa, C.; Matsuyama, T.; Toda, T. The effect of the labile organic fraction in food waste and the substrate/inoculum ratio on anaerobic digestion for a reliable methane

yield. ***Bioresour. Technol.*** 2014, ***157***, 174–180.

68. Labatut, R.A.; Angenent, L.T.; Scott, N.R. Biochemical methane potential and biodegradability of complex organic substrates. ***Bioresour. Technol.*** 2011, ***102***, 2255–2264.
69. Sawatdeenarunat, C.; Surendra, K.C.; Takara, D.; Oechsner, H.; Khanal, S.K. Anaerobic digestion of lignocellulosic biomass: Challenges and opportunities. ***Bioresour. Technol.*** 2015, ***178***, 178–186.
70. Chandra, R.; Takeuchi, H.; Hasegawa, T. Methane production from lignocellulosic agricultural crop wastes: A review in context to second generation of biofuel production. ***Renew. Sustain. Energy Rev.*** 2012, ***16***, 1462–1476.
71. Davidsson, Å.; Gruvberger, C.; Christensen, T.H.; Hansen, T.L.; Jansen, J.C. Methane yield in source-sorted organic fraction of municipal solid waste. ***Waste Manag.*** 2007, ***27***, 406–414.
72. Iroba, K.L.; Tabil, L.G.; Sokhansanj, S.; Dumonceaux, T. Pretreatment and fractionation of barley straw using steam explosion at low severity factor. ***Biomass Bioenergy*** 2014, ***66***, 286–300.
73. Sapci, Z. The effect of microwave pretreatment on biogas production from agricultural straws. ***Bioresour. Technol.*** 2013, ***128***, 487–494.
74. Xu, F.; Wang, Z.-W.; Li, Y. Predicting the methane yield of lignocellulosic biomass in mesophilic solid-state anaerobic digestion based on feedstock characteristics and process parameters. ***Bioresour. Technol.*** 2014, ***173***, 168–176.
75. Mayer, F.; Gerin, P.A.; Noo, A.; Foucart, G.; Flammang, J.; Lemaigre, S.; Sinnaeve, G.; Dardenne, P.; Delfosse, P. Assessment of factors influencing the biomethane yield of maize silages. ***Bioresour. Technol.***2014, ***153***, 260–268.
76. Kačík, F.; Šmíra, P.; Kačíková, D.; Veľková, V.; Nasswettrová, A.; Vacek, V. Chemical alterations of pine wood saccharides during heat sterilisation. ***Carbohydr. Polym.*** 2015, ***117***, 681–686.
77. Gao, J.; Chen, L.; Yuan, K.; Huang, H.; Yan, Z. Ionic liquid pretreatment to enhance the anaerobic digestion of lignocellulosic biomass. ***Bioresour. Technol.*** 2013, ***150***, 352–358.
78. Szczerbowski, D.; Pitarelo, A.P.; Zandoná Filho, A.; Ramos, L.P. Sugarcane biomass for biorefineries: Comparative composition of carbohydrate and non-carbohydrate components of bagasse and straw. ***Carbohydr. Polym.*** 2014, ***114***, 95–101.

79. Ji, W.; Shen, Z.; Wen, Y. Hydrolysis of wheat straw by dilute sulfuric acid in a continuous mode. ***Chem. Eng. J.*** 2015, ***260***, 20–27.
80. Bateni, H.; Karimi, K.; Zamani, A.; Benakashani, F. Castor plant for biodiesel, biogas, and ethanol production with a biorefinery processing perspective. ***Appl. Energy*** 2014, ***136***, 14–22.
81. Moshi, A.P.; Crespo, C.F.; Badshah, M.; Hosea, K.M.M.; Mshandete, A.M.; Elisante, E.; Mattiasson, B. Characterisation and evaluation of a novel feedstock, Manihot glaziovii, Muell. Arg, for production of bioenergy carriers: Bioethanol and biogas. ***Bioresour. Technol.*** 2014, ***172***, 58–67.
82. Gonela, V.; Zhang, J. Design of the optimal industrial symbiosis system to improve bioethanol production. ***J. Clean. Prod.*** 2014, ***64***, 513–534.
83. Taheripour, F.; Hertel, T.W.; Tyner, W.E.; Beckman, J.F.; Birur, D.K. Biofuels and their by-products: Global economic and environmental implications. ***Biomass Bioenergy*** 2010, ***34***, 278–289.
84. Latif, M.A.; Mehta, C.M.; Batstone, D.J. Low pH anaerobic digestion of waste activated sludge for enhanced phosphorous release. ***Water Res.*** 2015, ***81***, 288–293.
85. Riggio, V.; Comino, E.; Rosso, M. Energy production from anaerobic co-digestion processing of cow slurry, olive pomace and apple pulp. ***Renew. Energy*** 2015, ***83***, 1043–1049.
86. Maspolim, Y.; Zhou, Y.; Guo, C.; Xiao, K.; Ng, W.J. Comparison of single-stage and two-phase anaerobic sludge digestion systems–Performance and microbial community dynamics. ***Chemosphere*** 2015, ***140***, 54–62.
87. Tian, Z.; Mohan, G.R.; Ingram, L.; Pullammanappallil, P. Anaerobic digestion for treatment of stillage from cellulosic bioethanol production. ***Bioresour. Technol.*** 2013, ***144***, 387–395.
88. Mustafa, A.F.; McKinnon, J.J.; Christensen, D.A. Chemical characterization and in vitro crude protein degradability of thin stillage derived from barley- and wheat-based ethanol production. ***Anim. Feed Sci. Technol.*** 1999, ***80***, 247–256.
89. Wilkie, A.C.; Riedesel, K.J.; Owens, J.M. Stillage characterization and anaerobic treatment of ethanol stillage from conventional and cellulosic feedstocks. ***Biomass Bioenergy*** 2000, ***19***, 63–102.
90. Wang, W.; Xie, L.; Luo, G.; Zhou, Q. Enhanced fermentative hydrogen production from cassava stillage by co-digestion: The effects of different co-substrates. ***Int. J. Hydrog. Energy*** 2013, ***38***, 6980–6988.

91. Kim, Y.; Mosier, N.S.; Hendrickson, R.; Ezeji, T.; Blaschek, H.; Dien, B.; Cotta, M.; Dale, B.; Ladisch, M.R. Composition of corn dry-grind ethanol by-products: DDGS, wet cake, and thin stillage. ***Bioresour. Technol.***2008, ***99***, 5165–5176.

92. Wang, K.; Zhang, J.; Tang, L.; Zhang, H.; Zhang, G.; Yang, X.; Liu, P.; Mao, Z. Establishment and assessment of a novel cleaner production process of corn grain fuel ethanol. ***Bioresour. Technol.*** 2013, ***148***, 453–460.

93. Krzywonos, M.; Cibis, E.; Miśkiewicz, T.; Ryznar-Luty, A. Utilization and biodegradation of starch stillage (distillery wastewater). ***Electron. J. Biotechnol.*** 2009, ***12***.

94. Moraes, B.S.; Zaiat, M.; Bonomi, A. Anaerobic digestion of vinasse from sugarcane ethanol production in Brazil: Challenges and perspectives. ***Renew. Sustain. Energy Rev.*** 2015, ***44***, 888–903.

95. Kaparaju, P.; Serrano, M.; Angelidaki, I. Optimization of biogas production from wheat straw stillage in UASB reactor. ***Appl. Energy*** 2010, ***87***, 3779–3783.

96. Krzywonos, M.; Cibis, E.; Ryznar-Luty, A.; Miśkiewicz, T.; Borowiak, D. Aerobic biodegradation of wheat stillage (distillery wastewater) at an elevated temperature—Effect of solids separation. ***Biochem. Eng. J.***2010, ***49***, 1–6.

97. Cesaro, A.; Velten, S.; Belgiorno, V.; Kuchta, K. Enhanced anaerobic digestion by ultrasonic pretreatment of organic residues for energy production. ***J. Clean. Prod.*** 2014, *74*, 119–124.

98. Park, J.-H.; Yoon, J.-J.; Park, H.-D.; Lim, D.J.; Kim, S.-H. Anaerobic digestibility of algal bioethanol residue. ***Bioresour. Technol.*** 2012, ***113***, 78–82.

99. Park, J.-H.; Kim, S.-H.; Park, H.-D.; Lim, D.J.; Yoon, J.-J. Feasibility of anaerobic digestion from bioethanol fermentation residue. ***Bioresour. Technol.*** 2013, ***141***, 177–183.

100. Luo, G.; Xie, L.; Zou, Z.; Wang, W.; Zhou, Q.; Shim, H. Anaerobic treatment of cassava stillage for hydrogen and methane production in continuously stirred tank reactor (CSTR) under high organic loading rate (OLR). ***Int. J. Hydrog. Energy*** 2010, ***35***, 11733–11737.

101. Westerholm, M.; Hansson, M.; Schnürer, A. Improved biogas production from whole stillage by co-digestion with cattle manure. ***Bioresour. Technol.*** 2012, ***114***, 314–319.

102. Andalib, M.; Hafez, H.; Elbeshbishy, E.; Nakhla, G.; Zhu, J. Treatment of thin stillage in a high-rate anaerobic fluidized bed bioreactor (AFBR). ***Bioresour. Technol.*** 2012, ***121***, 411–418.

103. Eskicioglu, C.; Kennedy, K.J.; Marin, J.; Strehler, B. Anaerobic digestion of whole stillage from dry-grind corn ethanol plant under mesophilic and thermophilic conditions. ***Bioresour. Technol.*** 2011, ***102***, 1079–1086.

104. Eskicioglu, C.; Ghorbani, M. Effect of inoculum/substrate ratio on mesophilic anaerobic digestion of bioethanol plant whole stillage in batch mode. ***Process Biochem.*** 2011, ***46***, 1682–1687.

105. Alkan-Ozkaynak, A.; Karthikeyan, K.G. Anaerobic digestion of thin stillage for energy recovery and water reuse in corn-ethanol plants. ***Bioresour. Technol.*** 2011, ***102***, 9891–9896.

106. Nasr, N.; Elbeshbishy, E.; Hafez, H.; Nakhla, G.; Hesham El Naggar, M. Comparative assessment of single-stage and two-stage anaerobic digestion for the treatment of thin stillage. ***Bioresour. Technol.*** 2012, ***111***, 122–126.

107. Wang, Z.; Lv, Z.; Du, J.; Mo, C.; Yang, X.; Tian, S. Combined process for ethanol fermentation at high-solids loading and biogas digestion from unwashed steam-exploded corn stover. ***Bioresour. Technol.*** 2014, ***166***, 282–287.

108. Schmidt, T.; Pröter, J.; Scholwin, F.; Nelles, M. Anaerobic digestion of grain stillage at high organic loading rates in three different reactor systems. ***Biomass Bioenergy*** 2013, ***55***, 285–290.

109. Moestedt, J.; Nordell, E.; Schnürer, A. Comparison of operating strategies for increased biogas production from thin stillage. ***J. Biotechnol.*** 2014, ***175***, 22–30.

110. Schmidt, T.; Nelles, M.; Scholwin, F.; Pröter, J. Trace element supplementation in the biogas production from wheat stillage – Optimization of metal dosing. ***Bioresour. Technol.*** 2014, ***168***, 80–85.

111. Shakeri Yekta, S.; Lindmark, A.; Skyllberg, U.; Danielsson, Å.; Svensson, B.H. Importance of reduced sulfur for the equilibrium chemistry and kinetics of Fe(II), Co(II) and Ni(II) supplemented to semi-continuous stirred tank biogas reactors fed with stillage. ***J. Hazard. Mater.*** 2014, ***269***, 83–88.

112. Sun, Z.-Y.; Yamaji, S.; Cheng, Q.-S.; Yang, L.; Tang, Y.-Q.; Kida, K. Simultaneous decrease in ammonia and hydrogen sulfide inhibition

during the thermophilic anaerobic digestion of protein-rich stillage by biogas recirculation and air supply at 60 °C. ***Process Biochem.*** 2014, ***49***, 2214–2219.

113. Patni, N.; Pillai, S.G.; Dwivedi, A.H. Wheat as a Promising Substitute of Corn for Bioethanol Production. ***Procedia Eng.*** 2013, ***51***, 355–362.
114. Dererie, D.Y.; Trobro, S.; Momeni, M.H.; Hansson, H.; Blomqvist, J.; Passoth, V.; Schnürer, A.; Sandgren, M.; Ståhlberg, J. Improved bio-energy yields via sequential ethanol fermentation and biogas digestion of steam exploded oat straw. ***Bioresour. Technol.*** 2011, ***102***, 4449–4455.
115. Zhang, Q.-H.; Lu, X.; Tang, L.; Mao, Z.-G.; Zhang, J.-H.; Zhang, H.-J.; Sun, F.-B. A novel full recycling process through two-stage anaerobic treatment of distillery wastewater for bioethanol production from cassava. ***J. Hazard. Mater.*** 2010, ***179***, 635–641.
116. Demirel, B.; Yenigün, O. Two-phase anaerobic digestion processes: A review. ***J. Chem. Technol. Biotechnol.***2002, *77*, 743–755.
117. Kobayashi, T.; Tang, Y.; Urakami, T.; Morimura, S.; Kida, K. Digestion performance and microbial community in full-scale methane fermentation of stillage from sweet potato-shochu production. ***J. Environ. Sci.*** 2014, ***26***, 423–431.
118. Moestedt, J.; Påledal, S.N.; Schnürer, A.; Nordell, E. Biogas Production from Thin Stillage on an Industrial Scale—Experience and Optimisation. ***Energies*** 2013, ***6***, 5642–5655.
119. García-Gen, S.; Rodríguez, J.; Lema, J.M. Optimisation of substrate blends in anaerobic co-digestion using adaptive linear programming. ***Bioresour. Technol.*** 2014, ***173***, 159–167.
120. Hamawand, I. Anaerobic digestion process and bio-energy in meat industry: A review and a potential. ***Renew. Sustain. Energy Rev.*** 2015, ***44***, 37–51.
121. Mata-Alvarez, J.; Dosta, J.; Romero-Güiza, M.S.; Fonoll, X.; Peces, M.; Astals, S. A critical review on anaerobic co-digestion achievements between 2010 and 2013. ***Renew. Sustain. Energy Rev.*** 2014, ***36***, 412–427.
122. Wang, W.; Xie, L.; Chen, J.; Luo, G.; Zhou, Q. Biohydrogen and methane production by co-digestion of cassava stillage and excess sludge under thermophilic condition. ***Bioresour. Technol.*** 2011, ***102***, 3833–3839.

123. Andalib, M.; Elbeshbishy, E.; Mustafa, N.; Hafez, H.; Nakhla, G.; Zhu, J. Performance of an anaerobic fluidized bed bioreactor (AnFBR) for digestion of primary municipal wastewater treatment biosolids and bioethanol thin stillage. ***Renew. Energy*** 2014, ***71***, 276–285.
124. Sharma, D.; Espinosa-Solares, T.; Huber, D.H. Thermophilic anaerobic co-digestion of poultry litter and thin stillage. ***Bioresour. Technol.*** 2013, ***136***, 251–256.
125. John, R.P.; Anisha, G.S.; Nampoothiri, K.M.; Pandey, A. Micro and macroalgal biomass: A renewable source for bioethanol. ***Bioresour. Technol.*** 2011, ***102***, 186–193.
126. Kemppainen, K.; Ranta, L.; Sipilä, E.; Östman, A.; Vehmaanperä, J.; Puranen, T.; Langfelder, K.; Hannula, J.; Kallioinen, A.; Siika-aho, M.; Sipilä, K.; von Weymarn, N. Ethanol and biogas production from waste fibre and fibre sludge – The FibreEtOH concept. ***Biomass Bioenergy*** 2012, ***46***, 60–69.
127. Narra, M.; Balasubramanian, V. Utilization of solid and liquid waste generated during ethanol fermentation process for production of gaseous fuel through anaerobic digestion–A zero waste approach. ***Bioresour. Technol.*** 2015, ***180***, 376–380.
128. Capecchi, L.; Galbe, M.; Barbanti, L.; Wallberg, O. Combined ethanol and methane production using steam pretreated sugarcane bagasse. ***Ind. Crops Prod.*** 2015, ***74***, 255–262.
129. Torry-Smith, M.; Sommer, P.; Ahring, B.K. Purification of bioethanol effluent in an UASB reactor system with simultaneous biogas formation. ***Biotechnol. Bioeng.*** 2003, ***84***, 7–12.
130. Uellendahl, H.; Ahring, B.K. Anaerobic digestion as final step of a cellulosic ethanol biorefinery: Biogas production from fermentation effluent in a UASB reactor—pilot-scale results. ***Biotechnol. Bioeng.*** 2010, ***107***, 59–64.
131. Global Solutions. Scale up history. Available online: http://www.inbicon.com/en/global-solutions (accessed on 20 July 2015).
132. Maabjerg Energy Concept. Status Report. Moving towards realization 2013. Available online: http://www.maabjergenergyconcept.eu/ (accessed on 20 July 2015).
133. Alburquerque, J.A.; de la Fuente, C.; Ferrer-Costa, A.; Carrasco, L.; Cegarra, J.; Abad, M.; Bernal, M.P. Assessment of the fertiliser potential of digestates from farm and agroindustrial residues. ***Biomass***

Bioenergy2012, ***40***, 181–189.

134. Massaccesi, L.; Sordi, A.; Micale, C.; Cucina, M.; Zadra, C.; Di Maria, F.; Gigliotti, G. Chemical characterisation of percolate and digestate during the hybrid solid anaerobic digestion batch process. ***Process Biochem.*** 2013, ***48***, 1361–1367.
135. Abdullahi, Y.A.; Akunna, J.C.; White, N.A.; Hallett, P.D.; Wheatley, R. Investigating the effects of anaerobic and aerobic post-treatment on quality and stability of organic fraction of municipal solid waste as soil amendment. ***Bioresour. Technol.*** 2008, ***99***, 8631–8636.
136. Bustamante, M.A.; Restrepo, A.P.; Alburquerque, J.A.; Pérez-Murcia, M.D.; Paredes, C.; Moral, R.; Bernal, M.P. Recycling of anaerobic digestates by composting: effect of the bulking agent used. ***J. Clean. Prod.*** 2013, ***47***, 61–69.
137. Cesaro, A.; Belgiorno, V.; Guida, M. Compost from organic solid waste: Quality assessment and European regulations for its sustainable use. ***Resour. Conserv. Recycl.*** 2015, ***94***, 72–79.
138. Menardo, S.; Gioelli, F.; Balsari, P. The methane yield of digestate: Effect of organic loading rate, hydraulic retention time, and plant feeding. ***Bioresour. Technol.*** 2011, ***102***, 2348–2351.
139. Yue, Z.; Teater, C.; Liu, Y.; MacLellan, J.; Liao, W. A sustainable pathway of cellulosic ethanol production integrating anaerobic digestion with biorefining. ***Biotechnol. Bioeng.*** 2010, ***105***, 1031–1039.
140. Teater, C.; Yue, Z.; MacLellan, J.; Liu, Y.; Liao, W. Assessing solid digestate from anaerobic digestion as feedstock for ethanol production. ***Bioresour. Technol.*** 2011, ***102***, 1856–1862.
141. MacLellan, J.; Chen, R.; Kraemer, R.; Zhong, Y.; Liu, Y.; Liao, W. Anaerobic treatment of lignocellulosic material to co-produce methane and digested fiber for ethanol biorefining. ***Bioresour. Technol.*** 2013, ***130***, 418–423.
142. Gao, T.; Li, X. Using thermophilic anaerobic digestate effluent to replace freshwater for bioethanol production. ***Bioresour. Technol.*** 2011, ***102***, 2126–2129.

12

Biomethane: A Renewable Resource as Vehicle Fuel

Federica Cucchiella, Idiano D'Adamo , and Massimo Gastaldi

Department of Industrial and Information Engineering and Economics, University of L'Aquila, Via G. Gronchi 18, 67100 L'Aquila, Italy

ABSTRACT

The European Union (EU) has set a mandatory target for renewable fuels of 10% for each member state by 2020. Biomethane is a renewable energy representing an alternative to the use of fossil fuels in the transport sector. This resource is a solution to reach this target. Furthermore, it contributes to reducing carbon dioxide emissions, gives social benefits and increases the security supply. Sustainability is reached also when the economic opportunities are verified. This work studies the profitability of small plants of biomethane, which is sold as vehicle fuel using the Net Present Value (NPV) and Discounted Payback Time (DPBT). The paper shows in detail the method used for the economic assessment of two typologies of feedstock recovered: (i) municipal solid waste and (ii) agricultural

Citation: Cucchiella F, D'Adamo I, Gastaldi M. Biomethane: A Renewable Resource as Vehicle Fuel. Resources. 2017; 6(4):58.

waste. Detailed information about the various parameters that affect the profitability of biomethane is given, and several case studies are analyzed as a function of two variables: subsidies and selling price. The results support the commercialization of small-scale plants, reducing also several environmental issues. The role of subsidies is strategic, and the profitability is verified only in some case studies.

Keywords: Biomethane, economic analysis, subsidies, sustainability, waste management

INTRODUCTION

A circular economy is based on the principle of maintaining the value of products, materials and resources as long as possible, minimizing waste and resource use [1]. At the end of its life, a product can be recovered to create further value [2]. Biogas is the product of anaerobic digestion beginning from several feedstocks, such as agricultural residues (e.g., manure and crop residues), energy crops, organic-rich waste waters, the organic fraction of municipal solid waste (ofmsw) and industrial organic waste [3].

Biomethane is obtained from properly-treated biogas through the process of purification [4]. In the last few decades, biomethane has achieved a significant importance in the field of Renewable Energy (RE) [5]. The evolution of the numbers of upgraded plants in Europe has grown from 187 in 2011 to 435 in 2015. It is the most significant market in the world (90%) (Figure 1) [6]. However, biomethane production is highly localized in a few countries, and this represents a great limit for this market [7].

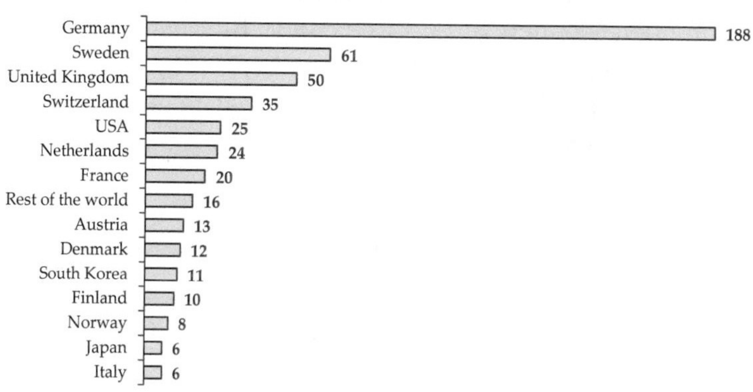

Figure 1: Number of upgraded plants worldwide in 2015 [6].

Biomethane has properties potentially equivalent to methane and can be used directly as vehicle fuel, or be injected into the natural gas grid, or be converted into electricity and heat in cogeneration units [8]. Current technologies of biomethane consume less than about 20% of biogas energy for upgrading and compression aims [9]. A review of this topic underlines innovative and highly effective technologies along the whole chain of biomethane production [10]. Among several feedstocks used, the application of a multi-criteria analysis underlines the advantageous linked to the use of organic waste [11]. A comprehensive analysis of agricultural waste highlights also its strategic role in a circular economy model [12].

The biogas-biomethane chain is able to tackle the environmental pollution as an alternative to the consumption of natural gas [13]. GHG emissions of vehicle powertrain systems are recorded from the Well To Wheel (WTW). According to the DENA (Deutsche Energie-Agentur) study, methane or Compressed Natural Gas (CNG) has a reduction potential of the order of around 21% and 24% in comparison to diesel and petrol, respectively. The use of biomethane is able to reduce emissions further. This value is equal to 24 gCO_2eq/km, if CNG is composed also by 20% of biomethane (also called BIO-CNG (20%)). Instead, when pure biomethane (also called BIO-CNG (100%)) is used, WTW emissions are equal to 5 gCO_2eq/km (Figure 2) [14].

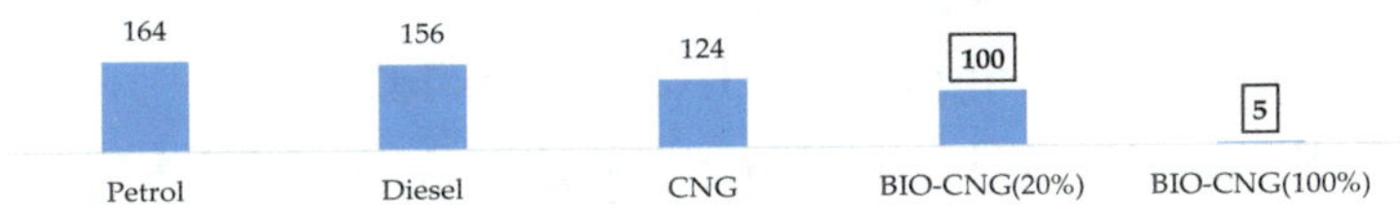

Figure 2: Well To Wheel (WTW) GHG emissions in gCO_2eq/km [14].CNG, Compressed Natural Gas.

For this reason, a policy support for biofuels is justified [15]. Changes in the legal framework had direct impacts on their development [16]. Furthermore, European countries can reduce their reliance on natural gas imports [17], reaching also renewable targets [18].

The economic feasibility of biogas plants is well defined in the literature [19], but also, the analysis of biomethane production is analyzed for several final uses (fed into the grid, destined for cogeneration or sold as vehicle fuel) [20]. However, there are several approaches.

In fact, the economic impact of upgrading can be evaluated varying the quantity of biogas processed and the technology used [21]. The biomethane cost of production is 0.54 €/m^3 injected into the grid and 0.73 €/m^3 as transportation fuel [22]. The discounted total cost for the organic fraction of municipal solid waste (ofmsw) substrate varies from 0.46–0.82 €/m^3, while it is equal to 0.49–0.76 €/m^3 for a mixed substrate (maize and manure residues) [20].

The EU project "Record Biomap" defines that it is relevant to propose solutions to make biomethane production profitable also at small- and medium-scale biogas plants. Literature analysis has shown attention to this topic [23]. A previous analysis underlining the profitability of ofmsw substrate is verified with the following configurations: a minimum size for the plant of 100 m^3/h if the biomethane is used for cogeneration and 250 m^3/h for the other two final destinations. Instead, the financial feasibility of mixed substrate is verified only for a 500 m^3/h plant if the biomethane is used as vehicle fuel [20].

This paper aims to evaluate the profitability of biomethane plants in a market with subsidies. One hundred and eighty case studies were analyzed as a function of four critical variables:

- Plant size (50 m^3/h, 100 m^3/h and 150 m^3/h).
- Feedstock (ofmsw and a mixture of 30% maize and 70% manure residues on a weight basis).
- Subsidies (varying from 0.162 €/m^3–0.487 €/m^3).
- Selling price of biomethane (varying from 0.1384 €/m^3–0.2397 €/m^3).

Italy is chosen as the case study, considering that it: (i) is the country with the highest number of Natural Gas Vehicles (NGVs) in Europe; (ii) has a widespread presence of biogas plants; and (iii) has high energy dependence on foreign imports [24]. Consequently, this country has great potentials, but the biomethane sector is not yet developed despite a decree of incentivization from 2013. This work can be useful for researchers, entrepreneurs and industry leaders since it provides an insight toward the commercialization of the technology and softens several environmental issues.

The paper is organized as follows. Section 2 presents the methodology used in this paper, and an economic model is defined to evaluate the profitability of a biomethane plant. Starting by input data, it is possible to calculate NPV and Discounted Payback Time (DPBT), and the results are shown in Section 3. A discussion of the role of biomethane in the transport sector and some concluding remarks are proposed in Section 4.

MATERIALS AND METHODS

Discounted Cash Flow (DCF) is a recognized economic assessment method used to evaluate the attractiveness of a project. Two financial indexes are considered: NPV is defined as the sum of the present values of the individual cash flows, and DPBT represents the number of years needed to balance cumulative discounted cash flows and initial investment [24].

The profitability of a project depends on the index used. Considering NPV, it is verified when the value obtained is greater than zero. Considering DPBT, it is verified when the value obtained is lower than the cut-off period. DPBT > 20 y (years) indicates that the investment cannot be recovered within this date, and this is coherent with an NPV < 0.

With a decree introduced in December 2013, Italy adopted a new system of incentives for biomethane, depending on its type of utilization [25]. In the case of biomethane used as vehicle fuel, the decree states that the subsidy is released through certificates called CIC (Certificates of Emission of Biofuel in Consumption). The obliged subjects are companies that market fuels in Italy and every year are responsible for entering into consumption a share of sustainable biofuels as a proportion of the amount of consumed fossil fuels. A pecuniary penalty is provided when this is not verified for each CIC. A single certificate is released when the biomethane quantity corresponding to 41,840 MJ (equivalent of 10 Gcal) is used as transportation fuel. This subsidy lasts twenty years.

The value of CIC is characterized by uncertainty, and a draft of the new decree of biomethane by the Italian Ministry of Economic Development provides a value of 375 €. Furthermore, this work proposes ten possible values of CICs starting in the range of 300–500 € [20] (Figure 3).

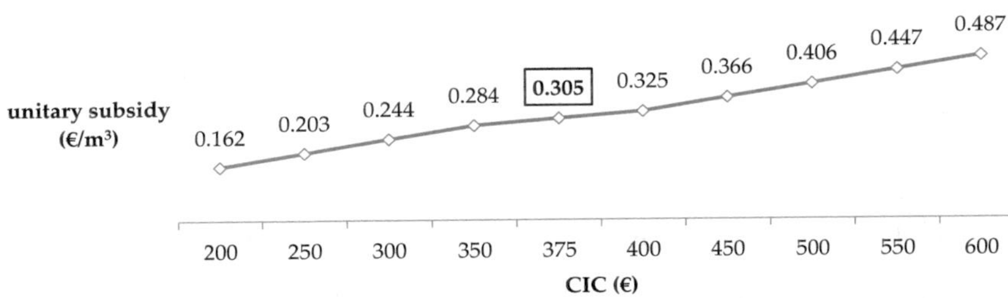

Figure 3: Certificates of Emission of Biofuel in Consumption expressed in i_{cic}^{u} . Adapted by [25].

For example, when one CIC is equal to 200 €, the transformation of this value in terms of cubic meters (i^{u}_{cic}) is obtained as the ratio between 200 € and 1231 m^3. In fact, one CIC corresponds to 0.837 t of biomethane, which is equivalent to 1231 m^3 of CH_4. This conversion is obtained considering that 1 m^3 CH_4 = 0.68 kg under normal conditions (standard temperature of 273.15 K and pressure of 101.325 kPa).

If the biomethane has been produced from specific products as reported in the Ministerial Decree, the number of released CIC doubles (double counting). For example, this is verified for ofmsw, while the corrective coefficient is equal to 1.7 with a mixture of 30% maize and 70% manure residues. Furthermore, there is a notable premium (another corrective coefficient equal to 1.5) when the producer firm is also the methane distributor. This paper did not analyze this scenario, considering that the benefits of this choice were defined by [24].

The revenues of a biomethane plant used for transportation fuel are basically two: (i) subsidies and (ii) the selling of biomethane. The first is determined by a Ministerial Decree, as defined above, while the second item is calculated as a function of the selling price of natural gas. It is calculated as a function of wholesale gas price defined in the virtual exchange point. Values proposed in Figure 4 were already decreased of 5%, in accordance with the draft of the new decree of biomethane.

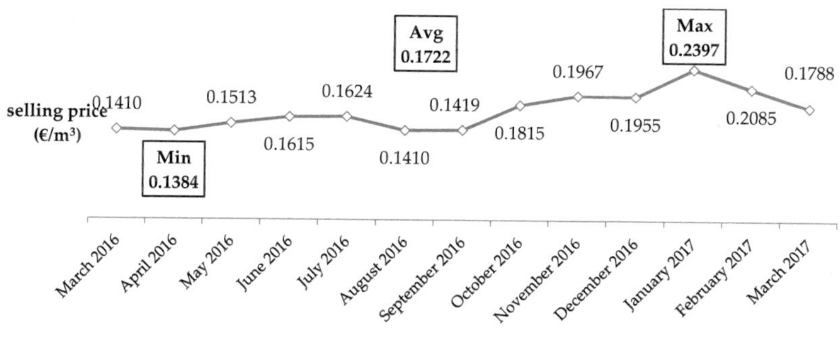

Figure 4: Selling price of biomethane expressed in P^{*}_{sng}. Adapted by [25].

This work proposes three possible values of selling price of biomethane to a vehicle-fuel distribution plant (P^{*}_{sng}). Considering the period March 2016–March 2017, three values during this interval are chosen: minimum, average and maximum.

The costs of biomethane production can be subdivided into three main phases: (i) biogas production; (ii) upgrading; and (iii) compression and distribution. Investment costs are lower than operational costs and range from 15%–20% [20].

The capital investment of a biogas facility is a function of feedstock and plant size [26], while the investment cost of upgrading phase depends on the technology used and the quantity of biogas processed [21]. This work analyzed the membrane separation (MEMS) technology that is typically used for small plants [27].

Operation costs associated with biogas and biomethane production are: (i) substrate; (ii) transport; (iii) labor; (iv) maintenance and overhead; (v) depreciation fund for mechanical and electrical elements; (vi) electricity consumption; and (vii) insurance. Among these items, maintenance and overhead is the most relevant cost varying from 46%–50% when ofmsw substrate is used and from 24%–29% with mixes. These considerations are obtained for 100−1000 m^3/h plants, and the results underline that transport costs become more relevant than maintenance and overhead costs when a 1000 m^3/h plant is considered for a mixed substrate [20].

Finally, the cost of the third phase is lower than the other two phases, if the production location is not far from the distribution grid [28]. The treatment of ofmsw as a substrate determines from one side the income because the process of urban waste is a paid service and from the other side also the cost for pre-treatment required. However, the net cash flow is positive [24].

The mathematical model used to evaluate the profitability of biomethane plant is reported below:

$$\mathrm{NPV} = \sum_{t=0}^{n} (I_t - O_t)/(1+r)^t \tag{1}$$

$$\sum_{t=0}^{DPBT} (I_t - O_t)/(1+r)^t = 0 \tag{2}$$

$$Q_{biogas}^{nom} = S_{biogas} * n_{oh} * \%CH_4 \tag{3}$$

$$Q_{feedstock} = Q_{biogas}^{nom} / \left(p_b^u * (\%vs/ts) * (\%ts/(ww + ts))\right) \tag{4}$$

$$Q_{biogas} = Q_{biogas}^{nom} * (1 - l_{bs}) \tag{5}$$

$$Q_{biomethane}^{nom} = S_{biomethane} * n_{oh} \tag{6}$$

$$Q_{biomethane} = Q_{biogas} * (\%CH_4) * (1 - l_{us}) \tag{7}$$

$$I_t = R_t^{subsidies} + R_t^{selling} + R_t^{ofmsw} \tag{8}$$

$$R_t^{subsidies} = Q_{biomethane} * i_{cic}^u * c_c \quad \forall t = 0 \ldots n \tag{9}$$

$$R_t^{selling} = Q_{biomethane} * p_{sng}^* \text{ with } p_{sng}^* = 0.95 * p_{sng} \quad \forall t = 0 \ldots n \tag{10}$$

$$R_t^{ofmsw} = Q_{ofmsw} * \left(R_{gross,t}^{ofmsw} - C_t^{ofmsw}\right) \quad \forall t = 0 \ldots n \tag{11}$$

$$0_t = C_{lcs,t}^{1^\circ s} + C_{lis,t}^{1^\circ s} + C_{lcs,t}^{2^\circ s} + C_{lis,t}^{2^\circ s} + C_{lcs,t}^{3^\circ s} + C_{lis,t}^{3^\circ s} + C_{l,t} + C_{s,t} + C_{ts,t} + C_{mo,t}^{1^\circ s} + \tag{12}$$

$$+C_{df,t}^{1^\circ s}+C_{e,t}^{1^\circ s}+C_{i,t}^{1^\circ s}+C_{mo,t}^{2^\circ s}+C_{df,t}^{2^\circ s}+C_{e,t}^{2^\circ s}+C_{i,t}^{2^\circ s}+C_{o,t}^{dis}+C_{o,t}^{com}+C_{tax,t}$$

$$C_{inv}^{1^\circ s}=C_{inv}^{u,1^\circ s}*S_{biogas} \quad (13)$$

$$C_{lcs,t}^{1^\circ s}=C_{inv}^{1^\circ s}/n_{debt} \quad \forall t=0\ldots n_{debt}-1 \quad (14)$$

$$C_{lis,t}^{1^\circ s}=\left(C_{inv}^{1^\circ s}-C_{lcs,t}^{1^\circ s}\right)*r_d \quad \forall t=0\ldots n_{debt}-1 \quad (15)$$

$$C_{inv}^{2^\circ s}=C_{inv}^{u,2^\circ s}*S_{biomethane} \quad (16)$$

$$C_{lcs,t}^{2^\circ s}=C_{inv}^{2^\circ s}/n_{debt} \quad \forall t=0\ldots n_{debt}-1 \quad (17)$$

$$C_{lis,t}^{2^\circ s}=\left(C_{inv}^{2^\circ s}-C_{lcs,t}^{2^\circ s}\right)*r_d \quad \forall t=0\ldots n_{debt}-1 \quad (18)$$

$$C_{inv}^{3^\circ s}=C_{inv}^{dis}+C_{inv}^{com} \quad (19)$$

$$C_{lcs,t}^{3^\circ s}=C_{inv}^{3^\circ s}/n_{debt} \quad \forall t=0\ldots n_{debt}-1 \quad (20)$$

$$C_{lis,t}^{3^\circ s}=\left(C_{inv}^{3^\circ s}-C_{lcs,t}^{3^\circ s}\right)*r_d \quad \forall t=0\ldots n_{debt}-1 \quad (21)$$

$$C_{l,t}=C_l^{u,a}*n_{op} \quad \forall t=0\ldots n \quad (22)$$

$$C_{s,t}=C_s^u*Q_{feedstock} \quad \forall t=0\ldots n \quad (23)$$

$$C_{ts,t}=C_{ts}^u*Q_{feedstock} \quad \forall t=0\ldots n \quad (24)$$

$$C_{mo,t}^{1^\circ s}=p_{mo}^{1^\circ s}*C_{inv}^{1^\circ s} \quad \forall t=0\ldots n \quad (25)$$

$$C_{df,t}^{1^\circ s}=p_{df}*C_{lcs,t}^{1^\circ s} \quad \forall t=0\ldots n \quad (26)$$

$$C_{e,t}^{1^\circ s}=c_e^{u,1^\circ s}*Q_{biogas}*p_e \quad \forall t=0\ldots n \quad (27)$$

$$C_{i,t}^{1^\circ s}=p_i*C_{inv}^{1^\circ s} \quad \forall t=0\ldots n \quad (28)$$

$$C_{mo,t}^{2^\circ s}=p_{mo}^{2^\circ s}*C_{inv}^{2^\circ s} \quad \forall t=0\ldots n \quad (29)$$

$$C_{df,t}^{2^\circ s}=p_{df}*C_{lcs,t}^{2^\circ s} \quad \forall t=0\ldots n \quad (30)$$

$$C_{e,t}^{2^\circ s}=c_e^{u,2^\circ s}*Q_{biogas}*p_e \quad \forall t=0\ldots n \quad (31)$$

$$C_{i,t}^{2^\circ s}=p_i*C_{inv}^{2^\circ s} \quad \forall t=0\ldots n \quad (32)$$

$$C_{o,1}^{dis}=C_o^{dis} \text{ with } C_{o,t+1}^{dis}=C_{o,t}^{dis}*(1+inf) \quad \forall t=0\ldots n \quad (33)$$

$$C_{o,1}^{com}=C_o^{com} \text{ with } C_{o,t+1}^{com}=C_{o,t}^{com}*(1+inf) \quad \forall t=0\ldots n \quad (34)$$

$$C_{tax,t}=p_{tax}^{unit}*ebt \quad \forall t=0\ldots n \quad (35)$$

This paper examines the financial feasibility of biomethane plants used as a vehicle fuel. One hundred and eighty case studies were proposed by the combination of these variables:

- plant size, in which three values are analyzed: 50 m^3/h, 100 m^3/h and 150 m^3/h.
- feedstock used, in which two typologies are considered: ofmsw and a mixture with 30% maize and 70% manure residues on a weight basis.
- unitary subsidy, in which ten values are considered (Figure 3), varying from 0.162 €/m^3–0.481 €/m^3.
- selling price of biomethane, in which three values are considered (Figure 4), varying from 0.1384 €/m^3–0.2397 €/m^3.

Other economic and technical inputs are proposed in Table 1, and the investment cost is covered by third party funds with the useful lifetime of the plant defined equal to the lifetime of the subsidies (20 years). The aggregation of cash flows is obtained through the opportunity cost of capital fixed equal to 5%. The definition of the optimal size of biogas plant is chosen in order to maximize the grade of saturation of the upgrading phase according to the approach used by [20].

Table 1: Input data

Variable	Value	Reference
c_c	1.7 [a]; 2 [b]	[25]
$c_e^{u,1°s}$	0.13 kWh/m^3	[27]
$c_e^{u,2°s}$	0.29 kWh/m^3	[26]
C_{inv}^{com}	53,000 €	[29]
C_{inv}^{dis}	237,500 €	[30]
$C_{inv}^{u,1°s}$	5100 €/kW [a,i]; 4800 €/kW [a,ii]; 4500 €/kW [a,iii] 5300 €/kW [b,i]; 5000 €/kW [b,ii]; 5300 €/kW [b,iii]	[20]
$C_{inv}^{u,2°s}$	6300 €/(m^3/h) [i]; 5800 €/(m^3/h) [ii]; 5300 €/(m^3/h) [iii]	[20]
$C_l^{u,a}$	25,000 €/y	[20]
C_o^{com}	47,000 €/y	[29]
C_o^{dis}	20,000 €/y	[30]
C_s^u	10 €/t [c]; 0 € [b,d]	[27]
C_t^{ofmsw}	49 €/t	[24]
C_{ts}^u	2 €/t	[31]
i_{cic}^u	Table 2	[25]
inf	2%	[20]
l_{bs}	6%	[30]
l_{us}	1.5%	[30]
N	20 y	[24]
n_{debt}	15 y	[24]
n_{oh}	8000 h	[24]
n_{op}	4	[24]
n_s	20 y	[24]
p_b^u	350 [d]–500 [b]—650 [c] m^{3}biogas/t(vs)	[20]
p_{df}	20%	[20]
p_e	0.13 €/kWh	[20]
p_i	1%	[20]
$p_{mo}^{1°s}$	10% [a]; 20% [b]	[26]
$p_{mo}^{2°s}$	10%	[24]
p_{sng}^*	Table 3	
p_{tax}^{unit}	27.5%	[24]
Q_{ofmsw}	5952 t [i]; 11,905 t [ii]; 17,857 t [iii]	[20]
Q_{maize}	1163 t [i]; 2363 t [ii]; 3525 t [iii]	[20]
$Q_{manure\ residues}$	19,579 t [i]; 39,789 t [ii]; 59,368 t [iii]	[20]
r	5%	[20]
r_d	3%	[24]
$R_{gross,t}^{ofmsw}$	70 €/t	[24]
S_{biogas}	155 kW [a,i]; 315 kW [a,ii]; 470 kW [a,iii]; 150 kW [b,i]; 300 kW [b,ii]; 450 kW [b,iii]	[20]
$S_{biomethane}$	50 (m^3/h) [i]; 100 (m^3/h) [ii]; 150 (m^3/h) [iii]	[20]
$\%CH_4$	57% [a]; 60% [b]	[24]
%ts/(ww + ts)	9.5% [d]; 27% [b]; 30.8% [c]	[32]
%vs/ts	80% [d]; 89.6% [b]; 95.9% [c]	[32]

Notes: a = mixed; b = organic fraction of municipal solid waste (ofmsw); c = maize; d = manure residues with a = 30%*c + 70%*d. These are % of total mass. i = 50 m^3/h; ii = 100 m^3/h; iii = 150 m^3/h.

RESULTS

The great challenges of climate change have increased the propensity towards sustainable choices. The automotive sector is characterized by new solutions from cradle to grave [33]. The substitution of fossil fuels with renewable sources is a choice adopted by several governments. Biomethane permits a reduction of emissions as defined in Section 1, while the economic opportunities are verified for some sizes and substrates.

This work follows this direction of research, and several case studies are proposed. The profitability is calculated as a function of two variables: unitary subsidy (i^{u}_{cic}) and selling price of biomethane (p^{*}_{sng}). NPV and DPBT for ofmsw substrate are reported in Table 2 and Table 3, respectively, while NPV and DPBT for mixed substrate are reported in Table 4 and Table 5, respectively.

Table 2: NPV (k€) for ofmsw substrate

	50 m³/h Plant			100 m³/h Plant			150 m³/h Plant		
CIC	Min p^*_{sng}	Avg p^*_{sng}	Max p^*_{sng}	Min p^*_{sng}	Avg p^*_{sng}	Max p^*_{sng}	Min p^*_{sng}	Avg p^*_{sng}	Max p^*_{sng}
200 €	−3897	−3729	−3392	−4517	−4180	−3507	−4618	−4112	−3103
250 €	−3489	−3320	−2984	−3699	−3362	−2689	−3391	−2886	−1876
300 €	−3080	−2911	−2575	−2882	−2545	−1872	−2165	−1660	−650
350 €	−2681	−2513	−2176	−2084	−1747	−1074	−969	−463	**546**
375 €	−2472	−2303	−1967	−1666	−1329	−656	−341	**165**	**1174**
400 €	−2272	−2104	−1767	−1267	−930	−257	**258**	**763**	**1772**
450 €	−1864	−1695	−1359	−449	−112	**561**	**1484**	**1989**	**2999**
500 €	−1465	−1296	−960	**348**	**685**	**1358**	**2680**	**3186**	**4195**
550 €	−1056	−888	−551	**1166**	**1503**	**2176**	**3906**	**4412**	**5421**
600 €	−657	−489	−152	**1963**	**2300**	**2973**	**5103**	**5608**	**6618**

Notes: Min p^*_{sng} = 0.1384 €/m³; Avg p^*_{sng} = 0.1722 €/m³; Max p^*_{sng} = 0.2397 €/m³. Profitable cases are shown in bold.

Table 3: Discounted Payback Time (DPBT) (y) for ofmsw substrate

	50 m³/h Plant			100 m³/h Plant			150 m³/h Plant		
CIC	Min p^*_{sng}	Avg p^*_{sng}	Max p^*_{sng}	Min p^*_{sng}	Avg p^*_{sng}	Max p^*_{sng}	Min p^*_{sng}	Avg p^*_{sng}	Max p^*_{sng}
200 €	>20	>20	>20	>20	>20	>20	>20	>20	>20
250 €	>20	>20	>20	>20	>20	>20	>20	>20	>20
300 €	>20	>20	>20	>20	>20	>20	>20	>20	>20
350 €	>20	>20	>20	>20	>20	>20	>20	>20	**3**
375 €	>20	>20	>20	>20	>20	>20	>20	**4**	**2**
400 €	>20	>20	>20	>20	>20	>20	**4**	**3**	**2**
450 €	>20	>20	>20	>20	>20	**3**	**2**	**2**	**1**
500 €	>20	>20	>20	**3**	**2**	**2**	**2**	**1**	**1**
550 €	>20	>20	>20	**2**	**2**	**1**	**1**	**1**	**1**
600 €	>20	>20	>20	**2**	**1**	**1**	**1**	**1**	**1**

Notes: Min p^*_{sng} = 0.1384 €/m³; Avg p^*_{sng} = 0.1722 €/m³; Max p^*_{sng} = 0.2397 €/m³. Profitable cases are shown in bold.

Table 4: NPV (k€) for mixed substrate

	50 m³/h Plant			100 m³/h Plant			150 m³/h Plant		
CIC	Min p^*_{sng}	Avg p^*_{sng}	Max p^*_{sng}	Min p^*_{sng}	Avg p^*_{sng}	Max p^*_{sng}	Min p^*_{sng}	Avg p^*_{sng}	Max p^*_{sng}
200 €	−5170	−5004	−4674	−7233	−6897	−6226	−8128	−7626	−6625
250 €	−4829	−4663	−4333	−6540	−6204	−5533	−7094	−6592	−5591
300 €	−4488	−4322	−3992	−5847	−5511	−4840	−6060	−5558	−4556
350 €	−4155	−3989	−3659	−5171	−4835	−4164	−5051	−4549	−3548
375 €	−3980	−3815	−3484	−4816	−4480	−3809	−4521	−4019	−3018
400 €	−3814	−3648	−3318	−4478	−4142	−3470	−4016	−3515	−2513
450 €	−3473	−3307	−2977	−3785	−3449	−2777	−2982	−2481	−1479
500 €	−3140	−2974	−2644	−3108	−2772	−2101	−1973	−1472	−470
550 €	−2799	−2633	−2303	−2415	−2079	−1408	−939	−437	**564**
600 €	−2466	−2301	−1970	−1739	−1403	−732	**70**	**571**	**1573**

Notes: Min p^*_{sng} = 0.1384 €/m³; Avg p^*_{sng} = 0.1722 €/m³; Max p^*_{sng} = 0.2397 €/m³. Profitable cases are shown in bold.

Table 5: DPBT (y) for mixed substrate

	50 m³/h Plant			100 m³/h Plant			150 m³/h Plant		
CIC	Min p^*_{sng}	Avg p^*_{sng}	Max p^*_{sng}	Min p^*_{sng}	Avg p^*_{sng}	Max p^*_{sng}	Min p^*_{sng}	Avg p^*_{sng}	Max p^*_{sng}
200 €	>20	>20	>20	>20	>20	>20	>20	>20	>20
250 €	>20	>20	>20	>20	>20	>20	>20	>20	>20
300 €	>20	>20	>20	>20	>20	>20	>20	>20	>20
350 €	>20	>20	>20	>20	>20	>20	>20	>20	>20
375 €	>20	>20	>20	>20	>20	>20	>20	>20	>20
400 €	>20	>20	>20	>20	>20	>20	>20	>20	>20
450 €	>20	>20	>20	>20	>20	>20	>20	>20	>20
500 €	>20	>20	>20	>20	>20	>20	>20	>20	>20
550 €	>20	>20	>20	>20	>20	>20	>20	>20	**3**
600 €	>20	>20	>20	>20	>20	>20	**5**	**3**	**2**

Notes: Min p^*_{sng} = 0.1384 €/m³; Avg p^*_{sng} = 0.1722 €/m³; Max p^*_{sng} = 0.2397 €/m³. Profitable cases are shown in bold.

The financial feasibility of biomethane production is verified in twenty-eight of nineteen case studies (31%) concerning of msw plants. These results are coherent with Section 1, and it can be underlined that:

- The 50 m³/h plant is always unprofitable.
- The 100 m³/h plant is profitable in ten scenarios; positive NPV varies from 3.48 k€/(m³/h)–29.73 k€/(m³/h), and DPBT ranges from 1–3 years. Starting from a value of CIC equal to 500 € the profitability is verified for all prices of selling of biomethane, while with a CIC of 450 €, only with a price of selling equal to 0.2397 €/m³.
- The 150 m³/h plant is profitable in eighteen scenarios; positive NPV varies from 1.72 k€/(m³/h)–44.12 k€/(m³/h) and DPBT ranges from 1–4 years. Starting from a value of CIC equal to 400 €, the profitability is verified for all prices of selling of biomethane, while with a CIC of 375 €, that is the supposed value by the new decree (see Section 2), the plant is profitable in the

Avg P^{*}_{sng} and Max P^{*}_{sng} scenarios. Finally, NPV is positive also with a value of CIC of 350 € and P^{*}_{sng} of 0.2397 €/m^3.

The role of subsidies is strategic, and the current incentive scheme does not provide specific supports to small plants. This choice could be incentivized by the structure of the incentive scheme adopted in the past for other REs.

A previous research work has defined that 250 m^3/h is the profitable minimum plant size in the baseline scenario, and also the 100 m^3/h plant has NPV > 0 in alternative scenarios with an increase of incentive (i^{u}_{cic} equal to 0.41 €/m^3) or a decrease of the maintenance and overhead cost of biogas production (p1°smo equal to 15%) [20]. Results obtained in Table 2 and Table 3 define that the 150 m^3/h plant is basically profitable, because CIC = 375 € is the more probable value. The analysis of several scenarios underlines that profits can be very relevant, but there also significant economic losses in some case studies.

The profitability of biomethane production is verified only in four of nineteen case studies (4%) concerning mixed plants. Furthermore, this part of the work is coherent with Section 1, and it can be highlighted that:

- The 50 m^3/h plant is always unprofitable.
- The 100 m^3/h plant is always unprofitable.
- The 150 m^3/h plant is profitable when a CIC equal to 600 € is present for all prices of selling of biomethane and in the case study with CIC = 550 € and P^{*}_{sng} = 0.2397 €/m^3. Positive NPV varies from 0.47 k€/(m^3/h)–10.49 k€/(m^3/h), and DPBT ranges from 2–5 years.

The current incentive scheme does not favor the use of substrates as energy crops, which could subtract agricultural land from the primary sector. A mixture plant with 30% maize and 70% manure residues on a weight basis presents a lower corrective coefficient than ofmsw (c_c is equal to 1.7 instead of two).

For this motive, NPV is lower in this typology of investment in addition to the absence of net revenues linked to the treatment of municipal solid waste. Certainly, plants with 100% of agricultural residues have a corrective coefficient equal to two, but in this case, there is a great risk of not maximizing the degree of saturation of the plant. In fact, agricultural residues have a lower potential of biomethane than one of the energy crops, and consequently, great quantities of these residues are required.

A previous work has defined 500 m^3/h as the minimum plant size in terms of profitability considering a baseline scenario. Furthermore, the 250 m^3/h plant has NPV > 0 when there is an increase of incentive (i^u_{cic} equal to 0.41 €/m^3) or a decrease of transport cost of the substrate (C^u_{ts} equal to 1 €/t) [20]. Table 4 and Table 5 highlight that the probability of positive NPV with mixed substrate is very low; in fact, CIC equal to 550 € or 600 € is not tracked in the current market.

DISCUSSION AND CONCLUSIONS

Italy presents a strong contradiction. This country has the highest number of NGVs in the global context (about equal to 950,000 units), and the number of biogas plants is very significant. In addition, the production of natural gas is very low, and consequently, Italy depends on foreign imports. Biomethane can be a valid solution, but its use is very limited [13]. This paper tries to offer new discussion items useful to researchers, firms and public administrators.

The first aim of this work was to define the environmental benefits linked to the use of biomethane. Initially, the quantity of biomethane required to satisfy the demand was calculated. Hypothesizing that the annual NGV's consumption was equal to 1100 m^3 of methane, two alternative scenarios were considered according to Section 1:

- A mixture composed of 20% biomethane and 80% methane (BIO-CNG (20%).
- Pure biomethane (BIO-CNG (100%).

The necessary resource is equal to 209 million m^3/y when BIO-CNG (20%) is considered, obtained by the product between the number of NGVs and the annual NGV's consumption. The reduction of GHG linked to this choice is 360 ktCO_2eq per year considering the unitary reduction of 24 gCO_2eq/km defined in Section 1 and hypothesizing that an NGV travels 15,000 km per year. Nine hundred fifty thousand NGVs powered by BIO-CNG (20%) save 6840 ktCO_2eq compared to those powered by fossil fuel during 20 years (Table 6).

Table 6: Environmental benefits of the biomethane used in the transport sector. NGV, Natural Gas Vehicle

	Scenario BIO-CNG (20%)	Scenario BIO-CNG (100%)
Biomethane demand (million m^3/y)	209	1045
Reduction of GHG (gCO_2eq/km)	24	119
1 NGV		
Reduction of GHG ($kgCO_2eq$/y)	360	1785
Reduction of GHG (tCO_2eq)	7.2	35.7
950,000 NGVs		
Reduction of GHG (tCO_2eq/y)	342,000	1,695,750
Reduction of GHG (tCO_2eq)	6,840,000	33,915,000

The annual amount of subsidies required to produce biomethane depends on two parameters: (i) the value of CIC (see Figure 3); and (ii) the quantity of biomethane demand (see Table 6) (Table 7). For example, when the CIC is equal to 375 € (0.305 €/m^3) and a use of 209 million m^3/y is hypothesized, the amount of subsidies is equal to 64 million €/y.

Table 7: Annual subsidies (million €/y)

CIC (€)	Scenario BIO-CNG (20%)	Scenario BIO-CNG (100%)
0.162	34	169
0.203	42	212
0.244	51	255
0.284	59	297
0.305	64	319
0.325	68	340
0.366	76	382
0.406	85	424
0.447	93	467
0.487	102	509

In 2015, the share of RE in Gross Final Energy Consumption (GFEC) is equal to 16.7% in EU 28, and Italy has achieved its national target (17%) in 2014 (Figure 5). Currently, this value is greater than the average of EU 28. This is verified also in electricity (+4.7%) and heating and cooling (+0.6%), and there is a reduction of this difference in comparison to previous years [20]. The EU has set a mandatory target for renewable fuels of 10% for each member state by 2020. Italy, but also EU 28, is far past this goal. Data presented in Figure 1 underline that biomethane is expanding in the

European context, while Italy presents only a number of upgraded plants equal to 1.4% and 1.2% of the European and global scenarios, respectively.

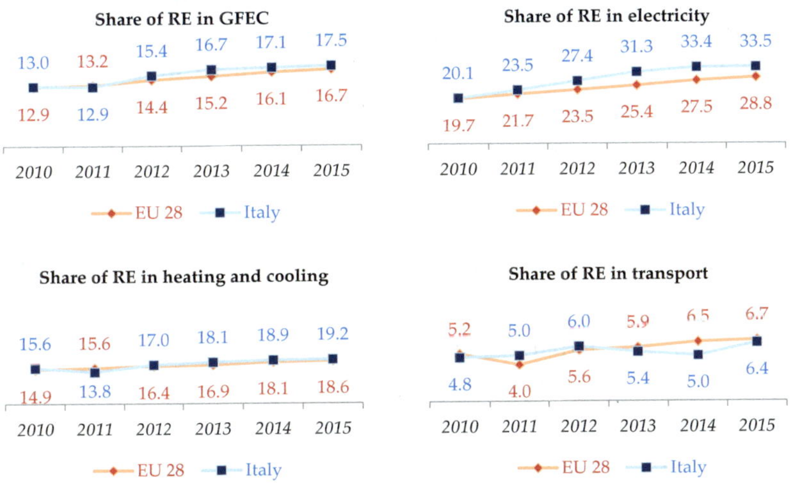

Figure 5: Share of renewable energy [34]. GFEC, Gross Final Energy Consumption.

Biomethane used as a vehicle fuel can be profitable, favoring the increase of the share of RE in the transport sector, and also, environmental improvements are obtained. These considerations permit biomethane to play a role in the development of the circular economy. Private and public firms operating in environmental services can use municipal solid waste to feed their own vehicle fleets. Agricultural residues are no longer a problem, but a resource for operators that are active in this sector.

Previous works have highlighted the key role of the plant size in the analysis of profitability. Furthermore, subsidies and the substrates used are critical variables. Results of this work define that the profitability of ofmsw substrate is verified also for a 150 m^3/h plant, while one with mixed substrate can be obtained starting with a 250 m^3/h plant in the scenario with a low cost of the transport of substrates. Alternatively, economic opportunities are provided by the incentive scheme when the producer of biomethane is also the distributor of methane. In this case, another corrective coefficient is recognized, and the pump price to the consumer is certainly higher than the price of selling of biomethane examined in this work. The excessive volatility of CICs influences in a negative way the investments in this sector, and consequently, new policy measures must be made to reduce this uncertainty.

REFERENCES

1. European Commission. Closing the Loop—An Eu Action Plan for the Circular Economy. Available online: http://ec.Europa.Eu/environment/circular-economy/index_en.Htm (accessed on 15 June 2017).
2. Capodaglio, A. Integrated, decentralized wastewater management for resource recovery in rural and peri-urban areas. ***Resources*** 2017, ***6***, 22.
3. Hanifzadeh, M.; Nabati, Z.; Longka, P.; Malakul, P.; Apul, D.; Kim, D.-S. Life cycle assessment of superheated steam drying technology as a novel cow manure management method. ***J. Environ. Manag.***2017, ***199***, 83–90.
4. Susmozas, A.; Iribarren, D.; Dufour, J. Assessing the life-cycle performance of hydrogen production via biofuel reforming in europe. ***Resources*** 2015, ***4***, 398.
5. Vassileva, I.; Thygesen, R.; Campillo, J.; Schwede, S. From goals to action: The efforts for increasing energy efficiency and integration of renewable sources in eskilstuna, sweden. ***Resources*** 2015, ***4***, 548.
6. European Biogas Association. Biogas and Biomethane Report 2015. Available online: http://european-biogas.Eu/ (accessed on 9 July 2017).
7. Parkes, R. Biomethane: Green gas rising? ***Renew. Energy Focus*** 2017, ***18***, 33–35.
8. O'Shea, R.; Wall, D.M.; Kilgallon, I.; Browne, J.D.; Murphy, J.D. Assessing the total theoretical, and financially viable, resource of biomethane for injection to a natural gas network in a region. ***Appl. Energy***2017, ***188***, 237–256.
9. Budzianowski,W.M.;Wylock,C.E.;Marciniak,P.A.Powerrequirements of biogas upgrading by water scrubbing and biomethane compression: Comparative analysis of various plant configurations. ***Energy Convers. Manag.*** 2017, ***141***, 2–19.
10. Miltner, M.; Makaruk, A.; Harasek, M. Review on available biogas upgrading technologies and innovations towards advanced solutions. ***J. Clean. Prod.*** 2017, ***161***, 1329–1337.
11. Billig, E.; Thraen, D. Renewable methane—A technology evaluation by multi-criteria decision making from a european perspective. ***Energy*** 2017, ***139***, 468–484.
12. Chen, L.; Cong, R.-G.; Shu, B.; Mi, Z.-F. A sustainable biogas model in china: The case study of beijing deqingyuan biogas project. ***Renew. Sustain. Energy Rev.*** 2017, ***78***, 773–779.

13. Patrizio, P.; Chinese, D. The impact of regional factors and new bio-methane incentive schemes on the structure, profitability and CO_2 balance of biogas plants in italy. ***Renew. Energy*** 2016, ***99***, 573–583.

14. DENA. The Role of Natural Gas and Biomethane in the Fuel Mix of the Future in Germany. Available online: https://www.Dena.De/en/home/ (accessed on 9 July 2017).

15. Petersen, A.M.; Melamu, R.; Knoetze, J.H.; Görgens, J.F. Comparison of second-generation processes for the conversion of sugarcane bagasse to liquid biofuels in terms of energy efficiency, pinch point analysis and life cycle analysis. ***Energy Convers. Manag.*** 2015, ***91***, 292–301.

16. Stürmer, B. Biogas—Part of austria's future energy supply or political experiment? ***Renew. Sustain. Energy Rev.*** 2017, ***79***, 525–532.

17. Foley, A.; Smyth, B.M.; Pukšec, T.; Markovska, N.; Duić, N. A review of developments in technologies and research that have had a direct measurable impact on sustainability considering the paris agreement on climate change. ***Renew. Sustain. Energy Rev.*** 2017, ***68 Pt 2***, 835–839.

18. D'Adamo, I.; Rosa, P. Current state of renewable energies performances in the european union: A new reference framework. ***Energy Convers. Manag.*** 2016, ***121***, 84–92.

19. Sgroi, F.; Foderà, M.; Trapani, A.M.D.; Tudisca, S.; Testa, R. Economic evaluation of biogas plant size utilizing giant reed. ***Renew. Sustain. Energy Rev.*** 2015, ***49***, 403–409.

20. Cucchiella, F.; D'Adamo, I. Technical and economic analysis of biomethane: A focus on the role of subsidies. ***Energy Convers. Manag.*** 2016, ***119***, 338–351.

21. Bauer, F.; Persson, T.; Hulteberg, C.; Tamm, D. Biogas upgrading—Technology overview, comparison and perspectives for the future. ***Biofuels Bioprod. Biorefin.*** 2013, *7*, 499–511.

22. Rotunno, P.; Lanzini, A.; Leone, P. Energy and economic analysis of a water scrubbing based biogas upgrading process for biomethane injection into the gas grid or use as transportation fuel. ***Renew. Energy***2017, ***102 Pt B***, 417–432.

23. Merlin, G.; Boileau, H. Eco-efficiency and entropy generation evaluation based on emergy analysis: Application to two small biogas plants. ***J. Clean. Prod.*** 2017, ***143***, 257–268.

24. Cucchiella, F.; D'Adamo, I.; Gastaldi, M. Profitability analysis for biomethane: A strategic role in the italian transport sector. ***Int. J. Energy Econ. Policy*** 2015, ***5***, 440–449.
25. MISE. Public Consultation on the Use of Biomethane and Biofuels. Available online: http://www.Sviluppoeconomico.Gov.It/index.Php/it/ (accessed on 28 June 2017).
26. Browne, J.; Nizami, A.-S.; Thamsiriroj, T.; Murphy, J.D. Assessing the cost of biofuel production with increasing penetration of the transport fuel market: A case study of gaseous biomethane in ireland. ***Renew. Sustain. Energy Rev.*** 2011, ***15***, 4537–4547.
27. Bortoluzzi, G.; Gattia, M.; Sognia, A.; Consonni, S. Biomethane production from agricultural resources in the italian scenario: Techno-economic analysis of water wash. ***Chem. Eng.*** 2014, ***37***, 259–264.
28. Bekkering, J.; Broekhuis, A.A.; van Gemert, W.J.T.; Hengeveld, E.J. Balancing gas supply and demand with a sustainable gas supply chain—A study based on field data. ***Appl. Energy*** 2013, ***111***, 842–852.
29. Uusitalo, V.; Soukka, R.; Horttanainen, M.; Niskanen, A.; Havukainen, J. Economics and greenhouse gas balance of biogas use systems in the finnish transportation sector. ***Renew. Energy*** 2013, ***51***, 132–140.
30. Smyth, B.M.; Smyth, H.; Murphy, J.D. Can grass biomethane be an economically viable biofuel for the farmer and the consumer? ***Biofuels Bioprod. Biorefin.*** 2010, ***4***, 519–537.
31. Sgroi, F.; Di Trapani, A.M.; Foderà, M.; Testa, R.; Tudisca, S. Economic performance of biogas plants using giant reed silage biomass feedstock. ***Ecol. Eng.*** 2015, ***81***, 481–487.
32. Schievano, A.; Scaglia, B.; D'Imporzano, G.; Malagutti, L.; Gozzi, A.; Adani, F. Prediction of biogas potentials using quick laboratory analyses: Upgrading previous models for application to heterogeneous organic matrices. ***Bioresour. Technol.*** 2009, ***100***, 5777–5782.
33. Schmid, A.; Batton-Hubert, M.; Naquin, P.; Gourdon, R. Multi-criteria evaluation of end-of-life vehicles' dismantling scenarios with respect to technical performance and sustainability issues. ***Resources*** 2016, ***5***, 42.
34. Eurostat. Statistics Database. Available online: http://ec.Europa.Eu/eurostat/data/database (accessed on 15 June 2017).

13

Process Disturbances in Agricultural Biogas Production—Causes, Mechanisms and Effects on the Biogas Microbiome: A Review

Susanne Theuerl [1], Johanna Klang [1], and Annette Prochnow [1,2]

[1]Leibniz Institute for Agricultural Engineering and Bioeconomy, Max-Eyth-Allee 100, 14469 Potsdam, Germany
[2]Humboldt Universität zu Berlin, Albrecht Daniel Thaer Institute for Agricultural and Horticultural Sciences, Hinter der Reinhardtstr. 6–8, 10115 Berlin, Germany

ABSTRACT

Disturbances of the anaerobic digestion process reduce the economic and environmental performance of biogas systems. A better understanding of the highly complex process is of crucial importance in order to avoid disturbances. This review defines process disturbances as significant changes in the functionality within the microbial community leading to unacceptable and severe decreases in biogas production and requiring an active counteraction to be overcome. The main types of process disturbances in agricultural biogas production are classified as unfavorable process temperatures, fluctuations in the availability of macro- and micronutrients

Citation: Theuerl S, Klang J, Prochnow A. Process Disturbances in Agricultural Biogas Production—Causes, Mechanisms and Effects on the Biogas Microbiome: A Review. Energies. 2019; 12(3):365.

(feedstock variability), overload of the microbial degradation potential, process-related accumulation of inhibiting metabolites such as hydrogen (H_2), ammonium/ammonia (NH_4^+/NH_3) or hydrogen sulphide (H_2S) and inhibition by other organic and inorganic toxicants. Causes, mechanisms and effects on the biogas microbiome are discussed. The need for a knowledge-based microbiome management to ensure a stable and efficient production of biogas with low susceptibility to disturbances is derived and an outlook on potential future process monitoring and control by means of microbial indicators is provided.

Keywords: anaerobic digestion, biogas, microbial biodiversity, process control

INTRODUCTION

One of the main challenges to optimize biogas production is to achieve high process stability and efficiency with low susceptibility to disturbances [1,2,3,4,5,6]. Process disturbances reduce the economic and environmental performance of biogas systems since they lead to decreased methane yields and hence reduce revenues, net energy yields and greenhouse gas emissions mitigation and result in additional expenditures for remedy. Consequently, there is the need to better understand the causes, mechanisms and effects of process instabilities and disturbances, especially the response of the biogas microbiome in order to avoid them.

Given that the anaerobic digestion plant is technically fully operational (e.g., regarding digester technology, mechanic feeding, temperature regulation, and stirring or pumping systems), all instabilities and disturbances of the digestion process are attributed to malfunctioning of the biogas microbiome induced by the operator's management measures. The objective of this review is to define process instabilities and disturbances while classifying different types, their causes, mechanisms and effects on the microbial community. Additionally, an outlook on potential future process monitoring and control strategies by means of using microbial indicators to deduce strategies for a sustainable microbial diversity management will be provided. The focus of this review is on agricultural biogas production where the plants are mainly operated by farmers and run on the basis of agricultural residues (e.g., livestock manure), energy crops (e.g., maize, grass, sugar beet) and partly residues from the food industry and biorefineries.

THE BIOGAS MICROBIOME

Biogas digesters are highly sensitive technological-biological systems with a diverse and complex interacting microbial community that converts anaerobically degradable biomass into biogas. The process can be roughly divided into the four main steps hydrolysis, acidogenesis, acetogenesis and methanogenesis whereby the metabolites of one step are the precursors for the next step (Figure 1). These steps run simultaneously as the involved microorganisms with their different physiological capacities depend on each other's degradation products and require a close spatial proximity.

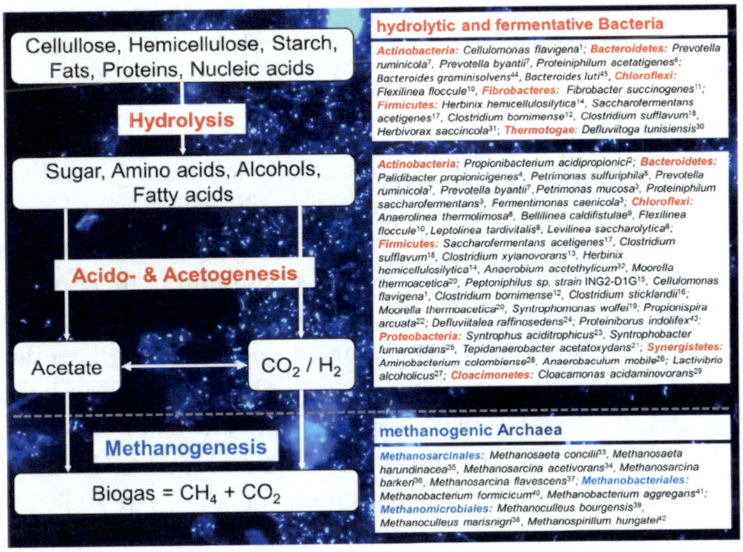

Figure 1: Compilation of the currently physiologically and/or genetically described microorganisms putatively involved in the different steps of the anaerobic digestion process. The corresponding reference list is given in Supplement 1. The microscopic image in the background shows a biogas microbiome stained with DAPI (4′,6-diamidino-2-phenylindole) of an anaerobic digester treating a mixture of energy crops and animal manure (photo by J. Klang). CO_2 = carbon dioxide, H_2 = hydrogen, CH_4 = methane.

So far, the most investigated species of the biogas microbiome are hydrolytic and fermentative bacteria and methanogenic archaea. Figure 1 shows a compilation of the physiologically and/or genetically described microorganisms involved in the different steps of the anaerobic digestion process.

The biogas microbiome does not only comprise bacteria and archaea but also viruses/phages [7] as well as eukaryotes including fungi and protists (mainly protozoa) which are less investigated [8,9,10]. While it is assumed that the functional characteristics of pro- and eukaryotes are closely related to each other in a cooperative way [11], viruses/phages are suspected to have regulatory effects on the microbial cell turnover [7,12].

The current state of knowledge on the biogas microbiome can be characterized by three general statements:

- Despite intensive research and growing knowledge on the microbiome most of the species and hence their ecological functions are still unknown [13,14,15].
- Each anaerobic digester develops its own specific microbiome [16,17,18].
- Most individuals of the biogas microbiome are generalists which are able to exist under various conditions, hence being found in most of the biogas plants and with regard to their abundance making the largest part of the community. In contrast, most species are specialists which occupy specific ecological niches, are detected only in few samples and are often specific to the digester. (In 20 anaerobic digestion plants a total number of 5938 bacterial operational taxonomic units (OUT) were detected. Only 2.5% (around 150 OTUs) were assigned to the biogas core microbiome with a median relative abundance of 70.3% ± 12.5% (the generalists), whereas 84.0% (around 5000 OTUs) were only found in a certain number of samples representing 3.5% ± 3.8% of the relative bacterial abundances (the specialists) [16]).

The microbial diversity in anaerobic digesters is affected by the plant operator's management as this influences the microbiome's living conditions (Figure 2). The nutritional basis for the microorganisms is provided by the feedstocks. Their chemical composition and physical characteristics affect the bioavailability/bioaccessibility of the anaerobically degradable compounds. The chosen type, amount and possibly pretreatment of feedstocks determine the available amount of macro- and micronutrients. The abiotic environment of the microorganisms is formed by the digester technology (e.g., continuously stirred/plug flow/container, one/two-phase) and the digester operation (e.g., loading rate, retention time or process temperature). When designing the abiotic environment of the biogas microbiome it has to be considered that each population (sum of individuals of a respective species in a habitat) of the biogas microbiome does not only have different optima regarding its living requirements, but also different tolerance levels against environmental factors. If these environmental factors exceed or fall below certain thresholds process disturbances might occur.

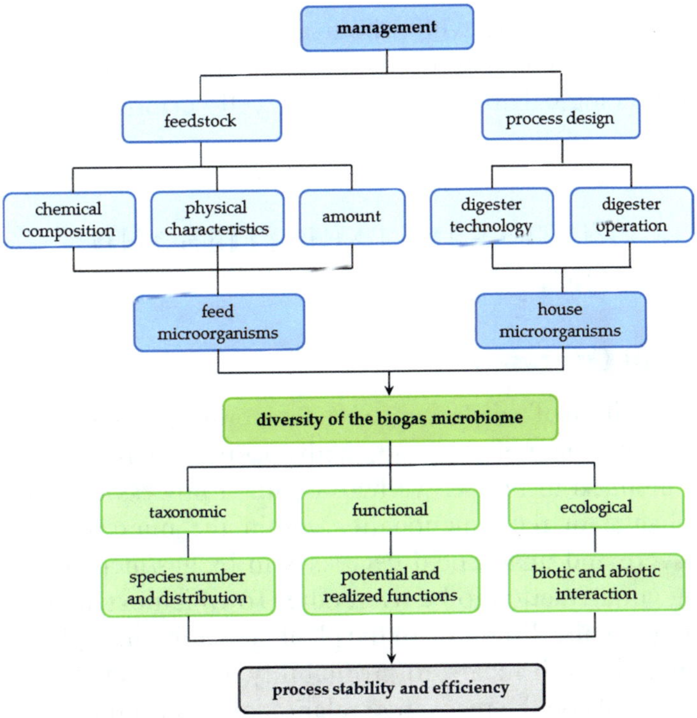

Figure 2: Management of the anaerobic digestion process and diversity levels of the biogas microbiome.

These management options shape the biogas microbiome which is characterized by its taxonomic, functional and ecological diversity (Figure 2). The taxonomic diversity covers the number and distribution of occurring species. The functional diversity defines the potential (i.e., genetically determined) and the actually realized functions and processes of the occurring species. The ecological diversity describes interactions of the microorganisms among each other and with their environment. The more we are able to understand and quantitatively describe the ecological functions of biogas microbiomes in their complexity, the more we will be able to deduce reliable and reproducible forecasts how systems react to varying external influences and how the performance of biogas microbiomes can be optimized and sustainably used for a knowledge-based process control. This in turn will enable us to better use the biogas microbiome technically to the benefit of human and environment (adapted from [19]).

In order to establish a sustainable microbial diversity management strategy for anaerobic digestions systems, the question that still has to be answered is "Who does what, when, with whom and why?" [5,8,14,20,21,22,23]. Moreover, there is an urgent need to better understand the taxonomic, functional and ecological difference between well-performing and disturbed microbiomes.

TYPES OF PROCESS INSTABILITIES AND DISTURBANCES

Definition and Overview

The general definition of a disturbance is a temporary event that significantly changes the normal state of a system. With specific regard to the anaerobic digestion of biomass to biogas we here define a process disturbance as a significant change in the functionality within the microbial community leading to severe and unaccepted decreases in biogas/methane generation and requiring counteraction to be overcome. To which extent a decrease is accepted or not is at the discretion of the plant operator. Since plant operators will decide mainly with regard to profitability, the accepted extent of the disturbance depends on the respective relation of costs and revenues [24].

We distinguish process instabilities from disturbances as temporary changes in the functionality within the microbial community that cause slight decreases in biogas/methane generation whereby the microbiome is able to reorganize itself without active counteraction of the plant operator.

The response of microbiomes or of populations within the microbiome to process disturbances can be described using three basic concepts [25]: (i) the microbiome can be resistant, meaning the microbiome withstands a disturbance without changing its taxonomic, functional and ecologic composition, (ii) the microbiome is resilient, which means that the microbiome composition returns back to its original composition after being disturbed, and (iii) members of the microbiome are functional redundant whereby inhibited community members can be replaced by others with similar function. Instabilities and disturbances of the anaerobic digestion process may have manifold causes with different underlying mechanisms

[26,27,28,29]. The main types occurring in agricultural biogas production can be classified as follows:

- unfavorable process temperatures,
- fluctuations in the availability of macro- and micronutrients (feedstock variability),
- overload of the microbial degradation potential,
- process-related accumulation of inhibiting metabolites such as hydrogen (H_2), ammonium/ammonia (NH_4^+/NH_3) or hydrogen sulphide (H_2S),
- inhibition by other organic and inorganic toxicants, e.g., light metal ions, heavy metals, chlorophenols, halogenated aliphatics, long chain fatty acids, nanomaterials, antibiotics, or viruses.

In practice it is nearly impossible to relate an instability or disturbance to one single cause/factor; mostly it is a combination of various factors that leads to process instabilities or disturbances (Figure 3). In the next sections, causes, mechanisms and effects on the biogas microbiome for the first four categories of the above mentioned disturbances are discussed. The last category is less intensively investigated in agricultural biogas production, especially regarding the response of the microbiome, and will not be discussed in detail here.

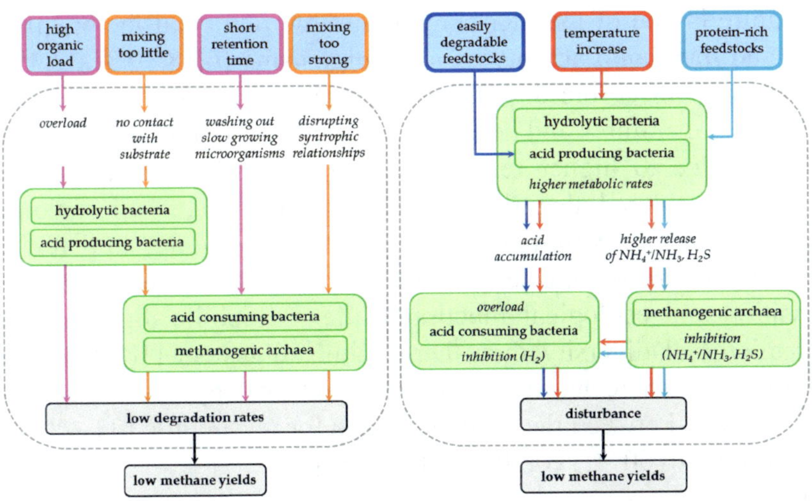

Figure 3: Causes, mechanisms and effects of process disturbances on the biogas microbiome (dashed line—microbial system; blue boxes—management

measures causing process disturbances; green boxes—microorganisms affected in the four steps of the digestion process; colored arrows—cause–effect chains of disturbance types).

So far, no studies are available on process disturbances in full-scale anaerobic digesters which investigate the dynamics of a process disturbance from the beginning to the end, distinguishing the "disturbed" microbiome from the "normal/healthy" or "recovered" microbiome in order to identify process-relevant microorganisms. This can mainly be attributed to the fact that process disturbances in full-scale biogas plants cannot be reasonably induced due to the high operation costs and potential economic losses. Moreover, process disturbances cannot be predicted. Consequently, studies on process disturbances in full-scale anaerobic digesters depend on randomly occurring events challenged by the "search" for instabilities/disturbances directly at the right time and place when and where the disturbances occur. Hence, such studies are case studies where the results cannot be validated or reproduced. Because of this, most studies and hence most knowledge on process disturbances originate from model digestion systems as they offer a broad spectrum of options to manipulate the process under controlled conditions.

Unfavorable Process Temperatures

Temperature is one of the most important environmental factors affecting growth and metabolic activity of microorganisms [6,18,30,31,32,33]. For the process of anaerobic digestion, biomass degradation accelerates with increasing temperatures and hence per time unit more biomass is converted into biogas due to higher reaction rates (Figure 3) [6,33]. However, it has to be considered that the members of the biogas microbiome are metabolically active within certain ranges of temperature, so that changes in the temperature regime lead to a change in the microbial community composition and ecological function while mesophilic digesters show a higher microbial diversity then thermophilic digesters [18,30,34]. Many agricultural anaerobic digesters (especially in Germany) are operated at 40–45 °C, which is between the optimal ranges for mesophilic (33–38 °C) and thermophilic (50–60 °C) conditions [6,18,31].

The temperature in anaerobic digesters should be changed slowly to enable the microbial community to adapt to the new environmental conditions. Strong fluctuations or fast increases or decreases carry the risk of process instabilities (Figure 3). This is well shown in an experiment which investigated how a change in the temperature regime influences the process stability, the potential for process optimization and the microbial community structure [6]. A temperature increase from 37 °C to 52 °C or a decrease from 52 °C to 37 °C was performed stepwise by 2 °C per week. The digester with decreasing temperature showed the most unstable process. During the temperature change acetate and propionate concentrations were high and methane yields low. The digester with increasing temperature operated more stable, but temporary disturbances still occurred when passing the range between 40 and 44 °C. These results are in accordance with previous findings [31]. At the microbial level a temperature sensitivity of members of the phylum ***Cloacimonetes*** was found. They disappeared when the temperature increase passed the 40–44 °C range [6]. Similar findings have been reported previously [35] indicating that this microbial group might be a suitable process indicator for upcoming process disturbances due to an inappropriate temperature regime.

As mentioned above, a temperature increase leads to an increased metabolic activity and hence to higher degradation rates with an accelerated release of organic acids and other process-inhibiting metabolites such as ammonium/ammonia (NH_4^+/NH_3) or hydrogen sulphide (H_2S) (see Section 3.6), making the process less stable. Hence, for a permanent or temporary thermophilic digester operation a knowledge-based management of the microbiome is necessary.

Fluctuation of Nutrient Availability

Feedstocks as Nutritional Basis

The feedstocks are the nutritional basis for the biogas microbiome. The macro- and micronutrients within the feedstocks affect microbial growth and activity and hence the degradation kinetics and the biogas/methane yield [10,36,37]. An optimal nutrient supply is essential for a stable digestion process. That means, the feedstock composition should (i) meet the nutritional requirements of the digester's microbiome, (ii) result in high biogas and methane yield and (iii) lead to a high quality digestate [10].

The availability of the macro- and micronutrients for the microbial community changes depending on the amount, composition and quality of the feedstocks as well as on the general process performance [18,30,38,39]. The main reasons for fluctuating nutrient availability are variation of the feedstock supply and insufficient mixing of the digester's content.

Chemical Composition of the Feedstocks

The chemical composition and the bioaccessibility of the feedstocks determine the potential amount of biogas/methane that can be produced [29]. The theoretical biogas yield from carbohydrates, fats and proteins are 750, 1390 and 800 norm liters per kg organic dry mass with theoretical methane amounts of 50%, 72% and 60% [33]. The amount and bioavailability of macro- and micronutrients varies strongly in different feedstocks. As different microorganisms have different nutrient requirements and also different degradation capacities to certain feedstock compounds, the chemical complexity of the supplied feedstock (either various compounds or complex molecule structures) directly affects the diversity of the biogas microbiome [18].

Anaerobic digesters operated with easily convertible feedstocks such as sugar beet silage or chemically stable/similar feedstocks (in function of time), such as wastewater sludge and industrial wastes show less diverse microbial communities. Several studies have shown that these communities are often dominated by members of the phylum ***Bacteroidetes***, especially from the order ***Bacteroidales*** [18,30,38,40,41,42,43]. Most of the so far described members of this phylum are known for rapid conversion of easily degradable sugars and alcohols into acids e.g., [44,45,46,47,48,49] which indicates that they play a crucial role in acidogenesis and acetogenesis.

In contrast, feedstocks such as biowaste, energy crops and livestock manure are chemically more complex and/or heterogeneous in function of time. Their conversion needs a structural more diverse community with members of various phyla (e.g., ***Firmicutes***, ***Bacteroidetes***, ***Chloroflexi***, ***Proteobacteria***, ***Spirochaetes***, ***Synergistetes***, ***Thermotogae***, ***Cloacimonetes*** as well as ***Euryarchaeota***) e.g., [18,30,38,43,50,51,52]. Such communities are enabled to successively, complementarily and efficiently degrade most of the bioavailable/bioaccessible feedstock compounds (except the anaerobically hardly-degradable lignin [51,53]) due to a high functional and ecological diversity (broad range of metabolic pathways, generalists, specialists, redundancy, resilience, concurrence, syntrophy, co-occurrence). As chemically more complex feedstocks often feature high lignocellulose contents, pretreatment techniques are applied in order to enhance rheology, degradability and hence biogas/methane production [52,54,55,56]. Some pretreatment techniques, mainly chemical ones, are known to generate toxic compounds [55,57] which have to be expected to cause inhibitory effects in the subsequent digestion process. Many studies deal with the impact of various pretreatment techniques on methane yields, but so far it has not been investigated how the pretreated feedstocks affect the diversity of the biogas microbiome and hence the process performance.

If the feedstock type changes, the microbial communities need to adapt to the new conditions, not only with respect to their taxonomic composition, but above all regarding their predominant functions and ecological behavior. Changing the feedstock does not necessarily have negative impacts on biogas production, but should be applied with caution. For example, a stepwise increase of poultry manure to a basic mixture of cattle slurry and solid manure led to altered nutrient supply and hence changing environmental conditions causing process instability. Without taking countermeasures, the microbiome reconstructed over a period of approximately 100 days [58] (see also Section 3.6). The new microbial community was functionally redundant as similar biogas/methane yields were recorded after the disturbance. From an ecological point of view, this can be called a naturally controlled microbial diversity management.

Nutrient Distribution

It is well known that a good stirring/mixing of the digester's content is of high importance to evenly spread fresh feedstock material within the digester, to ensure a close contact between the substrates (i.e., the nutrients supplied with the feedstocks) and the degrading microorganisms, to facilitate the removal of end products of metabolism and the upflow of gas bubbles, as well as to achieve even temperature conditions across the digester [33,59,60].

Depending on the kind of feedstock, insufficient mixing (low mixing intensity) might lead to stratification of the digester content. Light material tends to float on the surface (floating layer formation) while heavier material sinks to the bottom (sedimentation) [60]. Consequently, the contact area between microorganisms and their substrates is disrupted (Figure 3). In case of a floating layer formation, the contact area is restricted to the boundary zone of the solid phase (the biodegradable material) and the liquid phase where most of the microorganisms are located. Hence, only a small fraction of the nutrients is accessible to the microorganisms causing low degradation rates.

On the other hand, too high mixing intensities are detrimental to the biogas production as the required close contact between acid oxidation bacteria and their methanogenic partners is interrupted (Figure 3) [60,61].

Although stirring/mixing of the digester content is of high importance, little is known about the effects of stirring/mixing at the microbial level. Further research should investigate the effects of stirring on the microbiome at the different stages of anaerobic digestion [60], as this could be one option to manage the process efficiency by regulating the contact between the microorganisms and their nutrients and hence regulate the biogas flux.

Overload of the Microbial Degradation Potential

The digestion process becomes overloaded when the biomass input (amount of organic material added per active reactor volume and day) exceeds the degradation capacity of the biogas microbiome. This situation may occur for instance at too high organic loading rates in combination with too short retention times [10,40,62,63,64]. This leads to insufficient degradation rates and low biogas/methane yields in proportion to the amount of the supplied biomass (Figure 3). While the hydraulic retention time defines the medium duration of the feedstocks in the digester, the solids retention time describes the available time for the microorganisms to establish in the digester [65].

When the retention time is too short, which is typically the case with increasing the organic loading rate, microbial biomass cannot establish in required amounts to degrade the supplied biomass which might cause a wash-out of microorganisms resulting in a loss of essential functions for the ecosystem (Figure 3) [10,64,65]. Hence, organic load can only be increased when the retention time is long enough to ensure a certain reproduction (growth) and spreading time for the microorganisms to meet their function. This applies in particular to the slowly-growing acid-degrading and methanogenic members of the community. Thus, for a stable process performance, retention times of more than 10 days are generally recommended [66,67].

An overload of the process can also occur when the formation of one metabolic intermediate exceeds the degrading capacity of the microorganisms involved in the subsequent step. This can be caused by insufficient mixing of the digester's content, changing feedstock regimes, increasing temperature or inhibition of specific microbial populations by NH_3, H_2S or H_2.

Overloads often are related to acid accumulation (Figure 3), indicating an imbalance of microbial acid producers and consumers [40,63,68,69]. The underlying mechanism is based on different ways of obtaining energy and consequently on different growth rates: a thermodynamically more favorable energy production results in faster growth rates of the microorganisms [70,71,72,73].

While the energy balance of most hydrolytic and acidogenic reactions is negative —meaning a direct energy gain for the microorganisms— the reactions of acid-degrading microorganisms are thermodynamically unfavorable and require more energy than they yield [70,71,74]. Acid-degrading microorganisms often live in syntrophy with hydrogen scavengers, in the biogas process ideally hydrogenotrophic archaea, due to the fact that the conversion of most fatty acids is only possible via the formation of hydrogen (H_2). Without subsequent H_2 utilization by H_2scavengers such as hydrogenotrophic archaea or homoacetogens, the acid-degrading bacteria would be inhibited by their own metabolic end product [73]. However, the overall energy balance becomes negative and proceeds with a low energy gain for both partners (see Section 3.5).

With regard to an acid accumulation, this means that the biomass is relatively quickly converted into acids which can often be assigned to an increased relative abundance of members of the phylum ***Bacteroidetes***[40,63,69], but the acids cannot be converted into biogas in the same time frame due to the slower growth rates of the acid-degrading microorganisms [63]. This is of high importance and should be carefully considered while operating a biogas plant, especially under thermophilic conditions with a less diverse microbiome (Figure 3). From an ecological point of view, microbiomes with a high diversity level are assumed to be more stress resistant/resilient due to a higher metabolic potential and hence a high functional redundancy resulting in a higher stress tolerance potential according to increasing organic loads or even overloads [4,69].

Although it is well known that high loading rates (in combination with short retention times) have to be handled with caution, the entire regulatory processes—including the response of the microbiome or of its members— is still not fully understood, which hampers the recommendation of respective management strategies to efficiently control the process.

Hydrogen Inhibition

High amounts of H_2 may indicate a disturbance of the microbial community equilibrium. Methanogenic archaea are only capable to use acetic acid, H_2 and one-carbon (C1) compounds, such as carbon dioxide (CO_2) [75]. This means that C3-C6 fatty acids (propionic acid, butyric acid, succinic acid, and lactic acid) formed during acidogenesis, first have to be converted into utilizable compounds [70,74,76]. This is carried out by bacteria, such as members of the genera ***Syntophomonas***, ***Syntrophobacter***, ***Syntrophus***, ***Propionibacter***, ***Pelotomaculum***, ***Smithella*** or ***Clostridium***, who live in syntrophy with hydrogenotrophic archaea [70,74]. Syntrophy, in general terms, is an association of different microorganisms who mutually provide and use intermediates to and from the other partner and hence are interdependent for survival [73,74,75,76,77,78]. By the combined metabolic activity of microorganisms, endergonic reactions can become exergonic through the efficient removal of products and therefore enable a microbial community to survive with minimal energy resources [70,74].

Due to the lack of suitable electron acceptors, the conversion of fatty acids is often only possible via the formation of high-calorific H_2 [70,74,77,78]. As mentioned above, without subsequent utilization of the generated $H_2$by hydrogen scavengers, the syntrophic bacteria would be inhibited by their own end product since excessive H_2concentrations make the fatty acid oxidation thermodynamically impossible [73]. It follows that hydrogen scavengers (e.g., hydrogenotrophic archaea) depend on the H_2 which is produced by their neighboring bacteria who in turn can only grow, if the H_2 is consumed. Consequently, an inhibition of the function of hydrogen scavengers, such as hydrogenotrophic archaea by NH_4^+/NH_3 or H_2S inhibition, leads to a process disturbance which is reflected not only in reduced methane yields, but also in acid accumulation (Figure 3, see Section 3.4 and Section 3.6).

In this context syntrophic acetate oxidation is a particularly well investigated process that occurs under certain environmental conditions. The energy balance of acetate oxidation is quite unfavorable (free Gibb's energy ($\Delta \boldsymbol{G}^{\circ\prime}$) = +104.6 kJ·mol^{-1}). Together with hydrogenotrophic partners the energy balance improves to $\Delta \boldsymbol{G}^{\circ\prime}$ = −31 kJ·mol^{-1}($\Delta \boldsymbol{G}^{\circ\prime}$ = −15.5 kJ mol^{-1} per partner), whereby the H_2 partial pressure should be limited to the range 1.6 to 6.8 Pa under mesophilic conditions [79,80]. Due to the highly complex requirements of syntrophic microorganisms for their living conditions, they can hardly be cultivated and hence their genetic, physiologic and ecological characterization is limited. Bacteria currently known to be capable of syntrophic acetate oxidation are ***Thermacetogenium phaeum***, ***Pseudothermotoga lettingae***, ***Tepidanaerobacter acetatoxydans***, ***Clostridium ultunense*** and ***Syntrophaceticus schinkii*** [80].

Moreover, there are indications that high partial pressure of H_2 also negatively affects hydrolysis. An observed decrease in the degradation of wheat straw without acid accumulation (the metabolites from acidogenic bacteria) indicates that the hydrolytic activity (the first process step) was inhibited [81]. H_2 inhibition at the hydrolysis stage could be suppressed through the addition of CO_2 [81].

Ammonium/Ammonia and Hydrogen Sulphide Inhibition

The anaerobic digestion of feedstock with high nitrogen contents, such as grass silage, poultry manure or slaughterhouse waste, may negatively affect the function of the microbial community and hence biogas and methane generation (Figure 3). A critical issue is the formation of ammonium (NH_4^+) which is generated during the degradation of proteins, nucleic acids and/or uric acid/urea and cannot be further degraded under anaerobic conditions [80,82,83,84,85,86]. In aqueous solutions NH_4^+ is in equilibrium with NH_3, which shifts towards NH_3 with increasing temperature and pH. Generally, it is assumed that microorganisms, e.g., methanogenic archaea, might be affected "in two ways: (i) the ammonium ion (NH_4^+) may inhibit the methane producing enzymes directly and/or (ii) the hydrophobic ammonia molecule (NH_3) may diffuse passively into the microbial cells, causing proton imbalance or potassium deficiency" [84].

Methanogenic archaea generate methane either by converting acetic acid (acetoclastic methanogenesis) or by using hydrogen (H_2) and carbon dioxide (CO_2) (hydrogenotrophic methanogenesis) [75]. The obligate acetoclastic methanogens (members of the family ***Methanothrix***) have the narrowest ranges of tolerance to NH_4^+ and NH_3 e.g., [84,87,88,89]. Frequently described and generally accepted thresholds for concentrations of NH_4^+ range from 3 to 5 g L^{-1} and of NH_3 from 80 to 400 mg L^{-1} [80,84]. With rising concentrations of NH_4^+ and/or NH_3, changes in the taxonomic and hence the functional diversity within the archaeal community are recorded e.g., [58,83,90,91].

For example, in a long-term, lab-scale experiment the impact of increasing amounts of poultry manure and accordingly increasing NH_4^+/NH_3 concentration on the microbial diversity was investigated [58]. During the second experimental phase the addition of 50% poultry manure (based on VS) led to a serious process instability with a total acid concentration of 9.6 g L^{-1} at 5.9 g NH_4^+ L^{-1} and 500 mg $NH_3 \cdot L^{-1}$, which resulted in a decrease of the methane yield by 25%. Over the course of the instability, while operating the digester at constant conditions, the dominating bacterial order shifted from ***Bacteroidetes*** to ***Clostridiales***, combined with a change from acetoclastic to hydrogenotrophic methanogenesis. With similar overall process rates after the disturbance, the adapted microbiome was functional redundant [58]. Similar results regarding the microbiome reorganization were published previously e.g., [83,90,91,92,93]. However, another study investigated the impact of different ammonium sources (animal manure and ammonium carbonate) to reveal whether there is a relation between the microbial diversity level (structure and function) and the stress tolerance potential of the microbial community [90]. It was shown that a sugar beet silage digesting bacterial community with a pre-dominance of members from the phylum ***Bacteroidetes***associated with a functionally flexible archaeal community (mainly ***Methanotrix***, ***Methanosarcina***, ***Methanoculleus***), was more stress resistant under the certain experimental conditions, than a maize silage digesting community with a more diverse and evenly distributed bacterial community associated with a rigid archaeal community (mainly ***Methanotrix***). This indicates that a higher taxonomic and hence functional diversity at a certain community part, in the presented study the archaeal community, is one important factor for process stability.

Furthermore, members of the bacterial phylum ***Cloacimonetes*** were reported to disappear with increasing NH_4^+/NH_3 concentrations [6,90]. Especially in the study by Klang et al. [90], the relative abundance decreased more than 50 days prior to a process disturbance. In consequence, the disappearance of these microorganisms might be a potential indicator for upcoming process disturbances caused by increasing total ammonium nitrogen (sum of NH_4^+ and NH_3) concentrations.

As already mentioned and several times reported, obligate acetoclastic methanogens (members of the family ***Methanotrix***) are most sensitive against elevated NH_4^+/NH_3 concentrations. If acetic acid cannot directly be converted into methane due to NH_4^+/NH_3 inhibition of the respective microbial group, the community uses the pathway of syntrophic acetate oxidation (see Section 3.5). Here, the acetic acid is converted into CO_2 and H_2 via the inverse homoacetogenesis and further converted into methane by hydrogenotrophic archaea [79,80]. Based on the results of different studies, the threshold for the syntrophic acetate oxidation is limited to the range 140 to 250 mg NH_3 L^{-1} (approximately 3.0 to 3.3 g·NH_4^+ L^{-1}) at a temperature of 37 °C and a pH from 7.5 to 8.0 [80].

Another example for a process-related accumulation of inhibiting metabolites is the production of hydrogen sulphide (H_2S) (Figure 3). Since H_2S is produced by the degradation of proteins, or, more precisely, of sulphur-containing amino acids, such as cysteine and methionine, an increased amount of H_2S can be expected if protein-rich feedstocks, such as grass silage, slaughterhouse waste or biowaste, are used [18]. H_2S has a toxic effect on methanogenic archaea (Figure 3), as it diffuses through the cell membrane and denatures proteins within the cell, leading to inhibition of enzymes [65,94,95]. H_2S also reacts with metal ions. The precipitation of, for example, chromium (Cr) or nickel (Ni), which are essential as enzyme cofactors, is an indirect form of inhibition of methanogenesis by H_2S [37,95]. Although it is quite known that H_2S can lead to process disturbances due to the inhibition of the most important functional group of the biogas microbiome, the methanogenic archaea [28,82], little is known about the underlying microbial mechanisms. De Jonge et al. [94] investigated the reaction of a microbial community to a prolonged starvation time and the subsequent process reactivation under thermophilic conditions. A large shift within the acetogenic and methanogenic communities was observed, which was attributed to a strong increase in the volatile fatty acids (VFA), NH_4^+/ NH_3 as well as H_2S concentration. However, this topic still needs further research, especially regarding the causes, mechanisms and effects on the biogas microbiome.

Further Potential Causes for Process Instabilities/Disturbances

There are several other potential causes for process instabilities/disturbances which partly enter the system with the feedstock supply (e.g., heavy metals, organic toxicants, antibiotics), while others are by-products originated during the break down of biomass (e.g., light metals, trace elements, long chain fatty acids) or which are part of the native biogas microbiome, like viruses [7,12,37,53,82,96,97,98,99,100,101] Yet, little is known about the underlying mechanisms and how they affect the biogas microbiome.

For instance, is it well known that salts (or better the cations of salts, light metal ions such as Na^+, K^+, Mg^{2+}, Ca^{2+}, Al^{3+}) are required for microbial growth and activity like any other nutrient, but high salt levels can also lead to process instabilities/disturbances as they cause dehydration of microbial cells due to osmotic pressure [82]. In the case of potential salt stress, it has often been described that conductivity values above 30 mS cm^{-1} indicate unfavorable process conditions which are related to variations in the microbial diversity [18,82,102]. Also heavy metals (Cu, Zn, Pb, Hg, Cr, Cd, Fe, Ni, Co, Mo) are important trace elements and a certain amount is required for microbial growth and activity. Fe, Ni and Co, for example, are considered to be most essential in anaerobic digestion as they are important cofactors in enzyme systems in the methane formation pathway. Consequently, a certain amount of trace elements can have a stimulating effect on the anaerobic digestions, e.g., by increasing the volatile fatty acid utilization even at elevated TAN concentration of up to 7.2 g L^{-1} while mono-digesting chicken manure [103]. Moreover, trace elements also act as a binding component in sulfide precipitation and hence can be used to control the level of H_2S [37,104]. On the other hand H_2S has a negative impact on the trace element bioavailability since trace elements are often immobilized in sulfides, especially in anaerobic digestion with a redox potential of −300 mV and neutral pH [37]. This in turn indirectly inhibits the process (see Section 3.6). However, in practice heavy metals (trace elements) are often supplemented in excessive amounts to anaerobic digesters. As they are non-biodegradable, their accumulation can inhibit acetogens, acidogens as well as methanogens [37,96]. While methanogens are assumed to be most sensitive to changes in trace element speciation and bioavailability, knowledge on the effects of trace elements on the other anaerobic trophic groups is still lacking [37].

In addition, some recent studies show inhibitory effects of organic compounds such as lignin derivates [53], humic acids [99,101], or coumarin, a metabolite derived from the anaerobic conversion, e.g., of grassland biomass [105,106]. The underlying mechanisms of inhibition are still unknown. It is assumed that humic acids prevent cellulose degradation by binding to the corresponding hydrolytic enzymes [99,101].

Besides these examples for abiotic factors, there are also biotic factors which might cause process instabilities/disturbances. Significant correlations were shown between the occurrence of phages and prokaryotes over time at the α- and β-diversity level: 40.6% of total prokaryotic community variation was explained by the abundance of phages, in comparison to only 14.5% by abiotic factors [12]. This study showed that phages and the prokaryotic communities are highly interconnected, which might be in turn responsible for process stability or even instabilities as phages are one reason for prokaryotic cell lyses. This indicates that biotic factors should be much more considered for process or better microbiome control in the future.

PERSPECTIVES OF A FUTURE MICROBIAL PROCESS MONITORING AND CONTROL

To ensure a stable and efficient biogas production, regular process monitoring is standard nowadays. This does not only provide insights into the general process performance, it also provides the opportunity to detect/recognize process instabilities/disturbances [28,29]. Yet, today´s process control is still based on technical and chemical parameters and on the experience of the biogas plant operators (Table 1) [8,20] resulting in various process enhancing strategies such as pretreatment of feedstocks [52,55,56], optimization of the digester technology [107], optimization of operational parameters [29,108] as well as the supplementation of additives [107]. Although the production of biogas would be impossible without microorganisms, process control based on microbial diversity parameters is still missing.

Table 1: Overview on indicators commonly used to monitor process stability and efficiency (compiled from [27,28,29]). Given are the indicators, the direct or indirect information on the process as well as efforts which are required for their measurements. CH_4 = methane, VFA = volatile fatty acids, NH_4^+ – ammonium, NH_3 = ammonia

Indicator	Direct or Indirect Information on	Measuring Effort
feedstock composition	nutrient availability	high, laboratory
temperature	metabolic performance	low, thermometer
amount of produced biogas 1)	process performance	low, online-sensors
biogas composition	process performance, CH_4 content	low, online-sensors
hydrogen (H_2)	inhibition of methanogenesis	low, online-sensors
hydrogen sulphide (H_2S)	inhibition of methanogenesis	low, online-sensors
redox potential	reaction conditions	low, redox electrodes
conductivity	salt concentration	low, electrical sensors
pH value 2)	acid accumulation (yes/no)	low, pH electrodes
VFA concentration 3)	system overload	high, laboratory
VFA spectrum	concentration of single acids	high, laboratory
NH_4^+-N / NH_3	inhibition of methanogenesis	very high, laboratory

1 Significance of this parameter for the prediction of process instabilities/ disturbances is questionable, because the process is already disturbed when the amount of biogas decreases; 2 Useful indicator in digesters with low buffering capacity, but low informative value to predict process instabilities/ disturbances in well-buffered systems, 3 Nowadays several spectral sensors are available for on-line measurement of volatile fatty acids (VFA), but prices are still far too expensive [29].

Nearly all recent studies describe the effects of environmental parameters on the microbial community structure. One important research challenge is the reversed approach for exploring the influence of the microbial community on the digester functioning and stability [2,5]. Most chemical parameters which indicate process instabilities/disturbances (e.g., acid accumulation, elevated NH_4^+/NH_3, H_2S, or H_2 concentrations) result from previous microbial activities (see Figure 3). A proactive approach should be developed for process control and management, to detect upcoming instabilities/disturbances earlier in the cause–effect-chain, i.e., not only at the metabolic products that have been produced, but already at the microbiome.

To achieve this, it is essential to better understand the anaerobic microbiome, including the metabolic capacities of the occurring microorganisms and the fundamental mechanisms for biotic and abiotic interactions in a temporal dimension [8,10,20,109,110]. The application of modern sequencing tools, including 16S ribosomal ribonucleic acid (rRNA) gene amplicon libraries as well as metagenome, metatranscriptome, and metaproteome analyses has led to many new and deeper insights in the microbial diversity of anaerobic digestion plants e.g., [13,14,26,109,110,111,112,113]. However, similar as in other fields of microbiome research, 'omic' techniques only display snapshots of the microbial diversity at a certain time under certain conditions and provide limited information about community dynamics or ecological behavior [114], especially on the response of single species, groups of microorganisms or entire microbiomes to varying environmental factors over time. Hence, established techniques, such as the terminal restriction fragment length polymorphism (TRFLP), are still valuable for microbiome screening in full-scale biogas plants, in particular with respect to microbial process monitoring [64,80,115,116,117,118]. The disadvantage of this method, however, is a low phylogenetic resolution, so that the identification of indicator species is not completely possible. New sequencing technologies [119], such as Nanopore sequencing, are expected to overcome this limitation by the possibility to elucidate microbial diversity with high phylogenetic resolution down to the species level by full-length sequencing of the 16S rRNA gene in high temporal resolution [26,120,121].

Additionally, most of the occurring microorganisms are only known by their 16S rRNA gene sequence (or by a reconstructed whole genome sequence), but they are not yet cultivable and hence cannot be comprehensively described regarding their actually realized ecological functions [13,15,16,109,110]. This highlights that the metabolic potential of the anaerobic digestion process has not reached its limits yet, and that further optimization of the process is still to discover [4].

To optimize the process, the potentials and, above all, the limitations of the performance of the ecosystem "anaerobic digester" have to be defined and new methods, that assess the adaptability and resilience of microbial populations to specific environmental conditions, have to be developed. The fundamental challenge is to quantitatively describe and understand the biogas microbiome in its complexity and to predict its response to external influences. This can be made by co-occurrence network analyses which (i) provide an integrated vision of all the relationships existing between microorganisms in a given environmental matrix, (ii) show that in the case of disturbance, interactions are the first to be affected and thus alter the functions of the ecosystem even before the species disappear and (iii) deliver the opportunity to define keystone species within a complex environment [109,122,123].

Currently, the available tools for detecting the microbial diversity are highly complex regarding sample preparation and especially data evaluation and interpretation [109,110,111]. Identifying microbial indicators is a precondition for developing rapid and economical microbial detection methods to be used by plant operators and consultants. With regard to this, an efficient indicator needs to be (i) robust, reliable and accurate while it reflects the current process status and remains stable and coherent over time, (ii) sensitive and specific to reflect even weak variations (imbalances, disturbances) in the environment or system over time and space, (iii) straightforward to measure and understand, with easy interpretation by comparing with a suitable reference system, (iv) acceptable considering the cost-benefit ratio and (v) relevant to facilitate decision-making [27,109,123].

This is of particular importance for future process control and management strategies, as there are indications that specific microbiome members (e.g., members of the phylum ***Cloacimonetes***) probably can be used as indicators for upcoming process disturbances [6,90]. Further research is needed to identify potential microbial indicators and their ranges of tolerance regarding process conditions and to develop detection methods for practical application.

The concept to use microbial species and their abundances for monitoring, assessing and managing processes has already been applied in several fields such as hygiene [124], food safety [125] or ecology [126]. With the advances in microbiome research, the approach to use single species and/or microbial diversity parameters as indicators is also considered for other areas such as soil quality [127,128], crop health and productivity [129,130] or human health [131]. The conditions for successful identification, development and application of microbial indicators seem promising in anaerobic digestion since it is a closed system with a high potential to control environmental factors and reveal their interactions with the microbiome.

AUTHOR CONTRIBUTIONS

S.T.; A.P.; and J.K. developed the concept. S.T. wrote the major parts of the manuscript. A.P. wrote parts of the manuscript. A.P. and J.K. revised the manuscript.

ACKNOWLEDGMENTS

The authors thank Mareike Lausberg for language check and text editing. The authors kindly thank the Open Access Fund of the Leibniz Association for funding the publication of this article.

REFERENCES

1. Bensmann, A.; Hanke-Rauschenbach, R.; Heyer, R.; Kohrs, F.; Benndorf, D.; Kausmann, R.; Plöchl, M.; Heiermann, M.; Reichl, U.; Sundmacher, K. Diagnostic concept for dynamically operated biogas production plants. ***Renew. Energy*** 2016, ***96***, 479–489.
2. Carballa, M.; Regueiro, L.; Lema, J.M. Microbial management of anaerobic digestion: Exploiting the microbiome-functionality nexus. ***Curr. Opin. Biotechnol.*** 2015, ***33***, 103–111.
3. De los Reyes III, F.L.; Weaver, J.E.; Wang, L. A methodological framework for linking bioreactor function to microbial communities and environmental conditions. ***Curr. Opin. Biotechnol.*** 2015, ***33***, 112–118.
4. De Vrieze, J.; Verstraete, W. Perspectives for microbial community composition in anaerobic digestion: From abundance and activity to connectivity. ***Environ. Microbiol.*** 2016, ***18***, 2797–2809.
5. Venkiteshwaran, K.; Bocher, B.; Maki, J.; Zitomer, D. Relating anaerobic digestion microbial community and process function. ***Microbiol. Insights*** 2016, ***8***, 37–44.
6. Westerholm, M.; Isaksson, S.; Karlsson Lindsjö, O.; Schnürer, A. Microbial community adaptability to altered temperature conditions determines the potential for process optimisation in biogas production. ***Appl. Energy***2018, ***226***, 838–848.
7. Calusinska, M.; Marynowska, M.; Goux, X.; Lentzen, E.; Delfosse, P. Analysis of dsDNA and RNA viromes in methanogenic digesters reveals novel viral genetic diversity. ***Environ. Microbiol.*** 2016, ***18***, 1162–1175.
8. Castellano-Hinojosa, A.; Amato, C.; Pozo, C.; González-Martínez, A.; González-López, J. New concepts in anaerobic digestion processes: Recent advances and biological aspects. ***Appl. Microbiol. Biotechnol.*** 2018, ***102***, 5065–5076.
9. Dollhofer, V.; Callaghan, T.M.; Griffith, G.W.; Lebuhn, M.; Bauer, J. Presence and transcriptional activity of anaerobic fungi in agricultural biogas plants. ***Bioresour. Technol.*** 2017, ***235***, 131–139.
10. Schnürer, A. Biogas Production: Microbiology and Technology. In ***Anaerobes in Biotechnology***; Advances in Biochemical Engineering/ Biotechnology; Hatti-Kaul, R., Mamo, G., Mattiasson, B., Eds.; Springer: Cham, Switzerland, 2016; Volume 156, pp. 195–234. ISBN

978-3-319-45651-5.

11. Matsubayashi, M.; Shimada, Y.; Li, Y.-Y.; Harada, H.; Kubota, K. Phylogenetic diversity and in situ detection of eukaryotes in anaerobic sludge digesters. ***PLoS ONE*** 2017, ***12***, e0172888.
12. Zhang, J.; Gao, Q.; Zhang, Q.; Wang, T.; Yue, H.; Wu, L.; Shi, J.; Qin, Z.; Zhou, J.; Zuo, J.; et al. Bacteriophage–prokaryote dynamics and interaction within anaerobic digestion processes across time and space. ***Microbiome*** 2017, ***5***, 57.
13. Hassa, J.; Maus, I.; Off, S.; Pühler, A.; Scherer, P.; Klocke, M.; Schlüter, A. Metagenome, metatranscriptome, and metaproteome approaches unraveled compositions and functional relationships of microbial communities residing in biogas plants. ***Appl. Microbiol. Biotechnol.*** 2018, ***102***, 5045–5063.
14. Kundu, K.; Sharma, S.; Sreekrishnan, T.R. Influence of process parameters on anaerobic digestion microbiome in bioenergy production: Towards an improved understanding. ***Bioenergy Res.*** 2017, ***10***, 288–303.
15. Treu, L.; Kougias, P.G.; Campanaro, S.; Bassani, I.; Angelidaki, I. Deeper insight into the structure of the anaerobic digestion microbial community; the biogas microbiome database is expanded with 157 new genomes. ***Bioresour. Technol.*** 2016, ***216***, 260–266.
16. Calusinska, M.; Goux, X.; Fossépré, M.; Muller, E.E.L.; Wilmes, P.; Delfosse, P. A year of monitoring 20 mesophilic full-scale bioreactors reveals the existence of stable but different core microbiomes in bio-waste and wastewater anaerobic digestion systems. ***Biotechnol. Biofuels*** 2018, ***11***, 196.
17. Mei, R.; Nobu, M.K.; Narihiro, T.; Kuroda, K.; Muñoz Sierra, J.; Wu, Z.; Ye, L.; Lee, P.K.H.; Lee, P.H.; van Lier, J.B.; et al. Operation-driven heterogeneity and overlooked feed-associated populations in global anaerobic digester microbiome. ***Water Res.*** 2017, ***124***, 77–84.
18. Theuerl, S.; Klang, J.; Heiermann, M.; De Vrieze, J. Marker microbiome clusters are determined by operational parameters and specific key taxa combinations in anaerobic digestion. ***Bioresour. Technol.*** 2018, ***263***, 128–135.
19. German Federal Ministry of Food and Agriculture (Bundesministerium für Bildung und Forschung). ***National Research Stategy BioEconomy 2013—Our way to a Bio-Based Economy (Nationale***

Forschungsstrategie BioÖkonomie 2030—Unser Weg zu einer bio-basierten Wirtschaft); German Federal Ministry of Food and Agriculture (Bundesministerium für Bildung und Forschung): Berlin, Germany, 2010.

20. De Vrieze, J.; Christiaens, M.E.R.; Verstraete, W. The microbiome as engineering tool: Manufacturing and trading between microorganisms. ***New Biotechnol.*** 2017, ***39***, 206–214.
21. Marzorati, M.; Wittebolle, L.; Boon, N.; Daffonchio, D.; Verstraete, W. How to get more out of molecular fingerprints: Practical tools for microbial ecology. ***Environ. Microbiol.*** 2008, ***10***, 1571–1581.
22. Read, S.; Marzorati, M.; Guimarães, B.C.M.; Boon, N. Microbial Resource Management revisited: Successful parameters and new concepts. ***Appl. Microbiol. Biotechnol.*** 2011, ***90***, 861–871.
23. Verstraete, W.; Wittebolle, L.; Heylen, K.; Vanparys, B.; de Vos, P.; van de Wiele, T.; Boon, N. Microbial Resource Management: The Road to Go for Environmental Biotechnology. ***Eng. Life Sci.*** 2007, *7*, 117–126.
24. Theuerl, S.; Klang, J.; Prochnow, A. Prozessstörungen in landwirtschaftlichen Biogasanlagen—Ursachen, Mechanismen und Auswirkungen auf das Biogas-Mikrobiom. In Proceedings of the KTBL/FNR-Kongress: Biogas in der Landwirtschaft—Stand und Perspektiven, Bayreuth, Germany, 26–27 September 2017.
25. Allison, S.D.; Martiny, J.B. Resistance, resilience, and redundancy in microbial communities. ***Proc. Natl. Acad. Sci. USA*** 2008, ***105***, 11512–11519.
26. Amha, Y.M.; Anwarb, M.Z.; Brower, A.; Jacobsen, C.S.; Stadler, L.B.; Webster, T.M.; Smith, A.L. Inhibition of anaerobic digestion processes: Applications of molecular tools. ***Bioresour. Technol.*** 2018, ***247***, 999–1014.
27. Boe, K.; Batstone, D.J.; Steyer, J.P.; Angelidaki, I. State indicators for monitoring the anaerobic digestion process. ***Water Res.*** 2010, ***44***, 5973–5980.
28. Drosg, B. Process monitoring in biogas plants. In ***IEA Bioenergy Task 37—Energy from Biogas***; Frost, P., Baxter, D., Eds.; IEA Bioenergy: Paris, France, 2013; ISBN 978-1-910154-03-8.
29. Jimenez, J.; Latrille, E.; Harmand, J.; Robles, A.; Ferrer, J.; Gaida, D.; Wolf, C.; Mairet, F.; Bernard, O.; Alcaraz-Gonzalez, V.; et al. Instrumentation and control of anaerobic digestion processes: A review

and some research challenges. ***Rev. Environ. Sci. Biotechnol.*** 2015, ***14***, 615–648.

30. De Vrieze, J.; Saunders, A.M.; He, Y.; Fang, J.; Nielsen, P.H.; Verstraete, W.; Boon, N. Ammonia and temperature determine potential clustering in the anaerobic digestion microbiome. ***Water Res.*** 2015, ***75***, 312–323.
31. Kim, J.; Lee, C. Response of a continuous anaerobic digester to temperature transitions: A critical range for restructuring the microbial community structure and function. ***Water Res.*** 2016, ***89***, 241–251.
32. Luo, G.; De Francisci, D.; Kougias, P.G.; Treu, L.; Zhu, X.; Angelidaki, I. New steady-state microbial community compositions and process performances in biogas reactors induced by temperature disturbances. ***Biotechnol. Biofuels*** 2015, ***8***, 3.
33. Weiland, P. Biogas production: Current state and perspectives. ***Appl. Microbiol. Biotechnol.*** 2010, ***85***, 849–860.
34. Sundberg, C.; Al-Soud, W.A.; Larsson, M.; Alm, E.; Yekta, S.S.; Svensson, B.H.; Sørensen, S.J.; Karlsson, A. 454 pyrosequencing analyses of bacterial and archaeal richness in 21 full-scale biogas digesters. ***FEMS Microbiol. Ecol.*** 2013, ***85***, 612–626.
35. Regueiro, L.; Carballa, M.; Lema, J.M. Outlining microbial community dynamics during temperature dropand subsequent recovery period in anaerobic co-digestion systems. ***J. Biotechnol.*** 2014, ***192***, 179–186.
36. Li, Y.; Zhang, R.; Liu, G.; Chen, C.; He, Y.; Liu, X. Comparison of methane production potential, biodegradability, and kinetics of different organic substrates. ***Bioresour. Technol.*** 2013, ***149***, 565–569.
37. Thanh, P.M.; Ketheesan, B.; Yan, Z.; Stuckey, D. Trace metal speciation and bioavailability in anaerobic digestion: A review. ***Biotechnol. Adv.*** 2016, ***34***, 122–136.
38. Klang, J.; Theuerl, S.; Szewzyk, U.; Huth, M.; Tölle, R.; Klocke, M. Dynamic variation of the microbial community structure during the long-time mono-fermentation of maize and sugar beet silage. ***Microb. Biotechnol.*** 2015, ***8***, 764–775.
39. Zhang, W.; Werner, J.J.; Agler, M.T.; Angenent, L.T. Substrate type drives variation in reactor microbiomes of anaerobic digesters. ***Bioresour. Technol.*** 2014, ***151***, 397–401.
40. Goux, X.; Calusinska, M.; Lemaigre, S.; Marynowska, M.; Klocke, M.; Udelhoven, T.; Benizri, E.; Delfosse, P. Microbial community dynamics in replicate anaerobic digesters exposed sequentially to

increasing organic loading rate, acidosis, and process recovery. ***Biotechnol. Biofuels*** 2015, ***8***, 122–140.

41. Kirkegaard, R.H.; McIlroy, S.J.; Kristensen, J.M.; Nierychlo, M.; Karst, S.M.; Dueholm, M.S.; Albertsen, M.; Nielsen, P.H. The impact of immigration on microbial community composition in full-scale anaerobic digesters. ***Sci. Rep.*** 2017, *7*, 9343.
42. Ward, A.J.; Hobbs, P.J.; Holliman, P.J.; Jones, D.L. Optimisation of the anaerobic digestion of agricultural resources. ***Bioresour. Technol.*** 2008, ***99***, 7928–7940.
43. Werner, J.J.; Knights, D.; Garcia, M.L.; Scalfone, N.B.; Smith, S.; Yarasheski, K.; Cummings, T.A.; Beers, A.R.; Knight, R.; Angenent, L.T. Bacterial community structures are unique and resilient in full-scale bioenergy systems. ***Proc. Natl. Acad. Sci. USA*** 2011, ***108***, 4158–4163.
44. Chassard, C.; Delmas, E.; Lawson, P.A.; Bernalier-Donadille, A. ***Bacteroides xylanisolvens*** sp. nov., a xylan-degrading bacterium isolated from human faeces. ***Int. J. Syst. Evol. Microbiol.*** 2008, ***58***, 1008–1013.
45. Chen, S.; Dong, X. ***Proteiniphilum acetatigenes*** gen. nov., sp. nov., from a UASB reactor treating brewery wastewater. ***Int. J. Syst. Evol. Microbiol.*** 2005, ***55***, 2257–2261.
46. Grabowski, A.; Tindall, B.J.; Bardin, V.; Blanchet, D.; Jeanthon, C. ***Petrimonas sulfuriphila*** gen. nov., sp. nov., a mesophilic fermentative bacterium isolated from a biodegraded oil reservoir. ***Int. J. Syst. Evol. Microbiol.***2005, ***55***, 1113–1121.
47. Hahnke, S.; Langer, T.; Koeck, D.E.; Klocke, M. Description of ***Proteiniphilum saccharofermentans*** sp. nov., ***Petrimonas mucosa*** sp. nov. and ***Fermentimonas caenicola*** gen. nov., sp. nov., isolated from mesophilic laboratory-scale biogas reactors, and emended description of the genus ***Proteiniphilum***. ***Int. J. Syst. Evol. Microbiol.*** 2016, ***66***, 1466–1475.
48. Jabari, L.; Gannoun, H.; Cayol, J.-L.; Hedi, A.; Sakamoto, M.; Falsen, E.; Ohkuma, M.; Hamdi, M.; Fauque, G.; Ollivier, B.; et al. ***Macellibacteroides fermentans*** gen. nov., sp. nov., a member of the family ***Porphyromonadaceae*** isolated from an upflow anaerobic filter treating abattoir wastewaters. ***Int. J. Syst. Evol. Microbiol.*** 2012, ***62***, 2522–2527.

49. Ueki, A.; Akasaka, H.; Suzuki, D.; Ueki, K. ***Paludibacter propionicigenes*** gen. nov., sp. nov., a novel strictly anaerobic, Gram-negative, propionate producing bacterium isolated from plant residue in irrigated rice-field soil in Japan. ***Int. J. Syst. Evol. Microbiol.*** 2006, ***56***, 39–44.
50. Azman, S.; Khadem, A.F.; van Lier, J.B.; Zeeman, G.; Plugge, C.M. Presence and role of anaerobic hydrolytic microbes in conversion of lignocellulosic biomass for biogas production. ***Crit. Rev. Environ. Sci. Technol.*** 2014, ***45***, 2523–2564.
51. Li, W.; Khalid, H.; Zhu, Z.; Zhang, R.; Liu, G.; Chen, C.; Thorin, E. Methane production through anaerobic digestion: Participation and digestion characteristics of cellulose, hemicellulose and lignin. ***Appl. Energy***2018, ***226***, 1219–1228.
52. Shrestha, S.; Fonoll, X.; Khanal, S.K.; Raskin, L. Biological strategies for enhanced hydrolysis of lignocellulosic biomass during anaerobic digestion: Current status and future perspectives. ***Bioresour. Technol.*** 2017, ***245***, 1245–1257.
53. Mulat, D.G.; Horn, S.J. Biogas Production from Lignin via Anaerbic digestion. In ***Lignin Valorization: Emerging Approaches***; Beckham, G.T., Ed.; Energy and Environment Series; Royal Society of Chemistry (RSC) Publishing: Cambridge, UK, 2018; pp. 391–412. ISBN 978-1-78262-554-4.
54. Budzianowski, W.M. A review of potential innovations for production, conditioning and utilization of biogas with multiple-criteria assessment. ***Renew. Sustain. Energy Rev.*** 2016, ***54***, 1148–1171.
55. Carrère, H.; Antonopoulou, G.; Affes, R.; Passos, F.; Battimelli, A.; Lyberatos, G.; Ferrer, I. Review of feedstock pretreatment strategies for improved anaerobic digestion: From lab-scale research to full-scale application. ***Bioresour. Technol.*** 2016, ***199***, 386–397.
56. Montgomery, L.F.R.; Bochmann, G. Pretreatment of feedstock for enhanced biogas production. In ***IEA Bioenergy Task 37—Energy from Biogas***; Baxter, D., Ed.; IEA Bioenergy: Paris, France, 2014; ISBN 978-1-910154-05-2.
57. Jönsson, L.J.; Martín, C. Pretreatment of lignocellulose: Formation of inhibitory by-products and strategies for minimizing their effects. ***Bioresour. Technol.*** 2016, ***199***, 103–112.
58. Alsouleman, K.; Linke, B.; Klang, J.; Klocke, M.; Krakat, N.; Theuerl,

S. Reorganisation of a mesophilic biogas microbiome as response to a stepwise increase of ammonium nitrogen induced by poultry manure supply. ***Bioresour. Technol.*** 2016, ***208***, 200–204.

59. Kobayashi, T.; Wu, Y.-P.; Xu, K.-Q.; Li, Y.-Y. Effect of mixing driven by siphon flow: Parallel experiments using the anaerobic reactors with different mixing modes. ***Energies*** 2013, ***6***, 4207–4222.
60. Lindmark, J.; Thorin, E.; Bel Fdhila, R.; Dahlquist, E. Effects of mixing on the result of anaerobic digestion: Review. ***Renew. Sustain. Energy Rev.*** 2014, ***40***, 1030–1047.
61. McMahon, K.D.; Stroot, P.G.; Mackie, R.I.; Raskin, L. Anaerobic codigestion of municipal solid waste and biosolids under various mixing conditions—II: Microbial population dynamics. ***Water Res.*** 2001, ***35***, 1817–1827.
62. De Vrieze, J.; Verstraete, W.; Boon, N. Repeated pulse feeding induces functional stability in anaerobic digestion. ***Microb. Biotechnol.*** 2013, ***6***, 414–424.
63. Regueiro, L.; Lema, J.M.; Carballa, M. Key microbial communities steering the functioning of anaerobic digesters during hydraulic and organic overloading shocks. ***Bioresour. Technol.*** 2015, ***197***, 208–216.
64. Ziganshin, A.M.; Schmidt, T.; Lv, Z.; Liebetrau, J.; Richnow, H.H.; Kleinsteuber, S.; Nikolausz, M. Reduction of the hydraulic retention time at constant high organic loading rate to reach the microbial limits of anaerobic digestion in various reactor systems. ***Bioresour. Technol.*** 2016, ***217***, 62–71.
65. Gerardi, M.H. ***The Microbiology of Anaerobic Digesters***; Wiley: Hoboken, NJ, USA, 2003; ISBN 978-0-471206-93-4.
66. Ju, F.; Lau, F.; Zhang, T. Linking microbial community, environmental variables, and methanogenesis in anaerobic biogas digesters of chemically enhanced primary treatment sludge. ***Environ. Sci. Technol.*** 2017, ***51***, 3982–3992.
67. Nguyen, L.N.; Nguyen, A.Q.; Nghiem, L.D. Microbial community in anaerobic digestion system: Progression in microbial ecology. In ***Water and Wastewater Treatment Technologies***; Bui, X.T., Chiemchaisri, C., Fujioka, T., Varjani, S., Eds.; Energy, Environment, and Sustainability Book Series; Springer: Singapore, 2019; pp. 331–355, ISBN print 978-981-13-3258-6, ISBN online 978-981-13-3259-3.
68. Hori, T.; Haruta, S.; Sasaki, D.; Hanajima, D.; Ueno, Y.; Ogata, A.; Ishii,

M.; Igarashi, Y. Reorganization of the bacterial and archaeal populations associated with organic loading conditions in a thermophilic anaerobic digester. ***J. Biosci. Bioeng.*** 2015, ***119***, 337–344.

69. Li, Y.-F.; Abraham, C.; Nelson, M.C.; Chen, P.-H.; Graf, J.; Yu, Z. Effect of organic loading on the microbiota in a temperature-phased anaerobic digestion (TPAD) system co-digesting dairy manure and waste whey. ***Appl. Microbiol. Biotechnol.*** 2015, ***99***, 8777–8792.
70. Leng, L.; Yang, P.; Singh, S.; Zhuang, H.; Xu, L.; Chen, W.-H.; Dolfing, J.; Li, D.; Zhang, Y.; Zeng, H.; et al. A review on the bioenergetics of anaerobic microbial metabolism close to the thermodynamic limits and its implications for digestion applications. ***Bioresour. Technol*** 2018, ***247***, 1095–1106.
71. Montag, D.; Schink, B. Biogas process parameters—Energetics and kinetics of secondary fermentations in methanogenic biomass degradation. ***Appl. Microbiol. Biotechnol.*** 2016, ***100***, 1019–1026.
72. Roden, E.E.; Jin, Q. Thermodynamics of microbial growth coupled to metabolism of glucose, ethanol, short-chain organic acids, and hydrogen. ***Appl. Environ. Microbiol.*** 2011, *77*, 1907–1909.
73. Schink, B. Energetics of syntrophic cooperation in methanogenic degradation. ***Microbiol. Mol. Biol. Rev.*** 1997, ***61***, 262–280.
74. Morris, B.E.L.; Henneberger, R.; Huber, H.; Moissl-Eichinger, C. Microbial syntrophy: Interaction for the common good. ***FEMS Microbiol. Rev.*** 2013, ***37***, 384–406.
75. Liu, Y.; Whitman, W.B. Metabolic, phylogenetic, and ecological diversity of the methanogenic archaea. ***Ann. N. Y. Acad. Sci.*** 2008, ***1125***, 171–189.
76. McInerney, M.J.; Sieber, J.R.; Gunsalus, R.P. Syntrophy in anaerobic global carbon cycles. ***Curr. Opin. Biotechnol.*** 2009, ***20***, 623–632.
77. Baek, G.; Kim, J.; Kim, J.; Lee, C. Role and potential of direct interspecies electron transfer in anaerobic digestion. ***Energies*** 2018, ***11***, 107.
78. Stams, A.; Plugge, C. Electron transfer in syntrophic communities of anaerobic bacteria and archaea. ***Nat. Rev. Microbiol.*** 2009, *7*, 568–577.
79. Hattori, S. Syntrophic acetate-oxidizing microbes in methanogenic environments. ***Microbes Environ.*** 2008, ***23***, 118–127.
80. Westerholm, M.; Moestedt, J.; Schnürer, A. Biogas production through syntrophic acetate oxidation and deliberate operating strategies for

improved digester performance. ***Appl. Energy*** 2016, ***179***, 124–135.

81. Cazier, E.A.; Trably, E.; Steyer, J.P.; Escudie, R. Biomass hydrolysis inhibition at high hydrogen partial pressure in solid-state anaerobic digestion. ***Bioresour. Technol.*** 2015, ***190***, 106–113.
82. Chen, Y.; Cheng, J.J.; Creamer, K.S. Inhibition of anaerobic digestion process: A review. ***Bioresour. Technol.***2008, ***99***, 4044–4064.
83. Niu, Q.; Kubota, K.; Qiao, W.; Jing, Z.; Zhanga, Y.; Yu-Youa, L. Effect of ammonia inhibition on microbial community dynamic and process functional resilience in mesophilic methane fermentation of chicken manure. ***J. Chem. Technol. Biotechnol.*** 2014, ***90***, 2161–2169.
84. Rajagopal, R.; Massé, D.I.; Singh, G. A critical review on inhibition of anaerobic digestion process by excess ammonia. ***Bioresour. Technol.*** 2013, ***143***, 632–641.
85. Tian, H.; Fotidis, I.A.; Kissas, K.; Angelidaki, I. Effect of different ammonia sources on aceticlastic and hydrogenotrophic methanogens. ***Bioresour. Technol.*** 2018, ***250***, 390–397.
86. Wrońska, I.; Cybulska, K. Quantity and quality of biogas produced from the poultry sludge optimized by filamentous fungi. ***Ecol. Chem. Eng.*** 2018, ***25***, 395–404.
87. De Vrieze, J.; Hennebel, T.; Boon, N.; Verstraete, W. ***Methanosarcina***: The rediscovered methanogen for heavy duty biomethanation. ***Bioresour. Technol.*** 2012, ***112***, 1–9.
88. Meng, X.; Zhang, Y.; Sui, Q.; Zhang, J.; Wang, R.; Yu, D.; Wang, Y.; Wei, Y. Biochemical conversion and microbial community in response to ternary pH buffer system during anaerobic digestion of swine manure. ***Energies*** 2018, ***11***, 2991.
89. Schnürer, A.; Nordberg, A. Ammonia, a selective agent for methane production by syntrophic acetate oxidation at mesophilic temperature. ***Water Sci. Technol.*** 2008, ***57***, 735–740.
90. Klang, J.; Szewzyk, U.; Bock, D.; Theuerl, S. Nexus between the microbial diversity level and the stress tolerance within the biogas process. ***Anaerob*** 2019, in press.
91. Lv, Z.; Leite, A.F.; Harms, H.; Glaser, K.; Liebetrau, J.; Kleinsteuber, S.; Nikolausz, M. Microbial community shifts in biogas reactors upon complete or partial ammonia inhibition. ***Appl. Microbiol. Biotechnol.*** 2018.
92. Belostotskiy, D.E.; Ziganshina, E.E.; Siniagina, M.; Boulygina, E.A.;

Miluykov, V.A.; Ziganshin, A.M. Impact of the substrate loading regime and phosphoric acid supplementation on performance of biogas reactors and microbial community dynamics during anaerobic digestion of chicken wastes. ***Bioresour. Technol.*** 2015, ***193***, 42–52.

93. Gao, S.; Zhao, M.; Chen, Y.; Yu, M.; Ruan, W. Tolerance response to in situ ammonia stress in a pilot-scale anaerobic digestion reactor for alleviating ammonia inhibition. ***Bioresour. Technol.*** 2015, ***198***, 372–379.
94. De Jonge, N.; Moset, V.; Møller, H.B.; Lund Nielsen, J. Microbial population dynamics in continuous anaerobic digester systems during start up, stable conditions and recovery after starvation. ***Bioresour. Technol.*** 2017, ***232***, 313–320.
95. Paulo, L.M.; Stams, A.J.M.; Sousa, D.Z. Methanogens, sulphate and heavy metals: A complex system. ***Rev. Environ. Sci. Biol.*** 2015, ***14***, 537–553.
96. Chen, J.L.; Ortiz, R.; Steele, T.W.J.; Stuckey, D.C. Toxicants inhibiting anaerobic digestion: A review. ***Biotechnol. Adv.*** 2014, ***32***, 1523–1534.
97. Choong, Y.Y.; Norli, I.; Abdullah, A.Z.; Yhaya, M.F. Impacts of trace element supplementation on the performance of anaerobic digestion process: A critical review. ***Bioresour. Technol.*** 2016, ***209***, 369–379.
98. Du, L.; Liu, W. Occurrence, fate, and ecotoxicity of antibiotics in agro-ecosystems. A review. ***Agron. Sustain. Dev.*** 2012, ***32***, 309–327.
99. Azman, S.; Khadem, A.F.; Plugge, C.M.; Stams, A.J.M.; Bec, S.; Zeeman, G. Effect of humic acid on anaerobic digestion of cellulose and xylan in completely stirred tank reactors: Inhibitory effect, mitigation of the inhibition and the dynamics of the microbial communities. ***Appl. Microbiol. Biotechnol.*** 2017, ***101***, 889–901.
100. Geršl, M.; Kanduč, T.; Matýsek, D.; Šotnar, M.; Mareček, J. The role of mineral phases in the biogas production technology. ***Ecol. Chem. Eng.*** 2018, ***25***, 51–59.
101. Yap, S.D.; Astals, S.; Lua, Y.; Peces, M.; Jensen, P.D.; Batstone, D.J.; Tait, S. Humic acid inhibition of hydrolysis and methanogenesis with different anaerobic inocula. ***Waste Manag.*** 2018, ***80***, 130–136.
102. De Vrieze, J.; Christiaens, M.E.R.; Walraedt, D.; Devooght, A.; Ijaz, U.Z.; Boon, N. Microbial community redundancy in anaerobic digestion drives process recovery after salinity exposure. ***Water Res.*** 2017, ***111***, 109–117.

103. Molaey, R.; Bayrakdar, A.; Sürmeli, R.Ö.; Çalli, B. Influence of trace element supplementation on anaerobic digestion of chicken manure: Linking process stability to methanogenic population dynamics. ***J. Clean. Prod.***2018, ***181***, 794–800.

104. Seneesrisakul, K.; Sutabutr, T.; Chavadej, S. The effect of temperature on the methanogenic activity in relation to micronutrient availability. ***Energies*** 2018, ***11***, 1057.

105. Popp, D.; Schrader, S.; Kleinsteuber, S.; Harms, H.; Sträuber, H. Biogas production from coumarin-rich plants—Inhibition by coumarin and recovery by adaptation of the bacterial community. ***FEMS Microbiol. Ecol.***2015, ***91***.

106. Popp, D.; Plugge, C.M.; Kleinsteuber, S.; Harms, H.; Sträuber, H. Inhibitory effect of coumarin on syntrophic fatty acid-oxidizing and methanogenic cultures and biogas reactor microbiomes. ***Appl. Environ. Microbiol.***2017, ***83***, e00438-17.

107. Mao, C.; Feng, Y.; Wang, X.; Ren, G. Review on research achievements of biogas from anaerobic digestion. ***Renew Sustain. Energy Rev.*** 2015, ***45***, 540–555.

108. Hagos, K.; Zong, J.; Li, D.; Liu, C.; Lu, X. Anaerobic co-digestion process for biogas production: Progress, challenges and perspectives. ***Renew. Sustain. Energy Rev.*** 2017, ***76***, 1485–1496.

109. Bouchez, T.; Blieux, A.L.; Dequiedt, S.; Domaizon, I.; Dufresne, A.; Ferreira, S.; Godon, J.J.; Hellal, J.; Joulian, C.; Quaiser, A.; et al. Molecular microbiology methods for environmental diagnosis. ***Environ. Chem. Lett.*** 2016, ***14***, 423–441.

110. Zhang, L.; Loh, K.-C.; Lim, J.W.; Zhang, J. Bioinformatics analysis of metagenomics data of biogas-producing microbial communities in anaerobic digesters: A review. ***Renew. Sustain. Energy Rev.*** 2019, ***100***, 110–126.

111. Bozan, M.; Akyol, Ç.; Ince, O.; Aydin, S.; Ince, B. Application of next-generation sequencing methods for microbial monitoring of anaerobic digestion of lignocellulosic biomass. ***Appl. Microbiol. Biotechnol.*** 2017, ***101***, 6849–6864.

112. Cabezas, A.; Calabria de Araujo, J.; Callejas, C.; Galès, A.; Hamelin, J.; Marone, A.; Sousa, D.Z.; Trably, E.; Etchebehere, C. How to use molecular biology tools for the study of the anaerobic digestion process? ***Rev. Environ. Sci. Biol.*** 2015, ***14***, 555–593.

113. Vanwonterghem, I.; Jensen, P.D.; Ho, D.P.; Batstone, D.J.; Tyson, G.W. Linking microbial community structure, interactions and function in anaerobic digesters using new molecular techniques. ***Curr. Opin. Biotechnol.*** 2014, ***27***, 55–64.

114. Alivisatos, A.P.; Blaser, M.J.; Brodie, E.L.; Chun, M.; Dangl, J.L.; Donohue, T.J.; Dorrestein, P.C.; Gilbert, J.A.; Green, J.L.; Jansson, J.K.; et al. A unified initiative to harness Earth's microbiomes. ***Science*** 2015, ***350***, 507–508.

115. Bonk, F.; Popp, D.; Weinrich, S.; Sträuber, H.; Kleinsteuber, S.; Harms, H.; Centler, F. Intermittent fasting for microbes: How discontinuous feeding increases functional stability in anaerobic digestion. ***Biotechnol. Biofuels*** 2018, ***11***, 274.

116. De Vrieze, J.; Ijaz, U.Z.; Saunders, A.M.; Theuerl, S. Terminal restriction fragment length polymorphism is an "old school" reliable technique for swift microbial community screening in anaerobic digestion. ***Sci. Rep.***2018, ***8***, 16818.

117. Lim, J.W.; Ge, T.; Tong, Y.W. Monitoring of microbial communities in anaerobic digestion sludge for biogas optimisation. ***Waste Manag.*** 2018, ***71***, 334–341.

118. Prakash, O.; Pandey, P.K.; Kulkarni, G.J.; Mahale, K.N.; Shouche, Y.S. Technicalities and glitches of terminal restriction fragment length polymorphism (TRFLP). ***Indian J. Microbiol.*** 2014, ***54***, 255–261.

119. Shendure, J.; Balasubramanian, S.; Church, G.M.; Gilbert, W.; Rogers, J.; Schloss, J.A.; Waterston, R.H. DNA sequencing at 40: Past, present and future. ***Nature*** 2017, ***550***, 345–353.

120. Calus, S.T.; Ijaz, U.Z.; Pinto, A.J. NanoAmpli-Seq: A workflow for amplicon sequencing for mixed microbial communities on the nanopore sequencing platform. ***GigaScience*** 2018, *7*.

121. Kerkhof, L.J.; Dillon, K.P.; Häggblom, M.M.; McGuinness, L.R. Profiling bacterial communities by MinION sequencing of ribosomal operons. ***Microbiome*** 2017, ***5***, 116.

122. Berry, D.; Widder, S. Deciphering microbial interactions and detecting keystone species with co-occurrence networks. ***Front. Microbiol.*** 2014, ***5***, 219.

123. Karimi, B.; Maron, P.A.; Chemidlin-Prevost Boure, N.; Bernard, N.; Gilbert, D.; Ranjard, L. Microbial diversity and ecological networks

as indicators of environmental quality. ***Environ. Chem. Lett.*** 2017, ***15***, 265.

124. Browne, H.P.; Neville, B.A.; Forster, S.C.; Lawley, T.D. Transmission of the gut microbiota: Spreading of health. ***Nat. Rev. Microbiol.*** 2017, ***15***, 531–543.

125. Valero, A.; Rodríguez, M.-Y.; Posada-Izquierdo, G.D.; Pérez-Rodríguez, F.; Carrasco, E.; García-Gimeno, R.M. Risk factors influencing microbial contamination in food service centers. In ***Significance, Prevention and Control of Food Related Diseases***; Makun, H., Ed.; IntechOpen: London, UK, 2016; Chapter 2; ISBN 978-953-51-2277-7.

126. Siddig, A.A.H.; Aaron, M.; Ellison, A.M.; Ochs, A.; Villar-Leemand, C.; Lau, M.K. How do ecologists select and use indicator species to monitor ecological change? Insights from 14 years of publication in Ecological Indicators. ***Ecol. Indic.*** 2016, ***60***, 223–230.

127. Hermans, S.M.; Buckley, H.L.; Case, B.S.; Curran-Cournane, F.; Taylor, M.; Lear, G. Bacteria as emerging indicators of soil condition. ***Appl. Environ. Microbiol.*** 2017, ***83***, e02826-6.

128. Schloter, M.; Nannipieri, P.; Sørensen, S.J.; van Elsas, J.D. Microbial indicators for soil quality. ***Biol. Fertil. Soils*** 2018, ***54***.

129. Berg, G.; Köberl, M.; Rybakova, D.; Müller, H.; Grosch, R.; Smalla, K. Plant microbial diversity is suggested as the key to future biocontrol and health trends. ***FEMS Microbiol. Ecol.*** 2017, ***93***.

130. Dias, T.; Dukes, A.; Antunes, P.M. Accounting for soil biotic effects on soil health and crop productivity in the design of crop rotations. ***J. Sci. Food Agric.*** 2015, ***95***, 447–454.

131. Flandroy, L.; Poutahidis, T.; Berg, G.; Clarke, G.; Dao, M.-G.; Decaestecker, E.; Furman, E.; Haahtela, T.; Massart, S.; Plovier, H.; et al. The impact of human activities and lifestyles on the interlinked microbiota and health of humans and of ecosystems. ***Sci. Total Environ.*** 2018, ***627***, 1018–1038.

14

Perspectives of Biogas Conversion into Bio-CNG for Automobile Fuel in Bangladesh

M. S. Shah[1] , P. K. Halder[2,3] , A. S. M. Shamsuzzaman[2] , M. S. Hossain[1] , S. K. Pal[1], and E. Sarker[4]

[1]Department of Petroleum and Mining Engineering, Jessore University of Science and Technology, Jessore 7408, Bangladesh
[2]Department of Industrial and Production Engineering, Jessore University of Science and Technology, Jessore 7408, Bangladesh
[3]School of Engineering, Royal Melbourne Institute of Technology University, Melbourne, VIC 3001, Australia
[4]Hajee Mohammad Danesh Science and Technology University, Dinajpur 5200, Bangladesh

ABSTRACT

The need for liquid and gaseous fuel for transportation application is growing very fast. This high consumption trend causes swift exhaustion of fossil fuel reserve as well as severe environment pollution. Biogas can be converted into various renewable automobile fuels such as bio-CNG, syngas, gasoline, and liquefied biogas. However, bio-CNG, a compressed

Citation: M. S. Shah, P. K. Halder, A. S. M. Shamsuzzaman, M. S. Hossain, S. K. Pal, and E. Sarker, "Perspectives of Biogas Conversion into Bio-CNG for Automobile Fuel in Bangladesh," Journal of Renewable Energy, vol. 2017, Article ID 4385295, 14 pages, 2017. https://doi.org/10.1155/2017/4385295

biogas with high methane content, can be a promising candidate as vehicle fuel in replacement of conventional fuel to resolve this problem. This paper presents an overview of available liquid and gaseous fuel commonly used as transportation fuel in Bangladesh. The paper also illustrates the potential of bio-CNG conversion from biogas in Bangladesh. It is estimated that, in the fiscal year 2012-2013, the country had about 7.6775 billion m^3 biogas potential equivalent to 5.088 billion m^3 of bio-CNG. Bio-CNG is competitive to the conventional automobile fuels in terms of its properties, economy, and emission.

INTRODUCTION

Transportation system is the basement of the industrial and socioeconomic development of any country and predominantly depends on fossil fuel [1]. In the United States, about 28% of the total energy consumption is used for transportation system of which almost 86% comes from gasoline and diesel fuels [2]. Due to the rapid depletion and high cost of liquid fuel, natural gas is used in compressed form named compressed natural gas (CNG). Currently, it has become very popular alternative to liquid fuel for vehicles in the world due to its low price [3]. It is estimated that during 2013, approximately 18.09 million natural gas vehicles (NGV) have been run by CNG in the world. Nowadays some countries like the United States, Germany, Australia, Austria, India, and so forth already have been using bio-CNG as the vehicles fuel in place of CNG [4, 5].

In the United States, 8 renewable natural gas projects, namely, bio-CNG 50, bio-CNG 100, and bio-CNG 200, have been installed during 2011 to 2013 [5, 6]. These plants produce approximately 200 gallons of diesel equivalents (DGEs)/day, 399.77 DGEs/day, and 790 DGEs/day, respectively, which can be enough fuel for 25–30 vehicles, 5–8 trash trucks, or 40–50 passenger vehicles and 25 heavy duty vehicles or 40–50 light duty vehicles per day. In India, first indigenous bio-CNG plant with capacity of 85 cubic meters has been made at Rajasthan with the help from IIT, Delhi. The plant uses cow dung as feed material and the production cost of 1 kg of bio-CNG is about USD 0.23–0.24 which is much cheaper than the petro-based CNG [7]. During the year 2013, in Austria about 7 GWh energy equivalent vehicle fuel (biomethane) was produced from biogas and three filling stations were established at biogas upgrading plants. On the other hand, Germany produced almost 0.1% of total consumption of vehicle fuel from biogas in year 2012 equivalent to approximately 0.35 TWh of energy. At the end of 2012, about

120 biomethane feed plants were in operation with an installed capacity of 72,000 Nm^3/h in Germany that could be fueled for 0.1 million gas vehicles in 119 out of the 900 CNG filling stations [8]. Linköping, a city of Sweden is the promising model of bio-CNG integration into public transportation. The plant produces about 4.7 million m^3 of upgraded biogas per year that is used in 64 buses and a number of heavy and light duty vehicles through 12 public refueling stations. Besides, this is the pioneer in using biomethane in train. This plant reduces approximately 9,000 tons of CO2-emissions from urban transport as well as sulphur and nitrogen oxides per year [9].

Bangladesh is an energy starved country where natural gas and petroleum products are the main sources of energy. However, the country has only 0.3 trillion cubic meters of proved natural gas reserve at the end of 2014 which will be exhausted within next 10–12 years if the existing consumption rate continues. In addition, the petroleum reserve in Bangladesh is only about 8% of the total demand; it has to import about 1.2 million tons of crude oil and 2.6 million tons of refined petroleum products each year [10]. The country consumed almost 23.6 billion cubic meters of natural gas and 5.7 million tons of oil in the year 2014 [11]. The transportation sector consumes approximately 46.46% of total petroleum consumption and 6% of total natural gas consumption. Up to October 2015, the country has about 560 CNG refueling stations and 180 CNG conversion workshops which have converted approximately 245,372 vehicles into CNG vehicles. Currently, the country has total of 285,755 CNG run vehicles including imported vehicles. However, taking into account the future security of liquid and gaseous vehicle fuel, it is necessary to find out alternative renewable source for transport fuel immediately. Bio-CNG produced from the purification and then compression of biogas can be the most suitable alternative for traditional vehicle fuel in Bangladesh. Bangladesh has enormous amount feed material for biogas production which can be the most effective option for bio-CNG. However, most of the biogas plants in the country are small-scale domestic digester producing biogas mainly for cooking and lighting. The main issues for commercial biogas implementation in Bangladesh are technological knowledge gap, uncertainty of feed materials, and financial insufficiency. Although the Government of Bangladesh is financing domestic biogas digester through some organization, it has no specific policy regarding the commercialization of bio-CNG production from biogas. The study assesses the scope and potentiality of bio-CNG production from biogas and its technology for automobile in Bangladesh.

PRESENT STATUS OF TRANSPORTATION FUELS IN BANGLADESH

Petroleum Products

Petroleum products are the most usable liquid fuels which are derived from crude oil processing in oil refineries. The majority of crude oil is converted into petroleum products, which include several classes of fuels [12]. Petroleum fuel mainly contains kerosene, diesel, and petrol and is considered as the major sources of commercial energy in Bangladesh. The transportation sector is the primary consumer of petroleum fuels which accounts for about 46.46% of total consumption 5,321,423 tons as illustrated in Figure 1. In the fiscal year 2014-2015, transportation sector consumes approximately 2,472,486 tons of petroleum oil which is almost two times the consumption of power sector [13]. On the other hand, among the petroleum products, high speed diesel accounts for almost 63.82% of total consumption of which nearly 52.25% is consumed by vehicle. In fiscal year 2014-2015, Eastern refinery limited has produced about 386,449 tons of high speed diesel and 245,341 tons of superior kerosene oil. In addition to this, in the year 2014-2015, the country has imported about 5,398,789.20 tons of petroleum oil including crude oil and lube base oil of total cost of BDT 27,023.27 crores.

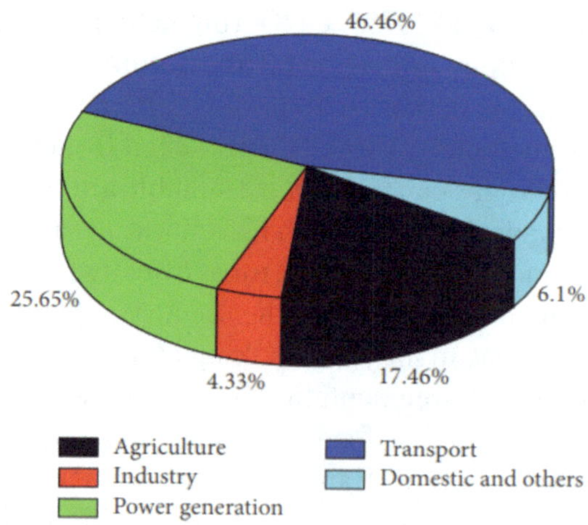

Figure 1: Sector-wise consumption of petroleum fuels in Bangladesh.

Compressed Natural Gas (CNG)

Compressed natural gas is produced by compressing the natural gas under a pressure between 21–25 kPa. CNG is a better eco-friendly fuel compared to gasoline/diesel; hence, its use as an alternative option in replacement of gasoline or diesel to run automobiles has attracted much more attention over the world. The use of CNG in vehicles mitigates the emission of nitrous oxide, hydrocarbons, carbon monoxide, and carbon dioxide by 40%, 90%, 80%, and 25%, respectively, compared to petrol or diesel fuel [14]. The need for CNG as vehicle fuel is growing very fast because of its low price and higher octane number (130) compared to petrol (93) indicating the high thermal efficiency and low emissions. In Bangladesh, CNG as a vehicle fuel was introduced around 1985. The CNG run vehicles are consuming about 100.43 million cubic meters of gas per month. The demand of natural gas for CNG in next 5 years will be nearly 110–127 million cubic meters [15]. On the other hand, the share of natural gas in power generation is increasing hastily. Therefore, due to lack of natural gas production, poor gas transmission, and network distribution in Bangladesh, almost 39 districts have not previewed CNG opportunities yet. Hence, to reduce the rapid depletion of natural gas, it is necessary to give emphasis for production of CNG from alternative sources.

Liquefied Petroleum Gas (LPG)

Liquefied petroleum gas (LPG) is a blend of propane (C_3H_8) and butane (C_4H_8) which is produced during the processing of natural gas and is sold in cylinder. It becomes liquid at atmospheric temperature when compressed to a pressure between 80 and 110 psi and returns into gases when the pressure is reduced adequately. The colorless and odorless LPG is easy to transport and store in liquid state and used as fuel in domestic cooking and commercial uses in the regions of insufficient pipeline. In addition to these, LPG is suitable for medium and small vehicles. Therefore, in Bangladesh, the annual demand for LPG is increasing hastily. However, the supply of LPG was only about 121 thousand tons against a demand of 500 thousand tons in year 2012 [16]. The sources of LPG in Bangladesh are presented in Table 1. Hence, the country has to import LPG to reduce the gap between demand and supply. The government has already taken initiatives to install import-based LPG storage and bottling plant in Mongla, Bagerhat, and in Chittagong [17]. In fiscal year 2014-2015, Eastern refinery limited has produced about 11,070 tons of LPG against the total consumption of 17,424 tons.

Table 1: Potentials of LPG in Bangladesh [16, 17]

Supply sources	Amount (ton per year)
Kailashtila fractionation plant, Golapgonj	7580
Eastern Refinery Ltd., Chittagong	13200
Import by private sectors	100000
Total supply	121050

Liquefied Natural Gas (LNG)

Liquefied natural gas (LNG) is produced from natural gas by condensing it artificially below a temperature of −162°C and is stored in very high pressure storage tank. LNG is mainly applicable as vehicle fuel for heavy duty transports. The storage and transportation system of LNG includes floating storage and regasification unit, subsea and overland gas pipelines to transport the gas from floating terminal to the consumers. To make LNG operation economically viable, Bangladesh has to dedicate at least 0.17 Trillion cubic meters gas with an investment cost of USD 6 billion. However, considering the present gas reserve in the country, it is necessary to give emphasis on alternative option immediately. Therefore, Bangladesh has already signed an agreement with Qatar for importing about 14.16 MMCM gas in the form of LNG to meet the country's increasing demand for LNG. However, the matter of concern is that whether the country can afford the costly LNG. Bangladesh has very limited transportation facilities as LNG storage and transportation require high draught. Hence, the country has set a plan to install LNG vessel at approximately 5-6 km offshore of Moheshkhali coast because of availability of required draught. The initial plan is to supply about 4 million tons of LNG annually. Global LNG production is expected to be 450 million tons in 2020 [18].

Biodiesel

Biodiesel is nonpetroleum oil generally produced from vegetable oil, animal fats, waste cooking oil, and so forth and used as renewable source in replacement of diesel because of its nontoxicity and environmental sustainability compared to diesel [19–21]. Transesterification is the chemical process to produce biodiesel by using base catalyst, acid catalyst, enzyme catalyst, and heterogeneously catalyst. Bangladesh has promising potential of edible (mustered, cottonseed, ground nut, sesame, rapeseed, sunflower, coconut, and soybean oil) oil and nonedible (jatropha curcas, karanja, castor, neem, and algae) oil crops for biodiesel feed stocks [22]. However, nonedible vegetable oil is the most suitable for biodiesel production in Bangladesh due to the food scarcity and lack of available land for oil crops production. On the contrary, the country has many unused marginal, road, and rail side land areas which can be used for commercial biodiesel plants. Therefore, it is possible to produce about 0.52 million tons of karanja biodiesel per year utilizing this unused land which reduce country's total diesel import approximately by 21.67% [23]. On the other hand, almost 1.19 million tons of Jatropha curcas, 0.15 million tons of castor, and 1.04 million tons of pithraj can be produced annually by considering the 50% use of the available land in Bangladesh [24]. However, there is no biodiesel generation plant in Bangladesh yet.

BIOGAS TO AUTOMOBILE FUELS

Bio-CNG

Bio-CNG, a methane rich compressed fuel, is also known as compressed biomethane. Bio-CNG is produced from pure biogas containing more than 97% methane at a pressure of 20–25 MPa. It is very similar to the regular CNG in terms of its fuel properties, economy, engine performance, and emissions. Like regular CNG, bio-CNG has high octane number which results in the high thermal efficiency. The performance of a constant speed internal combustion engine using CNG and bio-CNG was compared and it was noted that their engine performances were almost similar in terms of brake power output, specific gas consumption, and thermal efficiency [25]. A typical bio-CNG station comprises a biogas purification unit, a multistage compressor, and a high pressure storage system [26]. Because of the shorter driving range of bio-CNG compared to diesel fuel, the consumers have to install additional fuel cylinders to extend their driving range [27]. Bio-CNG can be injected into the CNG grid and blended with CNG. However, if the current CNG grid is inadequate, bio-CNG can be transported by trucks or in cylinder from the locations of production to the filling stations. From the literature, bio-CNG delivers greater environmental benefits than other traditional vehicle fuels as well as biodiesel and bioethanol [28–30]. Considering both technical and financial performance production of bio-CNG for vehicle fuel is at least feasible as it is produced from renewable wastes [31]. Therefore, the huge potentiality of different wastes in Bangladesh can make bio-CNG production a viable option.

Liquefied Biogas (LBG)

Biomethane from biogas can be liquefied to a fuel called liquefied biomethane (LBM) or LBG which has the similar characteristics to LNG. The conversion of LBM from biomethane requires a combination of high pressures and low temperatures and is a rather energy intensive process. LBM has the energy content of about 70% that of gasoline and can be used as a vehicle fuel in replacement of conventional vehicle fuel. Commonly used LBG production methods include cryogenic technology, liquefaction, and pressure letdown. During the liquefaction of biogas even a small amount of impurity can cause substantial difficulties and presence of oxygen can also cause danger of explosions. Therefore, biogas has to contain less than 25 ppm, 4 ppm, and 1 ppm of CO_2, H_2S, and H_2O, respectively, to produce liquefied biomethane [32]. Ignoring the energy input for liquefaction, it has been estimated that 1,000 cubic feet of gas yield about 10 gallons of LBM. On the other hand, assuming 10% losses, a plant producing about 70,000 cubic feet of biogas per day can generate approximately 500 gallons of LBG per day [33]. There is not much practical experience with this option because of its high capital and operating cost of liquefaction equipment though it has a much higher energy density in comparison to bio-CNG. Bangladesh needs to import the technology from Germany/Sweden for production of LBG from biogas available in the country.

Syngas

Syngas is a mixture of fuel gases which mainly contains hydrogen, carbon monoxide, and a small amount of carbon dioxide. Carbon containing fuels are the major sources of syngas generally produced by gasification process. Syngas can be produced from pure biogas through any of the three reforming process, for example, dry reforming, steam reforming, and partial oxidative reforming or any combination of these processes [34]. Steam reforming can produce high purity hydrogen as clean vehicle fuel commercially [35]. Dry reforming and steam reforming are endothermic process, while partial oxidative reforming is an exothermic process. In addition to this, syngas can be converted into methanol and dimethyl ether and further upgraded into transport fuel like diesel, jet fuel, gasoline, and so forth through Fischer–Tropsch synthesis method [26]. The calorific value of the syngas varies depending on the use of agent for gasification process. However, the use of oxygen or steam can increase the calorific value of syngas.

Biomethanol/Biogasoline

Gasoline is produced by upgrading methanol through methanol-to-gasoline (MTG) process. In recent years, the attention has been increased for biomethanol production from biogas through partial oxidation of methane, photocatalytic conversion, biological conversion, and biogas reforming to syngas and hence to methanol via FTS method [36]. Partial oxidation of methane is most attractive and commonly used technology for biomethanol. It has been estimated that biomethanol can mitigate greenhouse gas emissions by 25–40% compared to methanol produced from fossil fuels. In the year 2012, global biomethanol production was about 0.2 million tons and expected to be more than 1 million tons in the next few years [8]. Biomethanol is converted into aromatic hydrocarbon known as biogasoline by two steps of exothermic MTG process [37]. Biogasoline has almost twice the energy content of that of biomethanol and has higher vapor pressure compared to biomethanol. Biogasoline or blending of biogasoline and biomethanol can be directly used as transport fuel in vehicle.

PROSPECTIVE ANALYSIS OF BIO-CNG IN BANGLADESH

Potential of Biogas in Bangladesh

Biogas is a nonfossil, colorless, combustible gas containing about 40–70% methane. Biogas is produced by anaerobic digestion with anaerobic bacteria or fermentation of biodegradable materials such as manure, municipal waste and sewage, green waste, plant material, and crops [38, 39]. The biogas has combustion properties like natural gas as it burns at about 800°C with an ignition temperature of 650–750°C compared to 1000°C for natural gas [4, 17]. To cope with the present world, Bangladesh is going very fast by using biogas technology to produce biogas. Bangladesh has huge potential of biodegradable biomass resources including agricultural residues, animal manure, municipal solid waste, and human excreta to produce biogas. In the year 2012-2013, the country had almost 23.241 million cattle, 25.212 million goat, 246.60 million chicken, and 46.635 million duck [40]. These large numbers of livestock can produce huge amount of biomass residues. Additionally, field residues and process residues from agriculture are another promising candidate for biomass generation. It has been estimated that one ton of cattle excreta, sheep and goat excreta, poultry excreta, human excreta, crop residue, and organic fraction of MSW can generate almost 33 m^3, 58 m^3, 78 m^3, 50 m^3, 60 m^3, and 66 m^3 of biogas, respectively [41–44]. Taking into account the above biogas generation rate, it has been calculated that the country had about 213.81 million tons of biomass generation potential (Table 2) in the fiscal year 2012-2013 which could produce nearly 7.6775 billion m^3 biogas.

Table 2: Bioenergy potential in Bangladesh, 2012-2013 [40]

Biomass resources	Biomass generation (million tons)	Dry biomass recovery (million tons)	Energy content (PJ)
Agricultural residues	94.10	36.48	582.33
Animal manure	72.81	26.20	363.30
Poultry excreta	10.70	2.68	36.12
Human excreta	5.38	5.38	56.99
MSW	13.38	5.15	95.61
Forest residues	17.44	14.32	210.64
Total	213.81	90.21	1344.99

If only cow dung is brought under biogas production plant, then it could be possible to produce about 2.54 billion m^3 of biogas in 2012-2013 which is equivalent to 1.455 billion liters of diesel. On the other hand, poultry feces had the potential to produce approximately 0.749 billion m^3 of biogas in 2012-2013 as presented in Figure 2. However, if 50% of the total biomass is considered for biogas generation, then it is possible to generate almost 3.83875 billion m^3 of biogas which is equivalent to 2.20 billion diesel fuel. In Bangladesh, most of the biogas plants constructed are fixed dome type biogas plant based on cow dung and poultry litter of capacity between 1.6 and 4.8 m^3 as depicted in Table 3. Up to October 2014, almost 79,612 domestic biogas plants have been constructed by technical and financial assistance of various government and nongovernment organizations [45]. However, the country has a potential of 4 million small-scale biogas digesters. In Bangladesh, first commercial biogas plant was constructed by Paragon Poultry Ltd. in Mymensingh and Gazipur in 2010 which are producing 38.23 cubic feet and 76.46 cubic feet of biogas, respectively [46]. In addition to these, IDCOL has financed 8 biogas based power plants using poultry litters for gas production. Recently, Bangladesh has about 215,000 poultry farms and 15,000 cattle farms which may be an excellent source for biogas.

Table 3: Plant based daily biogas production data in Bangladesh [47]

Plant size (daily gas production), m^3	Operation hours per day	Cow dung, Kg	Poultry stool, Kg	Construction cost, (BDT)
1.6	3-4	43	23	26000
2.0	4-5	54	28	32000
2.4	5-6	65	34	36000
3.2	7-8	87	35	43000
4.8	10–12	139	68	52000

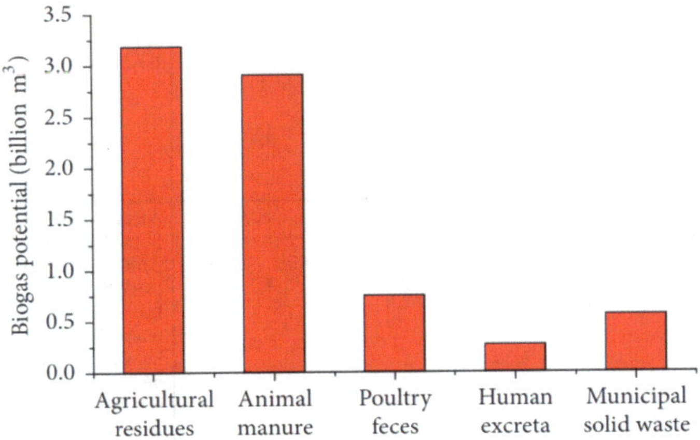

Figure 2: Biogas generation potential from different biomass wastes.

Biogas to Bio-CNG Conversion Technology/Process

Biogas produced from anaerobic digestion of biodegradable biomass contains significant amount of impurities like water, N_2, O_2, H_2S, NH_3, and CO_2 and so forth. Therefore, biogas has to be purified prior to the conversion into bio-CNG. Generally, pressurized water scrubbing, pressure swing adsorption, chemical absorption, membrane permeation, temperature swing adsorption, cryogenic approach, physical absorption, and biological filtration methods are used to purify the biogas before conversion [48]. However, pressurized water scrubbing is the most commonly used method as it offers several advantages and higher percentages of CH_4 purity compared to the other purifying methods [49]. Table 4 presents the process description of some common biogas cleaning methods. In context of Bangladesh, water scrubbing and membrane separation technology are the most feasible technology for biogas upgradation based on the technical availability and maintenance costs.

Table 4: Some commonly used biogas cleaning methods and their features

<table>
<tr><th>Biogas cleaning methods</th><th>Diagram</th><th>Characteristics</th><th>Reference</th></tr>
<tr><td>Water scrubbing</td><td>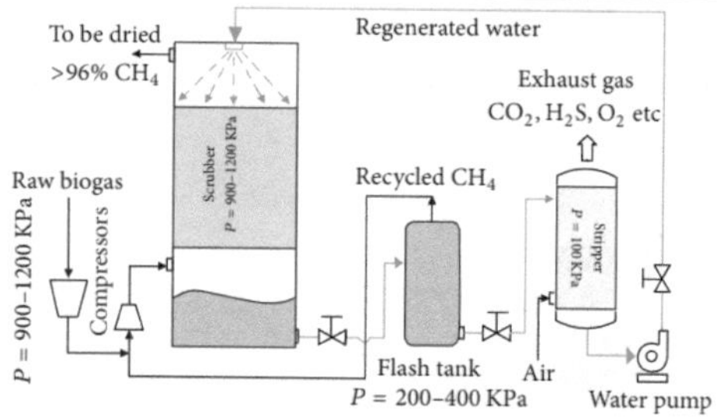
</td><td>(i) Eliminates CO_2, NH_3, dust, and so forth and provides up to 96.1% CH_4purity
(ii) Large quantity is required</td><td>[49]</td></tr>
<tr><td>Pressure swing adsorption</td><td>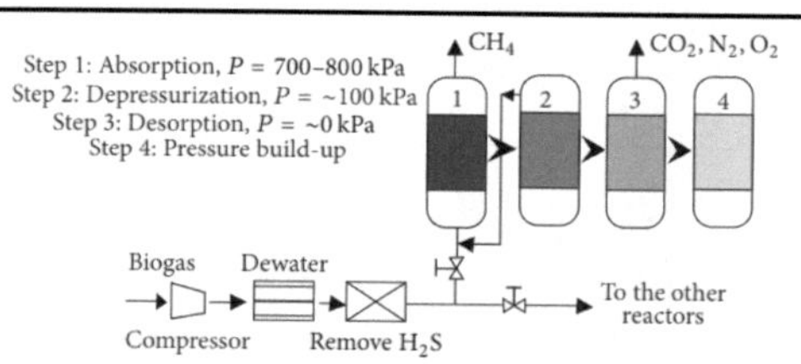
</td><td>(i) Eliminates CO_2, N_2, and O_2 and purifies biogas up to 95.8% of CH_4(ii) Low emission and power demand, however, relatively more expensive
(iii) Additional steps are required for removing H_2S prior to the process</td><td>[49]</td></tr>
<tr><td>Chemical absorption</td><td>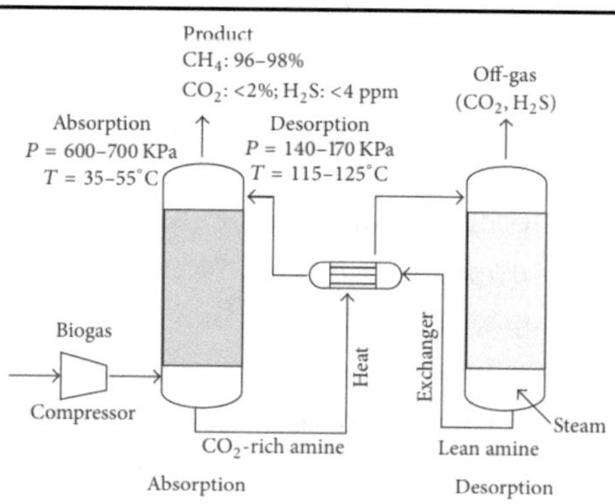
</td><td>(i) High efficiency, complete removal of H_2S, and 94.6% CH_4 containing biogas
(ii) Relatively high cost and additional chemical and energy input requirement</td><td>[49]</td></tr>
</table>

Membrane permeation	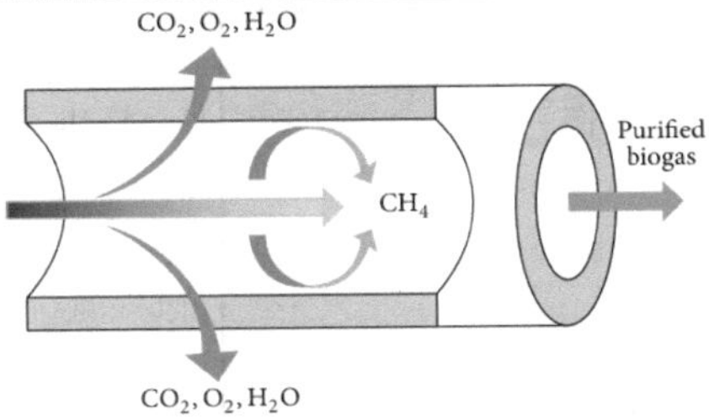	(i) Light in weight, simple in operation and maintenance (ii) Comparatively low CH_4 (90.3%) percentage in biogas (iii) High cost of membrane	[49]
Physical absorption	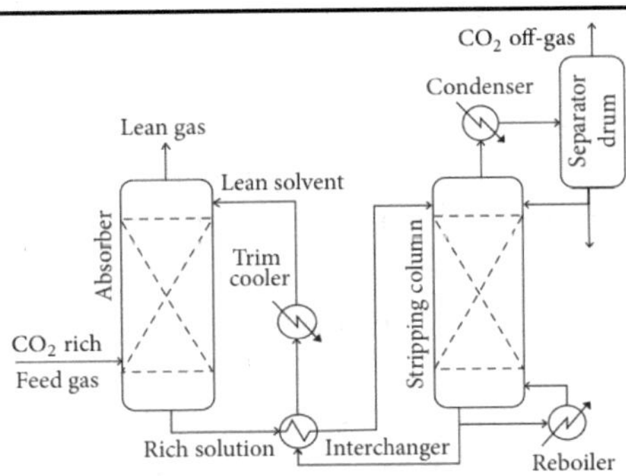	(i) Requires less energy input, about 93% CH_4 purity in biogas (ii) Requires more absorbent to capture CO_2	[50]
Cryogenic method	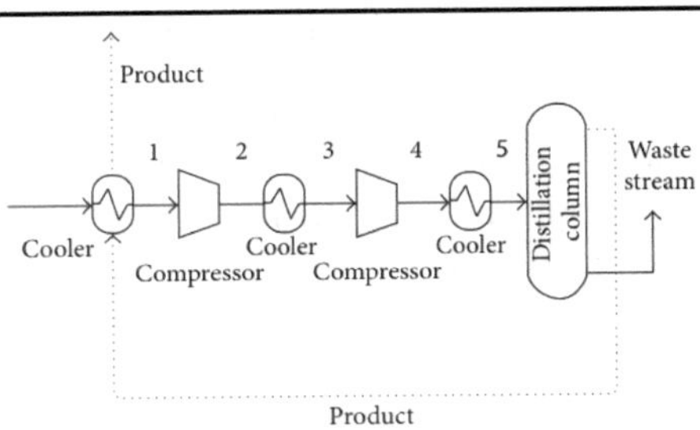	(i) Removes CO_2, H_2S, and all other impurities from biogas and contains 88% CH_4(ii) Additional equipment like turbines, compressors, and heat exchangers causes high investment cost	[51]
Biological filtration	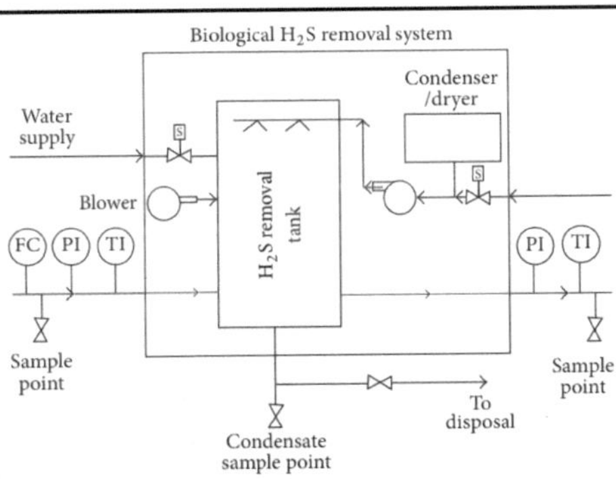	(i) Eliminates H_2S, and purifies biogas up to 94% of CH_4(ii) Low energy requirement (iii) Additional nutrients requirement and biogas may contain small amount of O_2 and N_2	[51]

Dynamic temperature swing adsorption	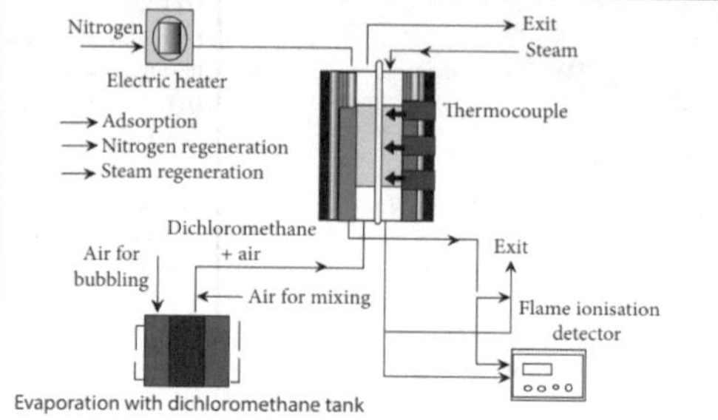	(i) Low operating cost with the lowest CH_4 (84%) percentage among the cleaning methods (ii) Requirement of high energy input	[52]

Cleaned biogas containing more than 97% CH_4 and less than 2% O_2 is considered for production of bio-CNG. Generally, two approaches named physical (compression and liquefaction) and chemical approach (catalytic reforming and Fischer–Tropsch synthesis) are applied for this conversion. Pure biogas then undergoes a high compression pressure between 20 and 25 MPa (Figure 3) and converts into bio-CNG which occupies less than 1% of its normal volume. It is required to store bio-CNG as it affects vehicle filling time, filling completeness, and energy consumption [53]. Typically, two storage systems such as buffer storage and cascade storage are used in filling station. The buffer storage system maintains the pressure of CNG in the range of 20–25 MPa and provides CNG with a maximum pressure of 20 MPa to a vehicle's on-board cylinders. In this case, all filling station reservoirs are connected and maintained at the same pressure [27].

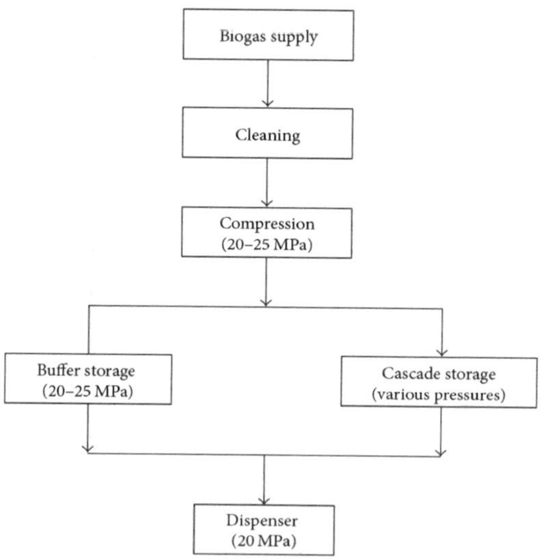

Figure 3: Outline of bio-CNG production, storage, and distribution [49].

On the contrary, the cascade storage system contains three reservoirs of low, medium, and high pressure and provides CNG in three steps to vehicle's on-board cylinders. In this case, the vehicle's cylinders are connected to the low pressure reservoir firstly and then to the medium pressure reservoir when the pressure is increased into the cylinder and finally to the high pressure reservoir for completing the filling process. In comparison with the cascade reservoir, the buffer reservoir offers fast filling and charges 80% more gas [54].

Bio-CNG Vehicle Technology

Bio-CNG technology consists of the refueling station and vehicle technology. Bio-CNG refueling station is much more complicated than the conventional diesel/petrol refueling station as all the components of bio-CNG vehicles and the refueling station are required to be maintained at a high pressure. On the other hand, slow filling refueling station is simpler than the fast filling refueling station which normally provides fuel overnight. Both the petrol (spark ignition) engine and diesel (compression ignition) engine driven vehicle can be converted to bio-CNG driven vehicle by the modification of some features through retrofitting. It offers the advantage of switching the option of using bio-CNG or conventional fuel. Figure 4 illustrates the process pathways of bio-CNG production and vehicle fueling in refueling station. In this case, a new fuel tank with a regulator, fuel lines, and new secondary injector are required to convert conventional fuel vehicle to bio-CNG vehicle. The regulator reduces tank pressure from 3,600 psi to 125 psi and the secondary injector injects the bio-CNG into the cylinder through fuel lines for combustion to produce energy.

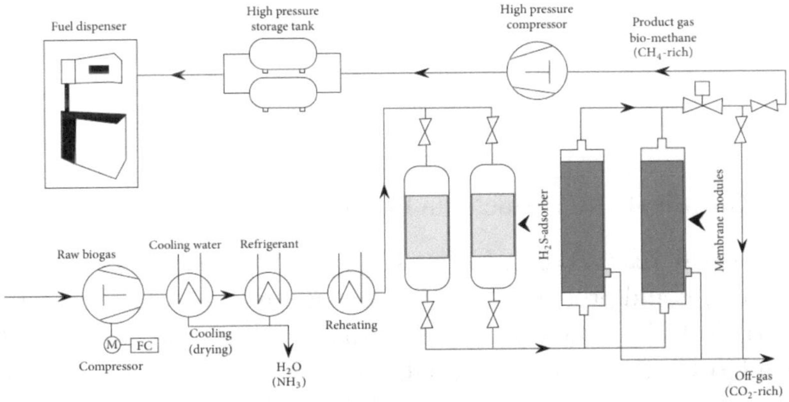

Figure 4: Schematic diagram of total bio-CNG process and vehicle fueling method [55].

Possibility of Bio-CNG in Bangladesh

Recently, in Bangladesh the consumption of CNG as transportation fuel and the number of CNG based vehicles are growing very fast because of its low cost and environmental sustainability. Figure 5 shows the cumulative growth of CNG vehicles including the number of converted CNG vehicles and the number of exported CNG vehicles. However, this high rate of consumption causes the threat for future reserve and forces us to harness the renewable and environmental friendly options for automobile fuel. It is obvious that some countries such as Germany, Austria, and the United States have proved that bio-CNG can take the place of fossil fuels as transportation fuel. However, in Bangladesh there are not so many industrial plants for biogas purification and bio-CNG production. Only, Effat BioCNG Limited has developed a technology based on water scrubbing for purifying the biogas by removing carbon dioxide and hydrogen sulphide and for producing bio-CNG with less than 2% methane loss.

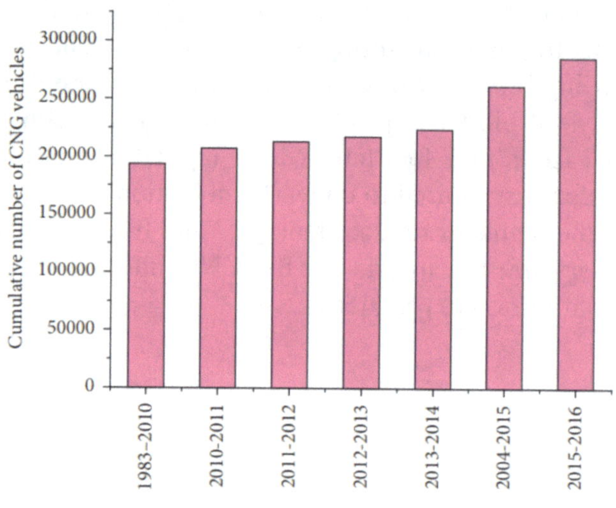

Figure 5: Growth of CNG vehicles in Bangladesh.

In the fiscal year 2012-2013, total biogas (7.6775 billion m^3) available in the country could produce about 3.4549 billion kg (1 m^3 biogas = 0.45 kg bio-CNG) bio-CNG which is equivalent to 5.088 billion m^3 ($\rho_{\text{bio-CNG}} = 0.679$ kg/m^3) of bio-CNG. Figure 6 shows the amount of bio-CNG produced from different biomass wastes. It is estimated that 1 liter of octane is equivalent to 0.81 m^3 of bio-CNG, 1 liter of petrol is equivalent to 0.80 m^3 of bio-CNG, and 1 liter of diesel is equivalent to 0.97 m^3 of bio-CNG. Accordingly, total bio-CNG potential in Bangladesh is equivalent to 5.25 billion liters of diesel, 6.36 billion liters of petrol, and 6.28 billion liters of octane. Therefore, Bangladesh shows the huge potentiality of bio-CNG production for alternative vehicle fuel. Sufficient feed material is the prerequisite for the continuous production of bio-CNG. It is difficult to collect feed materials from villages because of its high transportation cost. On the other hand, the six municipalities in Bangladesh produce about 7690 tons [60] of waste per day which can be available to generate fuel for vehicle running in the municipality. It has been estimated that the organic municipal solid waste available only in Dhaka can produce almost 41,830 m^3 per day of biogas equivalent to 27,721.4 m^3 of bio-CNG. Furthermore, the rate of migration of people to municipality is increasing very fast which can increase the quantity of organic waste as well. In addition, the inclusion of human manure for biogas production can be also a secure feed material in long run of bio-CNG plant in the municipality.

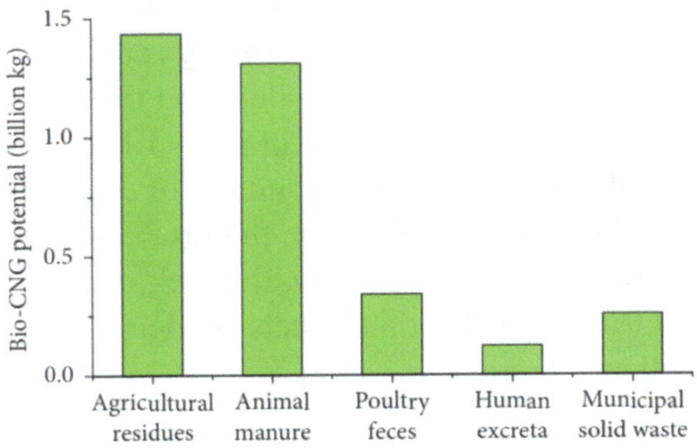

Figure 6: Bio-CNG potential from various biomass resources.

COMPETITIVE ANALYSIS OF BIO-CNG AS TRANSPORTATION FUEL

The feasibility of biogas conversion into bio-CNG as transportation fuel primarily depends on some key factors such as economic, technical, environmental, and safety. Fuel properties of bio-CNG are nearly the same as regular CNG and also competitive compared to the other automobile fuels such as diesel and petrol. The percentage of methane (>97%) in bio-CNG is higher than that of natural gas (93%) produced from different gas fields in Bangladesh. Besides this, the calorific value of bio-CNG is about 52 MJ/

kg which is higher than the calorific value of petrol (48 MJ/kg) and diesel (44.8 MJ/kg) as revealed in Table 5. The amount of energy cost in kJ/BDT is almost 3.5 times that of petrol and 2.25 times of diesel. Furthermore, in Bangladesh the cost of petrol and diesel fuel is significantly high as about BDT 96 per liter and BDT 68 per liter for petrol and diesel, respectively. On the other hand, the price of CNG and bio-CNG is approximately BDT 30 per m^3 and BDT 14.22 per m^3, respectively [13, 15, 61, 62]. Considering the equivalent ratio of bio-CNG to other transportation fuels, the use of 1 liter of bio-CNG as automobile fuel in replacement of 1 liter of petrol and 1 liter of diesel can save about BDT 84.62 and BDT 54.20, respectively. Additionally, bio-CNG contains negligible amount of impurities including less than 4% CO_2 and 8 ppm H_2S with no other impurities which are much lower than diesel and petrol fuel and responsible for up to 90% emission reduction compared to the conventional transportation fuel. Germany, Netherland, and Sweden are the most biomethane producing countries in the world. Sweden can be a role model for Bangladesh as the country is using 97% of bio-CNG for vehicle fuel, while Germany is using 1.4% of total bio-CNG successfully. Therefore, it is clearly depicted that bio-CNG is a rising candidate for automobile fuel. Although bio-CNG has several advantages over the conventional vehicle fuel, it has number of disadvantages. Because of decentralized feed material availability, the production cost of bio-CNG can be very high nowadays. Additionally, the lack of awareness about the suitability of bio-CNG over the other fuels hinders the support from the government as well as the other investors.

Table 5: Competitive analysis of bio-CNG among other fuels in Bangladesh [4, 13, 56–59]

Fuel	Calorific value (MJ/Kg)	Tariff/rate/cost (BDT/Kg)	Cost of energy (KJ/BDT)
CNG	52	42.10	1232.81
Bio-CNG	52	40.91	1271.08
LPG (commercial)	46	56	821.43
Petrol	48	133.44	359.71
Diesel	44.8	79.97	560.21

ISSUES AND CHALLENGES OF BIO-CNG TECHNOLOGY IN BANGLADESH

It is well evident that the use of bio-CNG as automobile fuel provides significant benefits in economic, emissions, and engine performance perspective. However, the successful implementation of this technology in developing country like Bangladesh is a great challenge. Requirement of sufficient amount of feed materials, upgradation of equipment and the cost, lack of technically sound man power, and refueling infrastructure are considered the significant barriers to the deployment of bio-CNG as vehicle fuel. Typically, a bio-CNG plant requires a biogas unit, purification unit, compression unit, and storage unit with other accessories maintained at high pressure. Lack of technology standardization is the main stumbling block for production of cost efficient renewable bio-CNG. Although Bangladesh has huge potential of biogas generation resources, there is insufficiency of industrial scale biogas plants with high level of automation to produce bio-CNG because of its high investment cost and lack of standard technology. In addition to this, almost 75% of total municipal waste in Bangladesh are unused and go for landfilling because of lack of proper technical knowledge and sufficient human resources to produce useful energy like bio-CNG. Besides this, in Bangladesh there are no specific guidelines and rules for renewable energy use. Additionally, there is no inclusion of legal, regulatory, and policy framework for bio-CNG conversion from waste in country's national energy policy.

CONCLUDING REMARKS

In Bangladesh, the energy shortage is the main hindrance for the development in every sector. Due to high price and crisis of crude oil, petroleum products and natural gas resources, the concept of using methane as a feed stock to produce transportation fuels has greatly increased. This paper has assessed that Bangladesh has about 7.6775 billion m^3 of biogas potential that can produce 5.088 billion m^3 of bio-CNG. If bio-CNG is used as automobile fuel in replacement of regular CNG or diesel fuel, it can save almost 5.088 billion m^3 of regular CNG or 5.25 billion liters of diesel fuel. Although the country has available raw materials for bio-CNG production, it requires special attention to utilize these resources properly. Therefore, the government should come forward with some concrete bio-CNG implementation strategies considering its necessity. Firstly, it is mandatory to prepare policy and regulatory framework for the production and utilization of bio-CNG instead of traditional vehicle fuel. Additionally, it is necessary to take initiatives for the development of bio-CNG distribution network right now and to motivate the nongovernment organizations for construction of bio-CNG plants.

REFERENCES

1. J.-P. Rodrigue, C. Comtois, and B. Slack, The Geography of Transport Systems, Routledge, Abingdon, UK, 4th edition, 2013.
2. The National Academies of Science, Engineering and Medicine, 2016, http://needtoknow.nas.edu/.
3. K. D. Patil, U. P. F. Pathan, P. S. M. Mahajan, and P. D. R. Satpute, "Alternative Fuel Public Transport Buses," International Journal of Engineering Research and Development, vol. 10, no. 12, pp. 29–35, 2014.
4. P. Jain, "Renewable Natural Gas-Bio CNG, in: Sustain, Outlook," 2012, http://sblf.sustainabilityoutlook.in/file_space/SBLF Summit Presentations 2012/Green Brick Eco Solutions_Prasun Joshi.pdf.
5. C. Voell, "Feasibility of Small Scale Biogas-to-CNG Vehicle Fuel Projects," 2016, http://biocng.us/.
6. Energy Vision, "Renewable Natural Gas (RNG): The Solution to a Major Transportation Challenge," 2016, http://energy-vision.org/ev-publications/EV-RNG-Facts-and-Case-Studies.pdf.
7. The Hindu, "New source for bio-CNG production," 2016, http://www.thehindu.com/todays-paper/tp-national/tp-otherstates/new-source-for-biocng-production/article1283967.ece.
8. "International Energy Agency (IEA)," 2016, http://www.iea.org/.
9. 100% Biogas for Urban Transport in Linköping , Sweden Biogas in Buses , Cars and Trains, 2005.
10. P. K. Halder, N. Paul, M. U. H. Joardder, and M. Sarker, "Energy scarcity and potential of renewable energy in Bangladesh," Renewable and Sustainable Energy Reviews, vol. 51, pp. 1636–1649, 2015.
11. British Petroleum, "Statistical Review of World Energy 2015," 2016, http://www.bp.com/en/global/corporate/energy-economics/statistical-review-of-world-energy.html.
12. W. W. Irion and O. S. Neuwirth, "Oil refining," in Ullmann's Encyclopedia of Industrial Chemistry, Wiley-VCH Verlag GmbH & Co. KGaA, Weinheim, Germany, 2000.
13. "Bangladesh Petroleum Corporation (BPC)," 2016, http://www.bpc.gov.bd/.
14. S. Wakdikar, "Compressed natural gas: A problem or a solution," Current Science, vol. 82, 2002.

15. "Rupantarita Prakritik Gas Company Limited (RPGCL)," 2016, http://www.rpgcl.org.bd/.
16. Government of the People of Republic of Bangladesh (GOB), "Ministry of Finance, Finance Division, Sector Road map: Second Update," 2016, http://mof.gov.bd/en/budget/12_13/power/power_energy_en.pdf.
17. B. Imam, Energy Resources of Bangladesh: Natural Gas, Oil, Coal, University Grants Commission of Bangladesh, Dhaka, Bangladesh, 2013.
18. D. A. Wood, "A review and outlook for the global LNG trade," Journal of Natural Gas Science and Engineering, vol. 9, pp. 16–27, 2012.
19. B. K. Barnwal and M. P. Sharma, "Prospects of biodiesel production from vegetable oils in India," Renewable and Sustainable Energy Reviews, vol. 9, no. 4, pp. 363–378, 2005.
20. J. Van Gerpen, "Biodiesel processing and production," Fuel Processing Technology, vol. 86, no. 10, pp. 1097–1107, 2005.
21. L. C. Meher, D. Vidya Sagar, and S. N. Naik, "Technical aspects of biodiesel production by transesterification—a review," Renewable and Sustainable Energy Reviews, vol. 10, no. 3, pp. 248–268, 2006.
22. M. Habibullah, H. H. Masjuki, M. A. Kalam et al., "Potential of biodiesel as a renewable energy source in Bangladesh," Renewable and Sustainable Energy Reviews, vol. 50, pp. 819–834, 2015.
23. P. K. Halder, N. Paul, and M. R. Beg, "Prospect of pongamia pinnata (karanja) in bangladesh: a sustainable source of liquid fuel," Journal of Renewable Energy, vol. 2014, pp. 1–12, 2014.
24. A. B. M. A. Islam, "Trees to Produce Biofuel For Combating The Energy Crisis In Bangladesh, University of Flensburg," 2008, http://slidegur.com/doc/3027232/bio-energy-utilization-in-developing-countries-and-future.
25. R. Chandra, V. K. Vijay, P. M. V. Subbarao, and T. K. Khura, "Performance evaluation of a constant speed IC engine on CNG, methane enriched biogas and biogas," Applied Energy, vol. 88, no. 11, pp. 3969–3977, 2011.
26. Y. Li and L. Yang, "Converting Biogas to Transportation Fuels, Ohio State University Extension," 2014.
27. S. Clarke, "Vehicle Conversion to Natural Gas or Biogas, Ministry of Agriculture, Food and Rural Affairs, Ontario," 2012, http://www.

omafra.gov.on.ca/english/engineer/facts/12-043.htm.

28. B. M. Smyth, J. D. Murphy, and C. M. O'Brien, "What is the energy balance of grass biomethane in Ireland and other temperate northern European climates?" Renewable and Sustainable Energy Reviews, vol. 13, no. 9, pp. 2349–2360, 2009.
29. T. Patterson, R. Dinsdale, and S. Esteves, "Review of energy balances and emissions associated with biomass-based transport fuels relevant to the United Kingdom context," Energy & Fuels, vol. 22, pp. 3506–3512, 2008.
30. J. D. Murphy and N. Power, "Technical and economic analysis of biogas production in Ireland utilising three different crop rotations," Applied Energy, vol. 86, no. 1, pp. 25 36, 2009.
31. T. Patterson, S. Esteves, R. Dinsdale, and A. Guwy, "An evaluation of the policy and techno-economic factors affecting the potential for biogas upgrading for transport fuel use in the UK," Energy Policy, vol. 39, no. 3, pp. 1806–1816, 2011.
32. D. Brown, J. Shi, and Y. Li, "Comparison of solid-state to liquid anaerobic digestion of lignocellulosic feedstocks for biogas production," Bioresource Technology, vol. 124, pp. 379–386, 2012.
33. K. Krich, D. Augenstein, J. Benemann, B. Rutledge, Salour, and D., "Biomethane from Dairy Waste: A Sourcebook for the Production and Use of Renewable Natural Gas in California," 2005, http://suscon.org/pdfs/cowpower/biomethaneSourcebook/Full_Report.pdf.
34. A. Serrano-Lotina and L. Daza, "Influence of the operating parameters over dry reforming of methane to syngas," International Journal of Hydrogen Energy, vol. 39, no. 8, pp. 4089–4094, 2014.
35. A. Boyano, T. Morosuk, A. M. Blanco-Marigorta, and G. Tsatsaronis, "Conventional and advanced exergoenvironmental analysis of a steam methane reforming reactor for hydrogen production," Journal of Cleaner Production, vol. 20, no. 1, pp. 152–160, 2012.
36. S. B. Jones and Y. Zhu, "Techno-economic Analysis for the Conversion of Lignocellulosic Biomass to Gasoline via the Methanol-to-Gasoline (MTG) Process," Tech. Rep. PNNL-18481, 2009.
37. W.-H. Cheng, Methanol Production and Use, CRC Press, Boca Raton, Fla, USA, 1994.
38. S. Kalambe, R. S. Sapkal, and V. S. Sapkal, "Low Pressure Separation Technique of Biogas Into Ch4 and Co2 Employing Pdms

Membrane," International Journal of Engineering and Advanced Technology, vol. 3, pp. 311–315, 2012.

39. National Non-Food Crops Centre (NNFCC), "Renewable Fuels and Energy Factsheet: Anaerobic Digestion," 2016, http://www.nnfcc.co.uk/.

40. P. K. Halder, N. Paul, and M. R. A. Beg, "Assessment of biomass energy resources and related technologies practice in Bangladesh," Renewable and Sustainable Energy Reviews, vol. 39, pp. 444–460, 2014.

41. L. Song, L. Deng, Y. Yin, X. Pu, and Z. Wang, "Biogas production potential and characteristics of manure of sheep, duck and rabbit under anaerobic digestion," Transactions from the Chinese Society of Agricultural Engineering, vol. 26, pp. 277–282, 2010.

42. A. Schunurer and A. Jarvis, "Microbiological Handbook for Biogas Plants," 2009.

43. F. F. Tatlidil, Z. Bayramoglu, and D. Akturk, "Animal manure as one of the main biogas production resources: Case of Turkey," Journal of Animal and Veterinary Advances, vol. 8, no. 12, pp. 2473–2476, 2009.

44. M. S. Rao, S. P. Singh, A. K. Singh, and M. S. Sodha, "Bioenergy conversion studies of the organic fraction of MSW: Assessment of ultimate bioenergy production potential of municipal garbage," Applied Energy, vol. 66, no. 1, pp. 75–87, 2000.

45. P. K. Halder, N. Paul, M. U. H. Joardder, M. Z. H. Khan, and M. Sarker, "Feasibility analysis of implementing anaerobic digestion as a potential energy source in Bangladesh," Renewable and Sustainable Energy Reviews, vol. 65, pp. 124–134, 2016.

46. Paragon Group, "Paragon Poultry Ltd," 2016, http://paragongroup-bd.com/.

47. Infrastructure Development Company Limited (IDCOL), 2016, http://www.idcol.org/.

48. O. W. Awe, Y. Zhao, A. Nzihou, D. P. Minh, and N. Lyczko, "A review of biogas utilisation, purification and upgrading technologies," Waste and Biomass Valorization, vol. 8, no. 2, pp. 267–283, 2017.

49. L. Yang, X. Ge, C. Wan, F. Yu, and Y. Li, "Progress and perspectives in converting biogas to transportation fuels," Renewable and Sustainable Energy Reviews, vol. 40, pp. 1133–1152, 2014.

50. F. Rossi, A. Nicolini, M. Palombo, B. Castellani, E. Morini, and M. Filipponi, "An innovative configuration for CO2 capture by high

temperature fuel cells," Sustainability (Switzerland), vol. 6, no. 10, pp. 6687–6695, 2014.

51. S. V. Khedkar and N. Bhautik, "Studies in Purification of Biogas Using Al Gas in Sewage Waste Water- A Review Paper," International Journal of Pure and Applied Sciences and Technology, vol. 3, pp. 95–108, 2015.
52. S. G. Ramalingam, P. Pré, S. Giraudet et al., "Recovery comparisons-Hot nitrogen Vs steam regeneration of toxic dichloromethane from activated carbon beds in oil sands process," Journal of Hazardous Materials, vol. 205-206, pp. 222–228, 2012.
53. M. Farzaneh-Gord, M. Deymi-Dashtebayaz, and H. R. Rahbari, "Studying effects of storage types on performance of CNG filling stations," Journal of Natural Gas Science and Engineering, vol. 3, no. 1, pp. 334–340, 2011.
54. M. Farzaneh-Gord and S. Branch, "Real and ideal gas thermodynamic analysis of single reservoir filling process of natural gas vehicle cylinders," Journal of Theoretical and Applied Mechanics, vol. 41, pp. 21–26, 2011.
55. M. Miltner, A. Makaruk, H. Bala, and M. Harasek, "Biogas upgrading for transportation purposes - Operational experiences with Austrias first Bio-CNG fuelling station," Chemical Engineering Transactions, vol. 18, pp. 617–622, 2009.
56. Green Brick Eco Solutions, "Financial Analyses of Biogas to Bio CNG projects in India; Projections based case study analyses," 2016, http://gbes.in/.
57. A. Petersson and A. Wellinger, "Biogas upgrading technologies–developments and innovations," IEA Bioenergy, vol. 20, pp. 1–19, 2009.
58. W. Papacz, "Biogas as Vehicle fuel," Journal of KONES Powertrain And Transport, vol. 18, pp. 403–410, 2011.
59. A. V. Nawade, "Biogas to BioCNG/Biopower-Solutions for Waste-to-Green-Fuel Projects: An Experience Sharing," in Proceedings of the World Futur Energy Summit, Abu Dhabi, UAE, 2013.
60. M. Alamgir and A. Ahsan, "Municipal solid waste and recovery potential: Bangladesh perspective, Iran," Journal of Environmental Health Science and Engineering, vol. 4, pp. 67–76, 2007.

61. K. A. Bashar and A. Rahman, "Increase Use of CNG as Public Transport & Reduce Emissions: A Comparative Study of the Benefits of CNG & Automobiles Fuel: Present," Journal of Business Management, vol. 4, pp. 131–141, 2012.
62. S. A. Iqbal, M. Iqbal, and A. F. M. Salauddin, "Present Scenario of Compressed Natural Gas (CNG) as a Vehicular fuel in Bangladesh," in Proceedings of the 2nd International Conference on Industrial Engineering and Engineering Management, pp. 222–228, 2011

INDEX

U

V

W

Printed in the United States
By Bookmasters